The Grand Canyon

Grand Canyon National Park

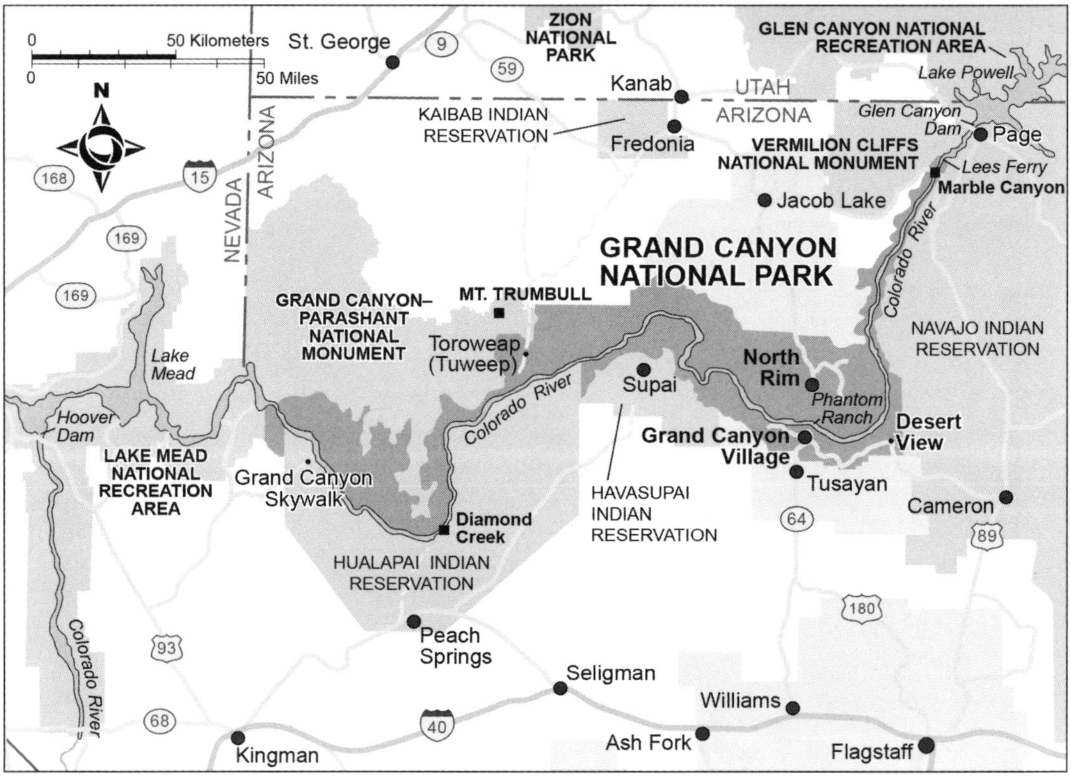

The Grand Canyon

An Encyclopedia of Geography, History, and Culture

RANDY MOORE AND KARA FELICIA WITT

ABC-CLIO™

An Imprint of ABC-CLIO, LLC
Santa Barbara, California • Denver, Colorado

Library of Congress Cataloging-in-Publication Data

Names: Moore, Randy, author. | Witt, Kara Felicia, author.
Title: The Grand Canyon : an encyclopedia of geography, history, and culture / Randy Moore and Kara Felicia Witt.
Description: Santa Barbara, California : ABC-CLIO, 2018. | Includes bibliographical references and index.
Identifiers: LCCN 2017055990 (print) | LCCN 2017060662 (ebook) | ISBN 9781610698405 (ebook) | ISBN 9781610698399 (hard copy : alk. paper)
Subjects: LCSH: Grand Canyon (Ariz.)—Encyclopedias. | Grand Canyon National Park (Ariz.)—Encyclopedias.
Classification: LCC F788 (ebook) | LCC F788 .M66 2018 (print) | DDC 979.1/32003—dc23
LC record available at https://lccn.loc.gov/2017055990

ISBN: 978-1-61069-839-9 (print)
978-1-61069-840-5 (ebook)

22 21 20 19 18 1 2 3 4 5

This book is also available as an eBook.

ABC-CLIO
An Imprint of ABC-CLIO, LLC

ABC-CLIO, LLC
130 Cremona Drive, P.O. Box 1911
Santa Barbara, California 93116-1911
www.abc-clio.com

This book is printed on acid-free paper ∞

Manufactured in the United States of America

For Jennifer, Oliver, Anita, and Marvin

Contents

Preface

Grand Canyon is the most internationally recognized landscape and symbol of nature in North America. People around the world have heard and seen books, postcards, documentaries, and televisions shows about Grand Canyon, and many of these people consider their visit to Grand Canyon a highlight of their lives. Whether at the rim, along an inner canyon trail, or at the bottom of Grand Canyon aboard a boat on the Colorado River, virtually everyone marvels at what they see and often struggles to understand the feelings that accompany the beauty and immensity of what is around them. Some people are scared by Grand Canyon, others are entertained, and almost all are inspired by what they see and feel.

But what is this place called Grand Canyon? What do Grand Canyon's millions of visitors per year come for, and what, exactly, do they see? For most visitors, Grand Canyon is a beautiful and famous place that they've heard much about and are anxious to see. Is it really 10 miles across the canyon? Is the river really a mile below the rim? Are the views really that stunning? Yes, yes, and yes. But there's more.

Although its landscapes are overwhelming, most visitors soon understand that Grand Canyon is more than just pretty views; this is why the experience of visiting Grand Canyon, unlike that of most other places, often lingers. And this, in turn, is why many people who come to Grand Canyon return to it. People are lured to Grand Canyon and Grand Canyon National Park (GCNP) for a multitude of reasons:

For many Native Americans, Grand Canyon is a sacred, revered, spiritual place where their ancestors originated and lived and to which their spirits will someday return. Today, some Native Americans continue to go on cultural and spiritual pilgrimages to and in Grand Canyon, as their families and ancestors have done for centuries.

For river runners, Grand Canyon is a place to experience challenging, world-class rapids in a unique setting unmatched anywhere else in the world. The violent, churning rapids, combined with the otherwise peaceful solitude at the bottom of Grand Canyon, inspire many people.

For geologists, Grand Canyon offers a deep look—literally—at Earth's history and the stunning power of erosion. Nowhere is Earth's past more obvious and accessible than at Grand Canyon.

For artists, Grand Canyon is a place for inspiration, challenges, and creative renewal. The ever changing colors and landscapes of Grand Canyon defy artists' attempts to capture what they see and even what they imagine.

For concessionaires, Grand Canyon and its throngs of tourists are an opportunity to build personal empires, design iconic buildings, and help visitors enjoy their time at the canyon. Grand Canyon is big business.

For conservationists and environmentalists, Grand Canyon remains a place worth fighting for as developers threaten to build dude ranches, massive theme parks, and polluting mines in and around the canyon.

For biblical literalists, Grand Canyon is powerful evidence of a worldwide flood described in the Bible. This is why Grand Canyon is featured prominently in virtually all of the growing number of "creation museums" in the United States and elsewhere. At Grand Canyon, businesses promoting this view on the rim and along the river are popular and growing fast.

For prospectors, Grand Canyon offered the promise of discovery and wealth. Indeed, many of the mines staked by Grand Canyon pioneers such as Peter Berry, Dan Hogan, and John Hance remain visible to visitors, and some mines near Grand Canyon remain operational. During GCNP's early years, more than half Arizona's tax revenues came from mining.

For many people, Grand Canyon is a cure for self-importance. As physician Mary E. Hart noted in 1895, "There is a certain malady, commonly termed 'big head,' which a large number of otherwise healthy people are afflicted. Prescription: Stand upon the brink of the Grand Cañon, gaze down, and still further down, into its awful depths, and realize for the first time your own utter insignificance" (Kinsey, 1998).

For NASA, Grand Canyon is a place that taught astronauts about vast, other-world-like landscapes and grandeur. This is why, in March 1964, the crew of Apollo 11 (Michael Collins, Buzz Aldrin, and Neil Armstrong) spent two days hiking to the bottom (and back) of Grand Canyon. NASA later brought two other groups of astronauts (and all but one of the 24 astronauts who went to the moon) to Grand Canyon for similar training. The training made an impression; Apollo 17 astronaut Harrison Hagan "Jack" Schmitt (b. 1935) likened being on the moon to "trying to describe what you feel when you're standing on the rim of the Grand Canyon" (Annerino, 2017).

Today, Grand Canyon remains culturally, economically, spiritually, and scientifically important to humanity, as well as an enduring symbol of America. In this book, you'll learn why.

This book discusses many places in Grand Canyon, but it is not meant to be a travel guide to accompany a visit to Grand Canyon. Several such guides are readily available elsewhere; moreover, the countless sites of Grand Canyon are far too diverse to summarize in a few essays here. Instead, this book describes what it's like to be at and experience Grand Canyon—what you'll see, what it means, where you'll go, and what many visitors overlook. Appreciating this experience at Grand Canyon is what this book is about. However, this book is also about the people, places, and events that have made Grand Canyon so famous. You'll learn about influential pioneers such as explorer John Wesley Powell, artist Thomas Moran, and geologist Clarence Dutton, and as well as sites such as Bright Angel Trail, Supai, the North Rim, Phantom Ranch, and historically significant sites along the river such as Separation Canyon. Together, these people, places, and stories create the lure and profound impact that Grand Canyon continues to have on so many people.

How to Use This Book

This book includes two major sections: (1) *Thematic essays* discuss the geology, history, first people, and other topics associated with Grand Canyon, and (2) *Topical entries* describe more than 160 people, places, events, and other aspects of Grand Canyon. An appendix provides additional information about geology, and a glossary helps with unfamiliar terms.

Throughout this book, you'll read about many people, places, stories, and issues associated with Grand Canyon, as well as the scientific, religious, and social contexts of a variety of Grand Canyon–related events. Many of these topics are interrelated; for example, John Wesley Powell—who led the first intentional expedition through Grand Canyon—is a towering figure whose name is common on the rims and river in Grand Canyon. Similarly, Theodore Roosevelt is associated with events ranging from the early federal protection of Grand Canyon to personal favors he granted several Grand Canyon pioneers. Mary Elizabeth Jane Colter designed several of Grand Canyon's iconic buildings, and mysteries such as the disappearance of honeymooning couple Glen and Bessie Hyde persist. Although some of these stories are well-known, many entries describe lesser known events, including the integration of Grand Canyon into American culture. Along the way, readers will get glimpses of a diverse group of people who have influenced Grand Canyon, ranging from courageous heroes to shameless profiteers.

Linkages of these people, places, and stories are listed as "See also" at the end of most entries. We have also included an extensive index to direct readers to other mentions of various people and places. Together, these people, places, and stories explain how we've viewed and imagined Grand Canyon and why it matters to do so.

We hope you enjoy learning about Grand Canyon.

Randy Moore
Kara Felicia Witt

Further Reading

Annerino, John. 2017. *Hiking the Grand Canyon: A Detailed Guide to More Than 100 Trails*. New York: Skyhorse Publishing.

Berger, Todd R. (Ed.). 2008. *Reflections of Grand Canyon Historians: Ideas, Arguments, and First-Person Accounts*. Grand Canyon, AZ: Grand Canyon Association.

Kinsey, John L. 1998. *The Majesty of the Grand Canyon: 150 Years in Art*. Petaluma, CA: Pomegranate Communications.

Lago, Don. 2014. *Canyon of Dreams: Stories from Grand Canyon History*. Salt Lake City: University of Utah Press.

Acknowledgments

Many people have helped us with our travel, research, and adventures in and around Grand Canyon. Moki Mac River Expeditions and Grand Canyon Expeditions expertly organized most of our 25 river trips through the canyon, during which we were educated, protected, pampered, and entertained by a variety of excellent boatmen and boatwomen, swampers, historians, rangers, and others. Of these river runners, we especially thank Lance Newman for making our first trips through Grand Canyon so interesting that we had to return, MaryLynn and Richard "Q" Quartaroli for never making things up, and Neal "Bear" Shapiro for conjuring many things with us. We also thank Common Good Books, Simone Stephenson, Greg Eastwood, Trevin Johnson, and Den McCormick for especially memorable river trips, as well as Stephanie Cox, Bette Englund, Jane Super, Little Socks, and others for their help at and in Grand Canyon. Andrew Witt, Matthew Witt, Marvin Witt III, Anne Witt, Petra Witt, Claudia and Jim Cameron, Gloria Lopez Franco and Scott O. Lilienfeld are family members and friends who were also helpful.

At the South Rim, we thank Robb Seftar, the GCNP rangers, and Grand Canyon Association officials who answered our questions, guided us to many sites, and helped us learn about the canyon and its surrounding area. We enjoyed excellent accommodations and advice at Red Feather Lodge. We also humbly thank the residents of Supai for their hospitality during our visits.

At the North Rim, we were often assisted by GCNP rangers, as well as by the staff of Kaibab Lodge and Grand Canyon Lodge. We especially thank Robert Pennell for his help, directions, advice, and early morning coffee and cinnamon rolls, which sustained us on hikes to and from Phantom Ranch, the South Rim, and the countless interesting points in between. We are also grateful to Grand Canyon historians Earle Spamer of the American Philosophical Society and Haley Johnson of the Grand Canyon Association. We could not have gotten to Toroweap without the help of NPS Ranger Todd Seliga, Pink Jeep Tours, and Will James and Meghann Kent of Dreamland Safari Tours. We also thank the many other people we met at Grand Canyon who gave us directions and answered our questions but whom we can thank only collectively and anonymously.

We especially thank Grand Canyon historian and river runner Richard Quartaroli for his suggestions and critical reading of our manuscript. Although any mistakes or other shortcomings that remain are ours alone, this does not lessen our gratitude to Richard and all the people named here for their help.

Finally, we thank Kaitlin Ciarmiello, Ellen Rasmussen, Shruti Chopta, and Fred Dahl for their help, guidance, and patience. All writers should be so fortunate.

Introduction

Grand Canyon is one of the most vast, beautiful, easily recognized, harsh, and inspiring places on Earth. It is one of our planet's few natural landmarks visible from outer space, and its magnificent panoramas, ecosystems, cultures, and history lure millions of visitors per year. Most of these visitors connect strongly with Grand Canyon from trails and overlooks on the rim. However, even these trails and overlooks can't encompass all of Grand Canyon. Although abundant and stretching for hundreds of miles, trails span less than 25 percent of the rims along Grand Canyon, and many parts of GCNP remain virtually unexplored.

Early visitors to Grand Canyon saw and interpreted the canyon in different ways. Native Americans were sustained by Grand Canyon, and they, in turn, protected and revered the land. The first known Europeans to see Grand Canyon arrived in 1540 (80 years before the Pilgrims landed at Plymouth Rock), when a detachment of Spanish soldiers led by Captain García López de Cárdenas reached the South Rim. Cárdenas and his men were astonished by the canyon's size; what appeared as small boulders were taller than the 342-foot-high Great Tower of Seville. Cárdenas considered the canyon little more than an obstacle to finding the Lost Cities of Gold that he sought. There was virtually no more interest in Grand Canyon for more than 300 years, when—in 1857—an expedition led by Lt. Joseph Christmas Ives rode a steamboat up the Colorado River into part of the lower canyon. Although the expedition's naturalist John Strong Newberry appreciated the canyon's geological significance, Ives declared that the area was "altogether valueless" and, after seeing it, that there was "nothing to do but leave."

Until the 1860s, the Colorado River through Grand Canyon remained unexplored because of its presumed dangers and remote location. The river's large drop in elevation was assumed to mean that there were violent rapids and high waterfalls that would make navigation dangerous and difficult, if not impossible. In 1869, despite these presumed dangers, Civil War hero John Wesley Powell decided to run the Colorado River through Grand Canyon. Along the way, Powell and his crew experienced many hardships, and their task became increasingly difficult. They knew the dangers ahead of them (Powell, 2013):

> And now we go on through this solemn, mysterious way. The river is very deep, the canyon very narrow, and still obstructed, so that there is no steady flow of the stream; but the waters reel and roll and boil, and we are scarcely able to determine where we can go. . . . We can neither land nor run as we please. The boats are entirely unmanageable; no order in their running

> can be preserved; now one, now another, is ahead, each crew laboring for its own preservation.

In August 1869, Powell and his crew arrived at the entrance to Grand Canyon's Inner Gorge. There, borrowing words used by famed explorer Alexander von Humboldt before he set out through South America's then unexplored Orinoco River in 1800, Powell wrote the most famous words ever written about the Colorado River in Grand Canyon (Powell, 2013):

> We are now ready to start on our way down the Great Unknown. Our boats . . . are chafing each other, as they are tossed by the fretful river. We have but a month's rations remaining. The flour has been resifted through the mosquito-net sieve; the spoiled bacon has been dried . . . the sugar has all melted and gone on its way down the river. We are three quarters of a mile in the depths of the earth, and the great river shrinks into insignificance, as it dashes its angry waves against the walls and cliffs, that rise to the world above; they are but puny ripples, and we are but pygmies, running up and down the sands, or lost among the boulders. We have an unknown distance yet to run; an unknown river yet to explore. What falls there are, we know not; what rocks beset the channel, we know not; what walls rise over the river, we know not. Ah, well! We may conjecture many things. The men talk as cheerfully as ever; jests are bandied about freely this morning; but to me the cheer is somber and the jests are ghastly.

Powell and most of his men made it through Grand Canyon. Their historic expedition made Powell a national hero and, more importantly, told the world of a new, seemingly indescribable place—a canyon more than a mile deep, defined by stunning views, geological wonders, and a famous river hundreds of miles long, often beset by rapids that could flip boats.

Powell's stories, combined with images provided by artists such as Thomas Moran and Louis Akin, introduced Grand Canyon to America. As people became increasingly curious about the fantastic place that Powell described, more and more people came to the canyon, with each visitor being uniquely awestruck by what he or she saw and experienced.

In 1901, the Santa Fe Railway began serving the South Rim and, in doing so, opened Grand Canyon to a growing flood of tourists that continues to this day. To accommodate these tourists, entrepreneurs began opening hotels, restaurants, tour facilities, photography studios, and souvenir shops. Two years later, President Theodore Roosevelt visited Grand Canyon and declared it "one of the great sights every American . . . should see" (Roosevelt, 1903). Tourists, heeding Roosevelt's advice, began coming to Grand Canyon in even greater numbers, and companies began featuring Grand Canyon in national advertising campaigns; for example, Coca-Cola ads told readers, "Even Bright Angel Trail leads to a soda fountain . . . [Coca-Cola has] to be good to get where it is." The canyon that had earlier been condemned and declared worthless had become a fabled, must-see destination.

As decades passed, Grand Canyon became embedded in American lifestyles and culture. For example, in 1931, then unknown actor John Wayne famously climbed out of Grand Canyon in *The Big Trail*, his Hollywood debut; in 1959, famous television star dog Lassie got her own room at Bright Angel Lodge when she visited Grand Canyon; and in 1969, television star Lucille Ball ran some of Grand Canyon's rapids in an episode of *Here's Lucy*. Grand Canyon has also inspired famous music. For example:

Roger Miller's (1936–1992) 1964 hit titled "King of the Road" recalled Miller's work in 1953 at Grand Canyon, during which time he—"a man of means, by no means"—was "pushin' broom" while working at Verkamp's Curios Store on the South Rim. Miller, who often performed in what is now the History Room of Bright Angel Lodge, earned 75¢ per hour at Verkamp's until he was fired (for spending too much time chatting with girls) and kicked out of his "trailer for rent." In 1964, Miller's "King of the Road" beat out the Beatles' "Yesterday" for the Grammy for Best Song.

In 1931, after three years of work, American composer Ferde Grofé (1892–1972) composed the jazz-styled classical *Grand Canyon Suite*. This five-movement suite (Sunrise, Painted Desert, On the Trail, Sunset, and Cloudburst) debuted that year in Chicago, and it remains the most famous orchestral composition about Grand Canyon. In 1958, Grofé's composition, which he claimed resulted from "an irresistible impulse" after a visit to Grand Canyon in 1926, accompanied the short film *Grand Canyon* by Walt Disney that won an Academy Award for Best Short Subject. (Grofé's famous sounds of clip-clopping mules were inspired by his hearing the rhythmic pile-drivers at a Los Angeles construction site.) American saxophonist Paul Winter's (b. 1939) 1985 composition *Canyon*, like Grofé's *Grand Canyon Suite*, is rooted in jazz. But whereas Grofé's music was from the perspective of a tourist, Winter's music is presented from that of a river runner on the Colorado River. Winter recorded his music during four trips through Grand Canyon.

Not surprisingly, Grand Canyon has also been associated with the unusual, notorious, and bizarre:

In the May 20, 1999, "Grand Canyon Death Jump"—broadcast live on television—motorcycle daredevil Robbie Knievel jumped over 2,500-foot-deep Quartermaster Canyon on the Hualapai Indian Reservation (31 years earlier, Robbie's father Evel had been denied a chance to jump over Grand Canyon aboard a rocket-powered motorcycle). The 228-foot jump broke one of Robbie's legs, but attracted worldwide attention to Grand Canyon. Not to be outdone, seventh-generation tightrope walker Nik Wallenda—in a live broadcast watched by 13 million people in 2013—walked untethered on a tightrope stretched across a 1,400-foot-wide, 895-foot-deep gorge of the Little Colorado River on Navajo land near Grand Canyon.

On May 12, 1992, the escape of convicted murderer Danny Ray Horning (disguised as a medical technician) from a Phoenix area prison triggered the largest manhunt in Arizona history. During some of his 54 days on the run, Horning hid out at Grand Canyon; at the height of the search for Horning, 385 police searched GCNP. Horning was caught after robbing a bank in Winslow, Arizona. As of early 2018, Horning remained on California's death row.

On April 5, 1909, the front-page headlines of Phoenix's *Arizona Gazette* announced that "Mysteries of Immense Rich Cavern Being Brought to Light" and claimed that "Remarkable Finds Indicate Ancient People Migrated from Orient" to Grand Canyon. The article claimed that a Smithsonian Institution explorer named G. E. Kinkaid, while on a solo boat trip through Grand Canyon, found a branched mile-long tunnel that had been dug by humans leading to a grand chamber filled with Egyptian treasures, mummies, golden urns, a Buddha-like statue, and a 700-foot-long dining table littered with cooking utensils. According to the article, "upwards of 50,000 people could have lived in the caverns comfortably" (Berger, 2007). Another Smithsonian explorer, S. A. Jordan, examined the cave and was convinced the inhabitants had been Egyptians, and he claimed that his findings would "stagger the wildest fancy of the fictionist" (Lago, 2009). Kinkaid said the cave was 42 miles upstream from "El Tovar Crystal Canyon." In 1962, these stories were revived, and writer David Childress claimed that GCNP and Smithsonian conspired to refuse him access to the area containing the cave. Soon thereafter, other conspirators used elaborate calculations to link the alleged Egyptian cave at Grand Canyon with cosmic secrets of the Great Pyramid of Giza and other wonders of the world. By the 1990s, there were stories linking Grand Canyon with the Egyptian cosmic matrix, Stonehenge, the Sphinx, the star Sirius, giant mummies, and UFO sightings at Area 51. Most people believe that the story was a hoax perpetuated by Joe Mulhatton (also spelled Mulhattan), a traveling salesman and prospector. Regardless, no one has found the cave.

In 1913, Alexander McAllan's eccentric *Ancient Chinese Account of the Grand Canyon* claimed that ancient Chinese texts described Grand Canyon and that Chinese explorers, not Columbus, had discovered America. Although these stories were soon forgotten, some were resurrected in the 1950s along with those describing the discovery of Atlantis, more space aliens at Area 51, vast governmental conspiracies, and a mummy called "The Thing" that, to this day, people pay to see in Dragoon, Arizona.

Many such stories continue. For example, in 2008, the SciFi Channel aired *Lost Treasure of the Grand Canyon*, a movie that revived tales about "lost treasures" in the canyon. Unlike the original stories from 1909, however, artifacts in this movie were Aztecan, not Egyptian, and explorers found Aztecs conducting human sacrifices. The next year, in Michael Cole's science fiction novel *Secrets of El Tovar Canyon*, an explorer found a gold tablet at the bottom of Grand Canyon, after which she was pursued by the military and unscrupulous collectors to Phantom Ranch, where she hiked to an Egyptian cave to learn about the dangers of global warming. From the inspiring to the absurd, Grand Canyon is omnipresent in American life.

The Ensemble of Grand Canyon

This book describes the people, places, and stories that collectively comprise the essence and experience of Grand Canyon. This experience is often far ranging and includes the language, culture, literature, art, history, and science of iconic sites such as Desert View Watchtower, Bright Angel Trail, and Phantom Ranch, and influential people such as John Wesley Powell, Clarence Dutton, and Mary Elizabeth Jane Colter. You'll learn the stories behind these people and places, and much more, in this book. These stories and landmarks are important because, among other things, they identify and honor the rich cultural history of Grand Canyon.

Grand Canyon can seem not only unknown but unknowable; its beauty and immensity impress everyone who stands at either rim and sees the canyon for the first time. Seeing Grand Canyon often leaves a lifelong impression, but these impressions vary; people see different things and have different impressions of Grand Canyon. For example:

Literary naturalist John Burroughs (1837–1921) was overwhelmed by Grand Canyon's immensity: "One's sense of the depths of the cañon is so great that it almost makes one dizzy to see the little birds fly over it, or plunge down into it. One seemed to fear that they, too, would get dizzy and fall to the bottom" (Burroughs, 1911). Burroughs ended an article about Grand Canyon by claiming "the remainder of our lives will be richer for having seen Grand Cañon" (Burroughs, 1911).

Conservationist John Muir (1838–1914), who first came to Grand Canyon in 1896, summarized Grand Canyon as "a gigantic statement for even nature to make" and that "it is a hard job to sketch it even in scrawniest outline . . . I cannot tell the hundredth part of the wonders of its features" (Muir, 1902).

Famed geologist Clarence Dutton (1841–1912) was "vexed at the inadequacy of our faculties to comprehend more . . . [I am] deeply moved by the splendor and grace of Nature's architecture" (Dutton, 1882).

Southwest advocate Charles Lummis (1859–1928) was impressed by the unexpected feeling that accompanies a visit to Grand Canyon: "I have seen people rave over it; better people struck dumb with it; even strong men who cried over it; but I have never yet seen the man or woman who *expected* it" (Lummis, 1902).

For clergyman and poet Henry Van Dyke (1852–1933), Grand Canyon's "fathomless abyss" was a symbol of his own meaninglessness as he became "a conscious grain of sand, lost in a desert of unconsciousness, thirsting for God and mocked by his own thirst" (Van Dyke, 1914). Another clergyman saw "Horror! Tragedy! Silence! Death! Chaos! There is the awful canyon in five words . . . It seemed to me as if it were the burying ground of the universe . . . There is a sense of terror that cannot be put into words." Newspaper editor William Allen White (1868–1944) felt "an awful spell that falls upon the soul like the lonesomeness of the grave" (Lago, 2015).

English writer and broadcaster John B. "J. B." Priestley (1894–1984) considered Grand Canyon "sort of a landscape Day of Judgment. It is not a show place, a beauty spot, but a revelation . . . The Colorado River made it, but you feel when you are there that God gave the Colorado River its instructions. It is all Beethoven's nine symphonies in stone and magic light." Priestley later noted, "If I were an American, I should make my remembrance of [Grand Canyon] the final test of men, art and politics. I should ask myself: Is this good enough to exist in the same country as the Canyon?" (Priestley, 1937).

Railroad executive George Reeve expressed the feelings of many visitors when he asked, "The Grand Canyon. What is it? I do not know. You do not know. Only God knows" (Santa Fe Railway, 1902).

Regardless of how writers, painters, and poets have tried, there is always more to Grand Canyon; as John Muir noted, "[O]f the many descriptions of [Grand Canyon], none seems adequate" (Muir, 1902). People see Grand Canyon in their own individual ways, but most are left with an impression elegantly stated by English novelist John Galsworthy (1867–1933): "This Grand Canyon of Arizona is the greatest sight in the world" (Morris, 1967). Dutton had similar feelings, proclaiming Grand Canyon to be "the most sublime and awe-inspiring spectacle in the world . . . [It is] a great innovation in modern ideas of scenery, and in our conceptions of the grandeur, beauty, and power of nature" (Dutton, 1882). The impressions made by Grand Canyon on Galsworthy, Dutton, and today's visitors are seldom forgotten and—combined with Grand Canyon's stunning vistas, geology, and history—produce an experience like no other. Famed Grand Canyon concessionaire Fred Harvey summarized the feelings of many visitors in 1905 when he simply noted, "No one can describe it to you" (Harvey, 1905).

Experiencing Grand Canyon begins with seeing Grand Canyon, and the essays in this book describe many of the famous places, rim-side overlooks, and unique ways that people see Grand Canyon. It is from these places—for example, the overlooks that showcase the canyon's compelling panoramas, the trails along which people have hiked for centuries, and the Colorado River that carved Grand Canyon and continues to sustain life in the canyon—that the Grand Canyon experience begins. Although these essays and stories cannot replace being at Grand Canyon, they will give you a sense of what's there, why it is important, and why it is so appealing and inspirational to visitors. It is these places and the experiences that accompany them that have changed so many people who have visited Grand Canyon.

Despite countless technological advances, Grand Canyon remains difficult to comprehend and even more difficult to describe. At the canyon's rim, as Dutton noted, "dimensions mean nothing . . . all that we are conscious of . . . is a troubled sense of immensity . . . Everything is superlative, transcending the power of the intelligence to comprehend it" (Dutton, 1882). As a result, different people see different things—including themselves—at Grand Canyon; as poet Carl Sandburg noted in *Many Hats* (1928), "For each man sees himself in the Grand Canyon—each one makes his own Canyon before he comes, each one brings and carries away

For centuries, people have been inspired by the immensity of Grand Canyon. In this photo, visitors admire the view from the South Rim's Mather Point, the most popular overlook in Grand Canyon National Park. In the distance, 10 miles away, is the North Rim. (Randy Moore)

his own Canyon" (Sandburg, 1928, 2003). The geology and colors of the gorges, trails, and buttes—which Sandburg likened to tumbled skyscrapers and wrecked battleships—are obvious and unavoidable, but the Grand Canyon experience also includes history, art, culture, literature, and other subjects. Countless writers, painters, explorers, poets, and photographers have admitted that this experience is indescribable. For example, pioneering river runner Robert Stanton noted, "It is impossible in a few pages to do justice, in the smallest degree, to the great gorge itself—'that sublimest thing on earth'—or to the perils and adventures of our journey through it. What then shall we write?" (cited in Compos, 2017). J. C. Martin, the editor of the Prescott *Journal Miner*, visited Bass Camp in 1894 and expressed his frustration this way:

> The artist stands aghast as he gazes on its cliffs and pinnacles, with a wealth of coloring the secret of whose blending is known only to the infinite . . . His pencil and brush are not trained to reproduce such perspectives. Nothing that the painter can picture can impart to one who has not seen it any idea of its depths and heights, to say nothing of its fantastic forms . . . The man of letters is appalled at its sight, as language fails him to tell of its wonders . . . One's powers of articulation are paralyzed. (Kinsey, 1998)

But each then tried to describe it. As American writer Irvin Cobb noted in 1913:

> Nearly everybody, on taking a first look at the Grand Canyon comes right out and admits its wonders are absolutely indescribable—and then proceeds to write anywhere from two thousand to fifty thousand words giving the full details . . . I do not know anybody who has yet succeeded in getting away with the job . . . When the Creator made it, He failed to make a word to cover it. (Cobb, 1913)

Today, Grand Canyon continues to lure millions of people to its rims and river. Artists come to Grand Canyon seeking beauty and inspiration; others visit Grand Canyon to see its immense panoramas, learn about Native American history and culture, and experience the thrills of running the Colorado River's famous rapids. Even others visit GCNP seeking the canyon's peace and solitude, or its challenging, beautiful trails. People have found all of these things and much more at Grand Canyon.

Whether viewing Grand Canyon from along the Colorado River a mile below the rim, from a crowded rim-side trail, or from a quiet, remote overlook such as Point Sublime, the overall experience is unique; as geologist Clarence Dutton (1882) noted:

> There are in the world valleys which are larger and a few which are deeper. There are valleys flanked by summits loftier than the palisades of the Kaibab. Still the Grand Canyon is the sublimest thing on earth. It is so not alone by virtue of its magnitude, but by virtue of the whole—its ensemble.

This book is about that ensemble.

Further Reading

Berger, Todd R. 2007. *It Happened at Grand Canyon*. Guilford, CT: TwoDot Publishers.

Burroughs, John. 1911. The Grand Canyon of the Colorado. *Century Illustrated Monthly Magazine*, January, 425–438.

Cobb, Irvin S. 1913. *Roughing It De Luxe*. New York: George H. Doran.

Compos, Jamie. 2017. Elves Chasm. Retrieved July 27, 2017, at http://elveschasm.com/grand-canyon-colorado-river-journal/day-19-book-worms/

Dutton, Clarence E. 1882. *Tertiary History of the Grand Canyon*. Washington, DC: U.S. Government Printing Office. (Reprinted 2001 by the University of Arizona Press, Tucson, Arizona, with an introduction by Wallace Stegner and a new foreword by Stephen J. Pyne.)

Harvey, Fred. 1905. *El Tovar: A New Hotel at Grand Canyon*. Chicago: Santa Fe Railway and W. H. Simpson.

Ives, Joseph C. 1861. *Report upon the Colorado River of the West, Explored in 1857 and 1858*. Washington, DC: Government Printing Office. (Republished in 2005 by University of Michigan Library.)

Kinsey, John L. 1998. *The Majesty of the Grand Canyon: 150 Years in Art*. Petaluma, CA: Pomegranate Communications.

Lago, Don. 2015. *Grand Canyon: A History of a Natural World and National Park*. Reno: University of Nevada Press.

Lago, Don. 2009. "Looks like a Mulhatton story": The origins of the Grand Canyon Egyptian cave myth. *The Ol' Pioneer*, 20 (2), 3–12.

Lago, Don. 2011. These infinities of beauty and terror: Poets and writers discover the Grand Canyon, Part 2. *The Ol' Pioneer* (3), pp. 6–16. Retrieved July 8, 2017, at http://www.grandcanyonhistory.org/Publications/TheOlPioneer/TOP_2011_3.pdf

Lummis, Charles. 1902. The greatest thing in the world. In *The Grand Canyon of Arizona*. Chicago: Santa Fe Railway.

Morris, Margaret. 1967. *My Galsworthy*. London: Peter Owen.

Muir, John. 1902. Grand Canyon of the Colorado. *Century Illustrated Monthly Magazine*, November, 107–116.

Powell, John Wesley. 2013. *The Exploration of the Colorado and Its Canyons*. New York: Simon & Brown.

Priestley, J. B. 1937. *Midnight on the Desert: Being an Excursion into Autobiography during a Winter in America, 1935–36*. New York: Harper & Brothers.

Roosevelt, Theodore. 1903. At Grand Canyon, Arizona, May 6, 1903. Retrieved June 29, 2017, at http://www.theodore-roosevelt.com/images/trenvpics/trgrandcanyonspeech.pdf

Sandburg, Carl. 1928. Many hats. *The American Mercury*, 14 (56). In Sandburg, Carl. 2003. *The Complete Poems of Carl Sandburg*. Boston: Houghton Mifflin Harcourt.

Santa Fe Railway. 1902. *The Grand Canyon of Arizona*. Chicago: Author.

Van Dyke, Henry. 1914. *The Grand Canyon, and Other Poems*. New York: Charles Scribner's Sons.

Chronology of Human History at Grand Canyon

11,000 BCE	The first peoples and, later, Archaic Indians are living at Grand Canyon and using stone spear points. We know relatively little about these people or their lifestyles.
1185 CE	Ancient Puebloan Indians are living at Grand Canyon, including at what is now Tusayan Ruins on the South Rim.
1540	Only 48 years after Christopher Columbus landed in the Americas, Hopi guides lead 25 Spanish soldiers to Grand Canyon in search of the Lost Cities of Gold. These were the first Europeans to see Grand Canyon.
1858	Joseph Ives and his crew become the first whites to reach the bottom of Grand Canyon from the rim.
1867	James White claims to be the first person to raft the Colorado River through Grand Canyon. Most people do not believe his story.
1869	John Wesley Powell leads the first successful and intentional expedition down the Colorado River through Grand Canyon. Powell, more than anyone else, made Grand Canyon a part of American lore.
1871	John Doyle Lee begins building a ferry across the Colorado River at what became Lees Ferry. Lee was the first person who catered to visitors at Grand Canyon.
1883	Two years after his first visit to Grand Canyon, prospector John Hance becomes the first permanent non-Native settler at the canyon when he moves to the South Rim.
1893	President Benjamin Harrison creates Grand Canyon Forest Reserve. This was the first federal protection for what became GCNP.
1891	Local entrepreneurs Ralph Cameron, Peter Berry, and their partners build and register what was later named the Bright Angel Trail. Today, this is Grand Canyon's most popular trail into and out of Grand Canyon. Berry's hotel at Grandview was the focus of tourism at Grand Canyon for the next decade.
1898	John G. Verkamp begins selling curios and Indian crafts from a tent just east of today's Hopi House. This was the first curio shop at Grand Canyon.
1901	Santa Fe Railway begins the first train service to Grand Canyon. This service transformed the South Rim into the bustling Grand

Canyon Village. Three months after the first train arrived, the first automobile arrived at Grand Canyon.

1904 Emery and Ellsworth Kolb build Kolb Studio at the trailhead of Bright Angel Trail on the South Rim. Hikers and mule caravans stopped at the studio to have their photos taken on their way into the canyon. Today, Kolb Studio houses a gallery, store, and exhibits.

1905 Mary Elizabeth Jane Colter—one of the few female architects in America at the time—completes Hopi House, her first of several iconic projects at Grand Canyon with the Atchison, Topeka and Santa Fe Railway and Fred Harvey Company.

1908 President Theodore Roosevelt uses the American Antiquities Act to create Grand Canyon National Monument.

1909 Wealthy industrialist Julius Stone hires Nathaniel Galloway to take him down the Colorado River through Grand Canyon simply for sport. This made Stone the first river runner to hire a guide.

1911 Santa Fe Railway builds and opens Hermit Trail, Hermit Camp, and Hermit Rim Road. Hermit Camp soon became a premier destination at Grand Canyon.

1912 The Kolb brothers complete a movie titled *The Grand Canyon Film Show* documenting their two-month-long boat trip through Grand Canyon. The Kolbs showed their movie in their studio every day for several decades.

1917 The Wylie Way Camp opens at the North Rim's Bright Angel Point. Elizabeth Wylie was the first woman to own and manage a tourist concession at Grand Canyon.

1919 President Woodrow Wilson signs Senate Bill 390, thereby creating GCNP, three years after the formation of the National Park Service (NPS). GCNP was the nation's 15th national park. In its first year, 44,173 visitors came to the park.

1921 The North Rim becomes part of Grand Canyon National Park.

1922 Phantom Ranch opens on the banks of Bright Angel Creek at the bottom of Grand Canyon. Today, Phantom Ranch—which is reachable only by mule, foot, or boat—is the only lodging facility at the bottom of GCNP.

1923 For the first time, more than 100,000 visitors come to Grand Canyon.

1925 To avoid the tolls charged at Ralph Cameron's Bright Angel Trail, NPS completes what is now called South Kaibab Trail, which stretches from Yaki Point to the Colorado River.

1928 Glenn and Bessie Hyde disappear while on their honeymoon voyage through Grand Canyon. Glenn and Bessie's bodies were never found. Bessie Hyde was the first of a growing number of women to operate a boat in Grand Canyon.

1928 Yavapai Observation Point opens on the South Rim. This was the first museum and formal interpretive structure in GCNP.

1929 The opening of Navajo Bridge at Marble Canyon speeds traffic between the North and South Rims and makes Lees Ferry obsolete for crossing the Colorado River.

1932 Mary Jane Colter's Desert View Watchtower opens at GCNP's east entrance on the South Rim. Desert View Watchtower, like all of Colter's buildings at Grand Canyon, harmonized design with the canyon's natural environment.

1932 President Herbert Hoover creates a new Grand Canyon National Monument that includes Toroweap to recognize the scientific value of western Grand Canyon.

1933 The Civilian Conservation Corps (CCC) arrives at Grand Canyon and begins nine years of work, including building phone lines, trails (North and South Kaibab), roads, and walls in Grand Canyon Village.

1935 Mary Jane Colter's Bright Angel Lodge opens on the canyon's South Rim as a cheaper alternative to the El Tovar Hotel.

1935 The completion of Hoover Dam forms the Lake Mead National Recreation Area.

1938 Botany professor Elzada Clover and Lois Jotter from the University of Michigan become the first women to successfully complete a boat trip through Grand Canyon when they join Norman Nevills's expedition down the Colorado River. This expedition, which was the first commercial trip through Grand Canyon, launched Nevills to national fame.

1954 A new entrance road from Williams, Arizona, to the South Rim of Grand Canyon opens.

1955 Bill Beer and John Daggett swim from Lees Ferry to Lake Mead. Their 26-day stunt made them celebrities, and the publicity generated by Beer and Daggett produced the largest single increase in the number of people running the Colorado River through Grand Canyon.

1956 For the first time, more than 1 million visitors come to Grand Canyon.

1956 The Colorado River Storage Project Act authorizes Glen Canyon Dam just upstream from Lees Ferry. The fight over Glen Canyon Dam was one of the first major battles of the conservation movement in the United States.

1956 Two commercial jets collide 20,000 feet above the eastern Grand Canyon over Chuar Butte, killing all 128 people aboard both planes.

This accident, which remains the worst of several disasters at Grand Canyon, led to the development of anticollision radar and the formation of the Federal Aviation Administration in 1958.

1963 Congress proposes to dam and flood part of Grand Canyon, thereby beginning a controversy that became one of the most bitter environmental debates in American history. As a result of this controversy, the Sierra Club gained a national following and political power.

1966 The completion of Glen Canyon Dam tames the Colorado River, changes the river's ecosystem, and ends the days of spring floods in GCNP.

1969 Just before leaving office, President Lyndon B. Johnson creates Marble Canyon National Monument, thereby bringing the proposed site of Marble Canyon Dam under the jurisdiction of NPS. In 1975, the Grand Canyon National Park Enlargement Act made this monument a part of GCNP and created today's boundaries of GCNP.

1969 More than 2 million people visit GCNP. By 1976, this number exceeded 3 million.

1969 Artist, photographer, and river runner Buzz Belknap produces the first waterproof, book-like guide to the Colorado River through Grand Canyon. Belknap's book, *The Grand Canyon River Guide*, became the industry standard.

1976 Emery Kolb, the last living pioneer of Grand Canyon, dies at age 95. During his career, Kolb photographed more than 3 million visitors at Grand Canyon. Kolb's 73-year residence at the South Rim exceeds that of any person at Grand Canyon Village.

1979 Grand Canyon is named a UNESCO World Heritage Site to recognize its cultural heritage, natural resources, and universal value to humanity.

1992 For the first time, more than 4 million visitors come to Grand Canyon.

1992 President George H. W. Bush signs the Grand Canyon Protection Act, which requires Glen Canyon Dam to protect GCNP. However, erosion of the canyon's beaches continued.

1995 Navajo Bridge is replaced by a wider, $15 million bridge near the same site.

2007 The Hualapai tribe on the so-called West Rim opens the Grand Canyon Skywalk, a horseshoe-shaped attraction on their reservation near River Mile 266.7 of the Colorado River. The popular Skywalk, which is 250 miles from the South Rim's Grand Canyon Village, is not part of GCNP.

2014 Almost 5 million visitors come to GCNP.

2014	In the latest of centuries-old attempts to commercialize Grand Canyon, developers seek permission to build Grand Canyon Escalade. The $1 billion development promises jobs for the Navajo but is opposed by environmentalists and many Native Americans.
2016	Conservationists begin a years-long restoration of Fred Kabotie's murals at Desert View Watchtower.
2016	GCNP attracts almost 6 million visitors.
2017	GCNP attracts more than 6 million visitors.
2018	A widely publicized and fatal crash of a tourist helicopter near Quartermaster Point reminds the public that the leading cause of death at Grand Canyon is aviation accidents.
2018	GCNP begins culling the herd of buffalo at the North Rim from about 600 to 200 individuals.
2018	GCNP's infrastructure problems continue, as broken pumps at Indian Garden force NPS to impose water restrictions on the South Rim.
2018	GCNP prepares to celebrate its upcoming 100th anniversary.

Grand Canyon at a Glance

The beauty and immensity of Grand Canyon can be stunning, inspiring, and even somewhat unsettling. When we stand at the rim, adjectives fail us as we realize that we have no standards for comparisons. This feeling is inevitably accompanied by predictable questions. For example, how big is Grand Canyon? How far is it to the other side? How wide and fast is the river at the bottom of Grand Canyon? Is the climate at the bottom of the canyon the same as along the rim? How old are the rocks in Grand Canyon? When did Grand Canyon become a national park? These and other basic questions are answered in this at-a-glance introduction to Grand Canyon.

GPS Coordinates of Primary Sites at Grand Canyon and Grand Canyon National Park

Area	Site	GPS Coordinates	
South Rim	Desert View	N36° 02.228	W111° 49.198
	Bright Angel Lodge	N36° 03.427	W112° 08.459
	Hermits Rest	N36° 03.443	W112° 12.404
North Rim	Grand Canyon Lodge	N36° 11.57	W112° 03.69
Grand Canyon West	Skywalk	N36° 0.735	W113° 48.682
Supai		N36° 22.996	W112° 69.260

National Parks and the U.S. National Park System

The first national park in the United States was Yellowstone, whose creation on March 1, 1872, began a worldwide national park movement. U.S. national parks created thereafter were managed by various governmental agencies (including the War Department) until President Woodrow Wilson created the National Park System (NPS) as part of the Department of Interior by signing the NPS Organic Act on August 25, 1916. (At the time, there were 14 national parks and 18 national monuments in the United States.) Today, 59 national parks (and many more monuments and associated areas) in NPS cover 210,000 square kilometers (2.18 percent) of the United States, an area about the same size as Kansas. In 2017—the 100th anniversary of NPS—NPS had 417 properties, a budget of $3.4 billion, and 22,914 full-time employees. Attendance in 2017 at national parks and monuments exceeded 331 million people who spent more than $18 billion at businesses in and around the parks.

On February 26, 1919—that is, 2.5 years after the formation of the NPS—President Woodrow Wilson signed "An Act to Establish the Grand Canyon National Park in the State of Arizona," thereby creating America's 15th national park. (The U.S. Geological Survey favored naming the park Powell National Park to honor John Wesley Powell. On the same day, Wilson also created LaFayette [Arcadia] National Park.) Most of Grand Canyon is owned by the federal government and managed by federal agencies. Grand Canyon National Park (GCNP) borders three Indian Reservations, two districts of a national forest, two national recreation areas, and two national monuments. GCNP, which some people wanted to name Carnegie National Park or Powell National Park, is one of Arizona's 22 national parks.

Grand Canyon–Parashant National Monument

Grand Canyon–Parashant National Monument is a remote, undeveloped area at the northern edge of Grand Canyon created by Presidential Proclamation 7265 by President Bill Clinton on January 11, 2000. The monument spans 1.05 million acres (i.e., it is larger than Rhode Island) and its elevations range from 1,230 feet (near Lake Mead) to 8,029 feet (Mount Trumbull). Grand Canyon–Parashant National Monument and Vermilion Cliffs National Monument (also created in 2000 by President Bill Clinton) are bigger than GCNP and protect parts of the North Rim watershed.

Grand Canyon–Parashant National Monument is jointly managed by NPS and the Bureau of Land Management. It has no developed campgrounds, paved roads, first-aid stations, or visitor services.

First Non-Natives to See Grand Canyon

Native Americans have lived in and around Grand Canon for thousands of years. The first recorded sighting of Grand Canyon by non-Natives occurred in 1540—that is, only 48 years after Christopher Columbus landed in the Americas—when Hopi guides led 13 Spanish soldiers searching for the mythical Seven Cities of Gold to Grand Canyon's South Rim, likely near Lipan Point. The soldiers considered the canyon an impassable obstacle to finding the treasures that they sought. After this visit, the Spanish lost interest in the canyon, and no other non-Native people entered Grand Canyon for more than 200 years.

Protection

President Abraham Lincoln made Arizona a territory in 1863. Thirty years later—that is, in 1893, just two years after Congress empowered U.S. presidents to create forest reserves (later renamed national forests) by taking unassigned land from the public domain—President Benjamin Harrison issued Presidential Proclamation 45 to create Grand Canyon Forest Reserve. This proclamation, which closed the area to homesteading, was the first federal protection of Grand Canyon. By 1897, however, other legislation had allowed grazing, mining, and lumbering to continue in the area. Administration of the Grand Canyon Forest Reserve was transferred from the General Land Office (of the Department of the Interior) to the U.S. Forest Service in 1905. Subsequent protections of Grand Canyon included making the area a game reserve in 1906 and a national monument in 1908 (the same year that Henry T. Ford sold his first cars).

Arizona became the 48th state (and last state in the contiguous United States) on February 14, 1912, and Grand Canyon became a national park in 1919. Grand Canyon became a World Heritage Site in 1979; a plaque describing this designation is outside the South Rim's Grand Canyon Visitor Center. In 1992, the canyon's protection was reaffirmed with the Grand Canyon Protection Act. Arizona is often referred to as The Copper State because it produces more copper than any other state, but its official nickname is The Grand Canyon State. The first closure of GCNP occurred in 1995 during a federal government shutdown.

Age

The age of Grand Canyon remains controversial, largely because so much of its history is missing (i.e., carried away by the Colorado River). According to recent studies, some western parts of Grand Canyon are up to 70 million years old, whereas some eastern parts may be less than 10 million years old.

Location and Size

Grand Canyon is in northwestern Arizona, near the border of Nevada and Utah. It stretches from the Paria River (just downstream from Lees Ferry) to the mouth of Grapevine Wash near Pearce Ferry.

To get to Grand Canyon, visitors cross the Colorado Plateau, a relatively flat, 130,000-square-mile bulge in the Earth's surface bordered by the Rocky Mountains, Arizona's Transition Zone, ancient volcanoes, and the stretched-apart Great Basin. The plateau, which has been uplifted nearly 2 miles over the ages, covers half of Utah and much of Colorado, New Mexico, and Arizona. This uplifting, combined with erosion and down-cutting by the Colorado River, created Grand Canyon.

Most of Grand Canyon is within GCNP. GCNP covers 4,950 square kilometers, which is 1.68 percent of Arizona. Most of this area is outside the river gorge. GCNP, which is slightly smaller than the state of Delaware, is the fourth largest national park in the United States outside of Alaska; only Yellowstone, Death Valley, and Everglades National Parks are larger. Although all of GCNP is in Arizona, the adjacent Glen Canyon National Recreation Area (at the northeastern end of Grand Canyon) is in Utah, and Lake Mead National Recreation Area (at the western end of Grand Canyon) is in Nevada.

Length of Grand Canyon

Most sources claim that Grand Canyon is 277 miles long; that's the distance you travel if you ride a raft on the Colorado River through GCNP from Lees Ferry (Mile 0, at the eastern end of GCNP) to Grand Wash Cliffs (Mile 277, at the western end of GCNP). The same two places are only 140 miles apart as the condor flies. If you walked along the canyon's rim between these two places, you would walk five times that distance. (The South Rim is 1,373 miles long, and the North Rim is 1,384 miles long.)

Width of Grand Canyon

The average width of Grand Canyon (i.e., the distance from rim to rim) is 10 miles. The narrowest point is 600 feet at Marble Canyon, and widest point (at several places) exceeds 15 miles. For comparison, the average commute of Americans going to work is 15 miles.

Volume of Grand Canyon

Grand Canyon has a volume of 4.7 trillion cubic meters. It would take the water of approximately 1.9 billion Olympic so-called long-course swimming pools to fill Grand Canyon.

Elevation

The South Rim averages 7,000 feet above sea level, and the North Rim averages 8,100 feet above sea level. The higher elevation of the North Rim produces a cooler and wetter environment than at the South Rim.

Deepest Point

Parts of Grand Canyon are more than a mile deep. For example, at Mather Point—the most visited site in GCNP—the rim is nearly a mile above the river. Over the ages, the surface of the rim has eroded a vertical mile.

Highest Point on Rim

The highest point along either rim in GCNP is the North Rim's Point Imperial, which is 8,803 feet above sea level. This popular, scenic site overlooks Mount Hayden and the Nankoweap Basin. Contrary to popular claims, this is not the highest point in GCNP. For example, the North Rim's entrance station, which is 11.7 miles from the rim, is 8,827 feet above sea level.

Lowest Point

The lowest point in Grand Canyon is at the head of Lake Mead. This area is only 1,200 feet above sea level.

Weather

Temperatures and precipitation at Grand Canyon vary according to season and elevation:

	South Rim	North Rim	Phantom Ranch
Average temperature, °F	63	56	82
January	30	26	46
April	47	41	68
July	69	61	93
October	51	46	73
Average annual rainfall, inches	15.6	25.3	8.5
Average annual snowfall, inches	60	144	0

Getting to GCNP

The South Rim of Grand Canyon is 220 miles from Phoenix, Arizona, 79 miles from Flagstaff, Arizona, and 278 miles from Las Vegas, Nevada. The North Rim of Grand Canyon is 275 miles from Las Vegas, 43 miles from Jacob Lake, Arizona, 193 miles from Flagstaff, and 130 miles from Page, Arizona. Although it is only 10 miles (as the condor flies) from the South Rim's Grand Canyon Village to the North Rim, it's a 216-mile drive.

Tourism and Visitation

During 2017, more than 6 million people visited Grand Canyon; this was the highest attendance in GCNP's history and, among U.S. national parks, was second only to Great Smoky Mountains National Park (more than 11 million visitors).

In 1904, when national parks began recording visitation, 120,000 people visited the 11 national parks. For comparison, more than 305 million people visited the NPS's more than 410 properties in 2017. Since the founding of NPS, more than 13 billion people have visited NPS properties. More than 200 million people have visited Grand Canyon since GCNP was created in 1919. In 2017, tourism at Grand Canyon brought in more than $590 million to local economies.

The South Rim, which is served by the Grand Canyon Railway and Grand Canyon National Park Airport, is visited by more than 90 percent of GCNP's tourists. The more secluded North Rim is visited by about 5 percent of Grand Canyon's tourists. On some days, as many people visit GCNP as came during the entire year that the park opened in 1919.

Only about 5 percent of visitors to GCNP venture below the rim, and fewer than 1 percent hike to the Colorado River at the bottom of Grand Canyon. The average stay at GCNP is approximately four hours.

In 1920, the first full year after GCNP became a national park, 16 percent of visitors came to Grand Canyon in cars. By 1926, more than 50 percent of the visitors came by car. When the park opened, the only paved road to or in the park was the road leading to Hermits Rest, which was built by the Atchison, Topeka and Santa Fe Railway. Since it opened in 1919, more than 200 million people have visited GCNP.

Oldest Rocks

Elves Chasm Gneiss in the Inner Gorge is 1.84 billion years old, which represents about 40 percent of Earth's history.

Youngest Rocks

The Kaibab Formation, which caps the rim along most of Grand Canyon, is 270 million years old.

Seeing Grand Canyon

Visitors to GCNP see Grand Canyon from boats, planes, helicopters, mules, and afoot.

By boat: More than 25,000 people ride rafts and boats through Grand Canyon every year. Although this seems like a large number of people, it is only 0.45 percent of Grand Canyon's total number of visitors. Most river running trips are provided by the 16 outfitters licensed by GCNP, but, in 2018, GCNP also sponsored a weighted lottery for 468 permits for noncommercial river trips. Individuals can make only one trip per year on the Colorado River in GCNP.

By air: In 2017, the FAA had capped air tours at GCNP at 93,971 flights per year; this number does not include the many flights per day that the Hualapai Tribe offers from Grand Canyon West adjacent to GCNP. Flights must stay at least 1,500 feet above the rim. Grand Canyon National Park Airport, which is just south of Tusayan, hosts more than 300,000 flight-passengers (i.e., people paying for a helicopter or airplane ride) per year, is Arizona's only state-owned airport and is the fourth-busiest commercial airport in Arizona. (The state owns the airport because Tusayan was not incorporated when the airport opened in 1967.) In 2017, the prices of air tours from the South Rim started at $149.

By mule: Tourists have ridden mules into and out of Grand Canyon since at least 1886. Today, thousands of people continue to do this every year from the North and South Rims. Grand Canyon's Mule Barn, which includes the longest continually operational blacksmith shop in the United States, is near the head of Bright Angel Trail on the South Rim.

By foot: There are 588 miles of trails in GCNP, of which 126 miles are maintained. Many of these trails connect rims to the river, and several others connect with the 235 campsites between Lees Ferry and Diamond Creek. The longest trail, the Tonto Trail, stretches more than 70 miles between Red

Canyon in the east and Garnet Canyon in the west (extensions of the Tonto Trail go even farther).

On average, the North Rim is twice as far from the Colorado River than is the South Rim. This is because the North Rim slopes to the south, causing runoff from the North Rim to flow into Grand Canyon and erode the landscape. The South Rim also slopes to the south, so runoff there doesn't flow into the canyon. As a result, erosion from the South Rim has been less extensive than erosion from the North Rim, meaning that the South Rim is closer to the river than is the North Rim.

Life Zones and Biodiversity

In 1889, ecologist C. Hart Merriam (1855–1942) showed that ecological diversity in Grand Canyon depends primarily on changes in elevation. Five of the seven so-called life zones of North America—that is, the Lower Sonoran, Upper Sonoran, Transition, Canadian, and Hudsonian life zones—lie between the Colorado River and North Rim in Grand Canyon. This means that you can see as many life zones in GCNP (more than you can see in any other national park) as you would if you traveled from Canada to north Mexico.

GCNP houses more than 1,750 known species of plants (including about 200 species of exotic plants), 355 species of birds, 89 species of mammals, 47 species of reptiles, 167 species of fungi, 64 species of mosses, 199 species of lichens, 9 species of amphibians, and 17 species of fish. Grand Canyon has the second greatest known diversity of plants in NPS (behind only New River Gorge National River in West Virginia).

School

GCNP is the only national park with a K–12 public school within its boundaries. (NPS employees living in other parks send their children on buses to schools outside the parks.) The school's mascot is The Phantom. In 2017, Grand Canyon High School graduated 18 students.

Colorado River

Grand Canyon is defined by the Colorado River, a 1,450-mile-long river that originates in the western Rocky Mountains. The river—the second longest in the United States—touches seven states and 11 national parks, drains 244,000 square miles (an area greater than Portugal and Spain combined), and provides water for more than 35 million people. In GCNP, the Colorado River flows 277 miles, from Lees Ferry to Grand Wash Cliffs; this is slightly longer than the drive from Grand Canyon Village to Las Vegas, Nevada. These 277 miles comprise 19.1 percent of the length of the Colorado River. The average width of the Colorado River in GCNP is 300 feet, with the narrowest point being 76 feet at Granite Narrows (River Mile 135). The river is widest (750 feet) at Mile 193. On average, the Colorado River is

40 feet deep in GCNP; the deepest site is at Mile 114.3, where the river is more than 85 feet deep. The Colorado River is one of the most storied rivers on Earth.

Today, the flow of the Colorado River through GCNP is determined not by nature but instead by power demands that regulate the release of water at Glen Canyon Dam. The speed of the Colorado River at particular places in GCNP varies, depending on location. The average speed of the Colorado River's smooth water is approximately 4 miles per hour; in steep rapids, the river's speed can exceed 10 miles per hour. Unimpeded, it would take about two weeks for melted snow to go from one end of the Colorado River to the other. However, today's flow of the Colorado River is slowed by several major dams (e.g., Hoover, Laguna, Imperial, Parker, and Davis Dams). There are also several diversions of water along the river's winding route.

The lowest documented flow rate of the Colorado River in GCNP was 1,000 cubic feet per second (cfs) in 1977, and the highest documented postdam flow exceeded 97,000 cfs in 1983. (Estimates from a flood in 1884 exceeded 250,000 cfs, an estimate from 1921 topped 180,000 cfs, and an estimate of an undated prehistoric flood exceeded 300,000 cfs.) In 2017, the median flow rate of the Colorado River was approximately 12,200 cfs. Contrary to popular opinion, high flow rates are not more dangerous for river runners than lower flow rates; indeed, medium flows are correlated with more drowning than are high or low flow rates.

The Colorado River is the lifeline of Grand Canyon. The most expansive views of the Colorado River are from Hermit Road (e.g., Hopi Point, Pima Point, Mohave Point) and the Desert View area (e.g., Desert View Watchtower, Lipan Point). Smaller parts of the river can be seen from other sites (e.g., Yavapai Point, Cape Royal).

The temperature of the Colorado River in GCNP averages 52°F; the water is this cold because it is released at Glen Canyon Dam from the bottom of Lake Powell. This is especially important for river runners who find themselves thrown in the river, for the cold water quickly lowers body temperature and reduces river runners' ability to hold their breath by 50 percent. In GCNP, the water of the Colorado River warms approximately 1°F every 25 miles.

Along its 1,450-mile length, the Colorado River drops an average of 9 feet per mile. (For comparison, the Mississippi River—the longest river in the United States at 2,340 miles—drops less than 1 foot per mile.) In GCNP, between Lees Ferry (elevation 3,117 feet at River Mile 0) and Lake Mead (elevation 1,221 feet at River Mile 277), the Colorado River drops 1,896 feet, which is 7.97 feet per mile; this is 25 times greater than that of the Mississippi River. (In its steepest 21 miles in GCNP, the Colorado River drops 45 feet per mile.) Almost two-thirds of the Colorado River's descent occurs in 161 recognized rapids that constitute 9 percent of the canyon's length, and 20 percent of this drop occurs in only 20 of these rapids. The number of rapids varies with the flow of the river: the lower the flow, the more rapids. High water flow diminishes or eliminates rapids.

More than 80 percent of the sediment that once entered Grand Canyon is now trapped in Lake Powell behind Glen Canyon Dam. Before this dam tamed the Colorado River in GCNP, the Colorado River carried as much as 380,000 tons of sand, mud, rocks, and dissolved minerals per day. Today, the river moves less than 50,000 tons of sand and rocks per day. This is why the water released into Grand

Canyon from Lake Powell is almost silt-free. As a result, there are fewer and smaller beaches throughout Grand Canyon.

Deaths and Injuries in Grand Canyon

More than 700 people have died in Grand Canyon. Most deaths have involved air mishaps, which have killed more people than drowning, flash floods, and heat-related problems combined. Every year, GCNP averages more than 1,000 medical emergencies, 15 deaths, and 280 rescues from the canyon.

Buildings and Development

GCNP includes four sewage treatment plants and 23 miles of trans-canyon pipeline (from Roaring Springs to the North and South Rims). There are approximately 1,210 buildings in GCNP, 58 percent of which belong to NPS. The remaining buildings are owned by concessioners such as Forever Resorts, Xanterra Parks & Resorts, and Delaware North. The park's more than 20 concessioners gross more than $200 million and pay more than $12 million in franchise and other fees per year.

Power and Water

All potable water in GCNP and at the adjoining North and South Rims (e.g., Tusayan) comes not from the Colorado River but instead from Roaring Springs, which is about 5 miles down North Kaibab Trail. Power in GCNP comes not from hydropower but instead from power plants on the rim.

Dams

There are no dams within GCNP, but dams bookend the park. At the upper end of Grand Canyon, the 710-foot-high Glen Canyon Dam (1964) forms Lake Powell; at the lower end of Grand Canyon, the 726-foot-high Hoover Dam (1936) holds back Lake Mead. These dams have had a profound impact on the Colorado River and Grand Canyon.

GCNP Budget

GCNP's budget is approximately $338 million, the fifth largest of all national parks. (For comparison, the 2017 budget for the NPS was $3.3 billion, while that for the Department of the Interior was $13.4 billion.) Despite spending $10.5 billion on maintenance from 2006 through 2015, NPS still has an $11.9 billion backlog of deferred maintenance. Of this amount, $372 million is at GCNP. The single biggest deferred project at GCNP (and NPS) is a new Trans-Canyon Pipeline and water system, which is projected to cost $163 million. GCNP is one of 24 NPS sites in Arizona.

Part I
Thematic Essays

Grand Canyon's Ancient Beginnings: The Proterozoic Eon

How long did it take for Grand Canyon to form? What forces stacked rocks in multicolored layers a full mile in height only to split them apart into a canyon more than 10 miles wide? And why did all this happen on the Colorado Plateau? These are among the questions that we will answer in this essay and the next.

The Grand Canyon includes three sets of rocks. The most ancient set is the Grand Canyon Metamorphic Suite, first laid down nearly 2 billion years ago and composed of metamorphic compounds. The second is the Grand Canyon Supergroup, one of the oldest, thickest, and rarest sedimentary sequences on Earth. The third and youngest set, the so-called postcard rocks, is renowned not only for their stunning beauty but also for cataloging the evolution of life from the beginning of the Paleozoic Era until its cataclysmic end.

In this essay, we tell the story of Grand Canyon's first two sets of rocks, from the initial crushing impact that created the Grand Canyon Metamorphic Suite, through the events that shaped the rare formations comprising the Supergroup. It is a tale of emerging life, from methane-exuding archaea to photosynthesizing bacteria that eventually produced sufficient atmospheric oxygen to support increasingly complex creatures. The stories within the rocks reveal their interconnectedness with air, water, weather, and life. As we learn how events in one Earth system influenced others, even the gaps in time between formations will gain import. In the next essay (The "Postcard Rocks" of Grand Canyon: The Paleozoic Era), we tell the story of the "postcard rocks," from the Cambrian Explosion that kicked off the Paleozoic Era to the Permian Extinction that brought it to its close.

The Proterozoic Eon and Grand Canyon's Oldest Rocks

A geologic event occurred nearly 2 billion years ago that would ultimately create Grand Canyon's bottom: two tectonic plates collided, one carrying a continent and the other, a sea surrounding an arc of volcanoes. The ocean crust, which was heavier, denser, yet thinner than that of the continent, slid underneath, grafting the volcanic islands onto the continent's edge, warping the earth's surface, mashing layers of sediment into the crust in vertical folds, and uplifting the region into a range of mountains. While this collision provided the force that initially pressed the sedimentary layers 10 to 15 miles into Earth's interior, the weight of the towering peaks prevented the strata from resurfacing.

Squeezed, folded, and driven downward by the weight of the overlying layers, the buried rocks gradually heated to the temperature of the mantle, a process that transformed them from sedimentary deposits into metamorphic compounds. The bottommost layers, buried so deeply that they melted entirely, seeped upward as liquid magma, lacing the overlying yet still submerged metamorphosing rock with salmon-colored ribbons.

Over the next hundreds of millions of years, the mountains eroded, gradually relieving the vertical pressure on the buried sediments. As the overlying weight lessened, the submerged layers floated upward, lifting the ever dwindling mountains and exposing their peaks to yet more weathering.

This cycle—that is, of the diminishing weight of the mountains allowing the basement rocks to rise and, in doing so, hastening the erosion of the surface rocks—continued until many millions of years of wind and water had worn the landscape down to sea level and exposed the long-buried metamorphosed strata. Meanwhile, molten magma rose periodically into fissures below the Earth's surface, cooling over millions of years into sashes of granite.

Levi Noble

Levi Fatzinger Noble (1882–1965) was a Yale University graduate and American geologist who produced pioneering studies of Grand Canyon. Noble first visited Grand Canyon in 1908, and returned to the canyon often because, as he noted in 1929, the canyon casts a spell from which an observer is never entirely free.

Noble, who was especially interested in Proterozoic and Paleozoic rocks in Grand Canyon, was the first person to produce a detailed geologic map (e.g., identifying the different sedimentary rocks) of a large part of GCNP. Noble, a friend of William and Ada Bass, was often assisted by John Waltenberg, a longtime employee of Bass. Noble died at age 82 on August 4, 1965, in Los Angeles, California.

Viewing Grand Canyon's Metamorphic Suite

Rising in steep walls above the river of the inner gorge, these earliest rocks, referred to by geologists as Grand Canyon's Metamorphic Suite, range from dark gray to black and sport curvy pink and white filaments that bear witness to the contortions that these buried layers endured early in their transformation. The metamorphic rock, which is made primarily of schist, gneiss, and granite, lies under almost 4,000 feet of younger rocks. Massive peach-colored granite intrusions form a stark contrast with the dark basement rocks and are visible in several places along the river, most notably near Phantom Ranch. Grand Canyon's oldest rock, Elves Chasm Gneiss (1.84 billion years old), can be found near Elves Chasm at River Mile 117.2.

On the South Rim, one can see the Metamorphic Suite at most viewpoints. From Grand Canyon Village, for example, these ancient rocks are clearly perceptible when one looks westward toward Hermits Rest. Alternatively, the South Kaibab Trail, leading to the bottom of the Canyon, cuts through the Vishnu Schist, offering an up close perspective.

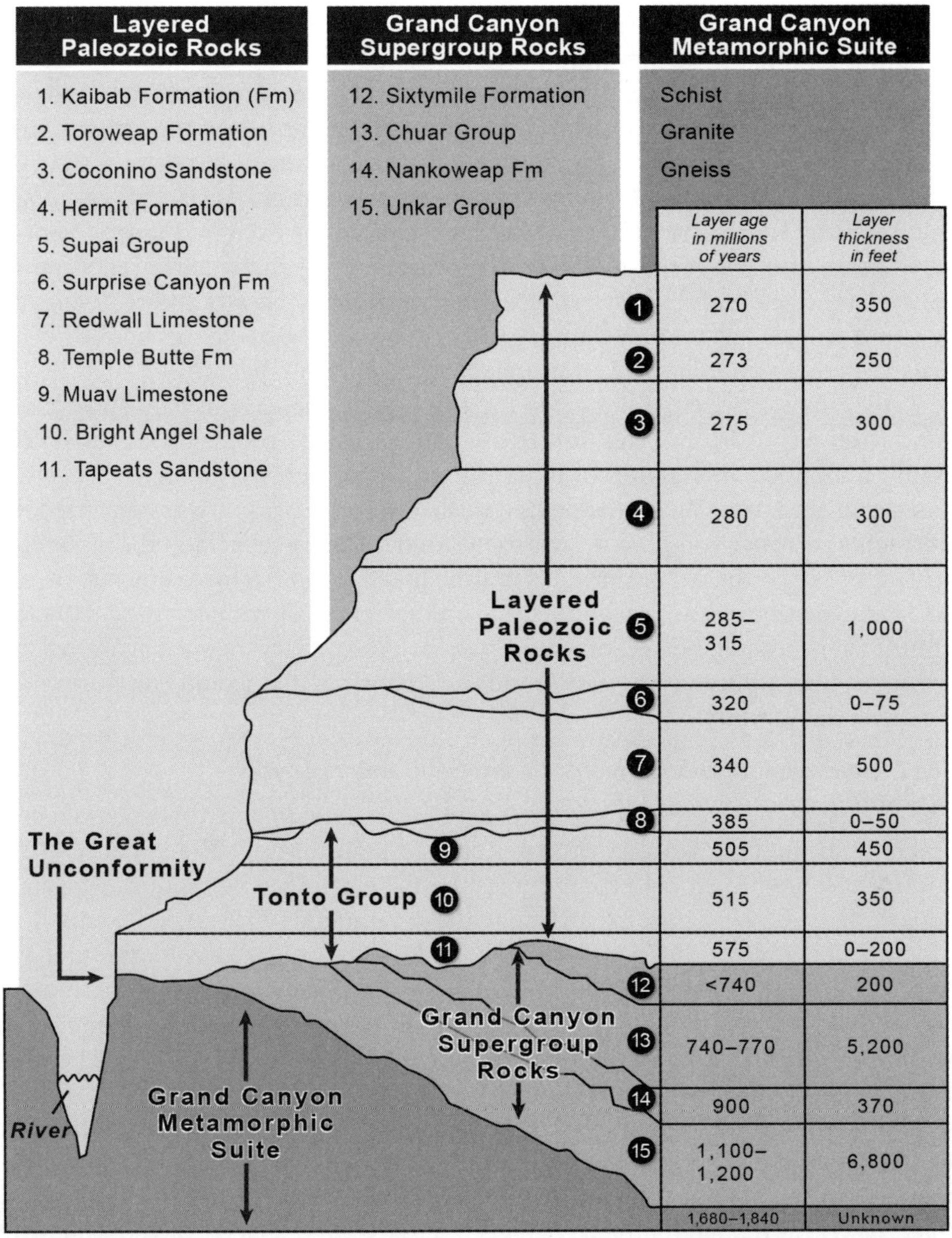

Grand Canyon includes three sequences of rocks, separated by vast expanses of time and deposited in distinctly different environments. At the bottom of the canyon lies the Grand Canyon Metamorphic Suite, which includes the canyon's oldest rocks; in the middle is the tilted Supergroup, among the rarest sedimentary strata in Grand Canyon; and on top are the famously photogenic Paleozoic layers, the horizontal strata that stretch as far as the eye can see. From the 1.84-billion-year-old rocks at the bottom to the 270-million-year-old rocks on which visitors stand at the rim, Grand Canyon displays more than 1.5 billion years of Earth's history.

First Life

Earth's emerging continents were desolate islands when the Grand Canyon Metamorphic Suite formed. Indeed, the atmosphere was so oxygen-poor that life on dry land had no chance. Deep within the superheated vents below the bottom of the ocean, however, primitive organisms called archaea thrived. And on the surface of the sea floor, an entirely different form of life, the cyanobacteria, had also begun to proliferate. Unlike the archaea, who flourished in the extreme environments of early Earth and produced methane as a waste gas, cyanobacteria relied on the energy of the sun for fuel, converting carbon dioxide into sugars and releasing oxygen into the air, all while arranging themselves in sheets at the bottom of the ocean. As these cyanobacteria multiplied, atmospheric oxygen rose.

Rather than supporting life, however, the sudden increase in atmospheric oxygen nearly wiped it out. The immediate cause was the profusion of sulfate that resulted from the interaction of oxygen with sulfur-rich pyrite, better known as fool's gold, and iron. When the sulfate washed into the seas, it was converted into the highly corrosive compound, hydrogen sulfide. The so-called Sulfidic Ocean that ensued poisoned the simple forms of marine life that had begun to emerge, thereby stalling evolution while also impeding the accretion of rock-forming minerals. Hundreds of millions of years would pass before conditions once again reached a balance that supported new life—and the sediments that would compose Grand Canyon's Supergroup.

The Supergroup: Grand Canyon's Oldest Sedimentary Layers

The Supergroup was the product of yet another tectonic pileup, this one even more severe than the first, since both plates carried land masses. The violence of this continent-to-continent collision whiplashed the earth in slow motion, alternately squeezing and pulling on its surface while leaving a down-warped basin at the juncture of the plates that provided a favorable environment for Supergroup layers to accumulate over the next 450 million years. The Grand Canyon Supergroup is one of the oldest and thickest sedimentary rock records on Earth, cataloguing the joining and tearing apart of the supercontinent of Rodinia. It is also one of the rarest, occurring in isolated sections in eastern Grand Canyon.

Unlike the unbroken horizontal sedimentary layers of the Paleozoic Era that would begin to form above them hundreds of millions of years later, the rocks of the Supergroup are not merely uncommon and discontinuous; they are curiously tilted as well, as though remnants had been selectively protected from erosion, while leaving the bulk to wear away. In fact, this is exactly what happened. The remainders of the Supergroup that have survived are those that fell into faults when Rodinia eventually tore apart, giving the strata their uniquely sloped and tilted orientation.

The Supergroup includes four primary layers: the Unkar Group, the Nankoweap Formation, the Chuar Group, and the Sixtymile Formation.

The Unkar Group

The five formations of the Unkar Group tell of the dramatic geologic events of the times, from gigantic waves flooding the land to Rodinia's rifting and uplifting, which, in turn, spawned massive volcanic eruptions extending across the newly formed supercontinent. The first and oldest layer, the Bass Limestone, reveals the encroachment of the Bass Sea; the Hakatai Shale, Shinumo Quartzite, and Dox Formations atop the Bass Limestone tell of the ocean's vacillating retreat, and the Cardenas Lava recalls the epic eruptions that the stretched earth ignited across Rodinia's vast interior. The time between the oldest Unkar layer and the Nankoweap that follows it offers a geologic testament to the creation of Rodinia.

Bass Limestone. When the Bass Sea encroached from the west, it left behind a layer of gray limestone interspersed with shale. Much of the limestone later became dolomite. The Bass Limestone is the first Grand Canyon formation in which a fossil record has survived. Indeed, the stromatolites that shape it reveal the relatively primitive nature of life prior to the Cambrian Explosion that marked the beginning of the Paleozoic Era to follow. Stromatolites are limestone structures that formed from the remains of massive colonies of cyanobacteria—the primitive bacteria that used the energy of the sun for fuel. Because they arranged themselves in sheets to best intercept light, their fossilized remains give the rock a laminated appearance—although some structures are round, domed or even mushroom-shaped. More varied fossils would not reveal themselves until life had evolved beyond single-celled organisms during the times of the Chuar Group, six layers into the Supergroup's future.

Hakatai Shale. When the Bass Sea retreated, it left mud flats that created a layer of shale intermixed with mudstone and sandstone. Salt-crystal casts, cross-beds, and cracks reflect an exceedingly arid climate, while the shale's rare orange-red color tells of the oxidation of the iron in the rock. This is evidence not only that Earth was exceedingly dry, but also of the progressive increase in atmospheric oxygen.

Shinumo Quartzite. The Shinumo Quartzite has a checkered past, having begun as flat sedimentary layers before eroding into islands surrounded by the Cambrian Sea that the Dox Formation (see the next paragraph) later covered. With its distinctive colors ranging from rich purple to chocolate brown, combined with its hardness and resistance to erosion, the Shinumo Quartzite, originally beach sand, is one of the most prominent layers of the Unkar Group.

Dox Formation. The Dox Formation consists of hills and valleys of interbedded sandstone, mudstone, and shale of orange, red, and greenish brown, replete with ripple marks and mud cracks, indicating a shoreline environment. Stromatolites and the fossilized remains of primordial algae occur here and there.

Cardenas Lava. Dark basaltic cliffs interbedded with sandstone and laced with dikes and sills comprise the Cardenas Lava, the youngest (i.e., uppermost) formation in the Unkar Group. This layer formed when the rifting of Rodinia summoned magma from below the surface, setting off a series of volcanic eruptions across the supercontinent. A gap in the rock record follows the outpourings of the Cardenas Lava—a breach in time that extends across the planet. A global glaciation had intervened, spawned by the massive volcanic eruptions of which the Cardenas Lava was but one small part. This all began with enormous quantities of volcanic ash and dust spewing from the belt of exploding volcanoes that engirdled the globe, dimming the sun and blocking its warming rays. The abrupt glut of ejected basalt drew the warming layer of carbon dioxide out of the atmosphere while increasing the concentration of atmospheric oxygen.

Earth cooled dramatically as a consequence. Ice crept from the poles to the tropics, boomeranging the sun's warming energies back into space and lowering temperatures further still. Once the frost had closed in at the equator, Earth was transformed into a shimmering snowball. With the sunlight continuing to bounce off the glowing ice, the planet grew colder still and the ice became ever thicker. Trapped in a positive feedback loop from which there was no apparent escape, life's future looked bleak.

Earth was spared an icy death, however, thanks to the warming effects of carbon dioxide released by a resurgence of volcanic eruptions. Life was granted a reprieve—but only briefly. With the volcanoes continuing to vent, yet another positive feedback loop arose, this time generating so much heat that all of Earth's ice and snow melted, even at the poles, draining the planet's water supply into the ocean basins. Huge tides, pulled by a much nearer moon, flooded the Grand Canyon region, despite its inland position on the rifting landmass. It was the water from these deluges that finally provided conditions fostering the formation of the Nankoweap Formation.

The Nankoweap Formation

The Nankoweap Formation consists of cross-bedded, mud-cracked, coarse-grained cliffs of sandstone in purple, white, and brown. It formed in the newly created sea, which had washed over the basaltic layers of the Cardenas Lava.

The Chuar Group

The Chuar Group includes the Galeros and the Kwagunt Formations, following a gap in the rock record during which sediments either failed to form or did not stick as Rodinia ripped apart. The fossiliferous formations of the Chuar Group reflect the initial presence of the sea, followed by its gradual retreat.

The Galeros Formation. Alternating beds of limestone, shale, and olive-green sandstone comprise the Galeros Formation. It contains four formal members, with the lower (i.e., older) layers of dolomite and limestone reflecting marine conditions and the upper members, of shale and sandstone, representing terrestrial environments. Ranging from green to purple, the Galeros Formation is one of the most colorful layers of the Supergroup.

The Kwagunt Formation. The Kwagunt Formation is a mixture of dark red to purplish shale, limestone, mud, and sandstone that forms a series of ledges and slopes. Up close, one can see the ancient evidence of raindrops, ripples, mud cracks, and stromatolites.

Fossils of the Chuar Group. A rare collection of microfossils documenting the life forms that preceded the proliferation of multicellular organisms during the Cambrian Period—still hundreds of millions of years in the future—have been found in the layers of the Chuar group. Acritarchs, which are round fossils that were once bacterial colonies, range from microscopic to 0.06 inch in diameter.

Three varieties of microfossils shaped like vases have also been found; these may be the remains of colonies of cyanobacteria. Of special interest are several unusual filamentous fossils that may reflect colonies of unspecified single-celled organisms.

The Sixtymile Formation

As Rodinia gradually pulled apart at its margins, stretching the crust from northwest to southeast, segments of the Supergroup elongated and slid into faults. Giant landslides ensued, with rubble piling up and hardening over geologic time into breccia, a sedimentary layer containing large chunks of other types of rock. This breccia, embedded in tan sandstone and shale, forms the Sixtymile Formation, the topmost layer of the nine-tiered Supergroup. The blocks of sedimentary rock comprising the breccia consist primarily of chunks of the Chuar Group that had not yet hardened prior to breaking apart. This suggests that the fault moved relatively soon after the rock layers had formed.

Viewing the Supergroup Strata

One can recognize the layers of the Supergroup by their unique colors and curious slope relative to the flat straight lines of the Paleozoic layers above them. From

Along Grand Canyon's Great Unconformity (marked here by a ruler), more than a billion years worth of eroded rock—that is, more than 12,000 feet of rock spanning more than one-fourth of Earth's history—is missing. At this eye-level view of the Great Unconformity in Blacktail Canyon at River Mile 120.6, the layered 550-million-year-old Tapeats Sandstone sits atop the 1.6-billion-year-old Vishnu Schist. The ruler is 4 inches long. (Randy Moore)

the South Rim, one can best see the Supergroup from easterly outlooks, including Desert View, Moran Point, and Lipan Point on Grand Canyon's South Rim.

The End of the Proterozoic: The Great Unconformity

Missing periods of time, called unconformities, occur in most rock sequences due to the continuously changing geologic forces that shape and erode the surface of our planet—and Grand Canyon is a textbook case. In fact, 14 major unconformities have been confirmed in Grand Canyon, and more will likely be discovered in the future. When the Bass Limestone of the Supergroup formed atop the metamorphic basement rocks, 450 million years came between them; when Grand Canyon's famous postcard rocks (from the Paleozoic Era see The "Postcard Rocks" of Grand Canyon: The Paleozoic Era) formed atop the Supergroup, hundreds of millions of years again went unaccounted for.

Yet, wherever the Supergroup strata are absent—which is in most of Grand Canyon—the interval between the Metamorphic Suite and the Paleozoic layers resting above them is so vast that the famed Grand Canyon explorer John Wesley Powell dubbed it The Great Unconformity. Scientists have since marveled that this 1.2-billion-year gap in the rock record, spanning more than a quarter of Earth's history, dwarfs the amount of time to which the existing layers testify.

During this time—between the creation of the crystalline rocks and the first Paleozoic formation—continents had migrated around the globe, joined together into a single supercontinent, and broken apart. Earth had frozen into a giant snowball and then overheated, simmering away its ice and leaving behind a parched planet, only to inch its way back to a temperate equilibrium. The moon had receded in its orbit, shrinking the ocean tides from the size of mega-tsunamis to waves that were merely large. The gradually increasing oxygen in the atmosphere had turned the sky from pink to blue. And, perhaps most dramatically, life had evolved from single-celled organisms into multicellular creatures of astonishing complexity and variety.

Further Reading

Abbott, Lon, and Cook, Terri. 2008. *Hiking the Grand Canyon's Geology*. Seattle: Mountaineers Books.

Blakey, Ron, and Ranney, Wayne. 2008. *Ancient Landscapes of the Colorado Plateau*. Grand Canyon, AZ: Grand Canyon Association.

Price, L. Greer. 1999. *Grand Canyon Geology*. Grand Canyon, AZ: Grand Canyon Association.

Thayer, Dave. 2009. *An Introduction to Grand Canyon Fossils*. Grand Canyon, AZ: Grand Canyon Association.

The "Postcard Rocks" of Grand Canyon: The Paleozoic Era

This essay gives an account of Grand Canyon's most visible and famous layers, the so-called Postcard Rocks. Conspicuous for miles from all points along both North and South Rims, these are the iconic strata that have lured photographers, artists, and scientists—not to mention millions of visitors—since before GCNP opened in 1919. Indeed, these rocks are so renowned that even people who have never seen Grand Canyon instantly recognize the view from postcards. These strata are much more than a mere wonder to behold, however. The 14 formations that comprise them, neatly stacked in horizontal layers, catalogue the evolution of complex life from the beginning of the Paleozoic Era until its catastrophic close. The stories within the rocks reveal their interconnectedness with air, water, weather, and life. Their tales tell us why, for example, the formations have different colors and shapes. The topmost layer, the one on which visitors now stand, even forecasts the Permian extinction, the greatest mass death that Earth has ever known.

If the story were to end here, however, there would be no Grand Canyon, for the layers would remain buried under a surface of soil. The carving of the Canyon required a separate force. That force was erosion, a Rubik's Cube of uplift, water, gravity, and time. We give that final part of the story an essay of its own, entitled "How the Colorado River Carved Grand Canyon."

The Paleozoic Era dawned on an Earth that was very different from the one in which the Grand Canyon Metamorphic Suite and the Grand Canyon Supergroup had formed. Oxygen levels had continued to rise, thanks both to the carbon dioxide and water emitted by the ongoing volcanic eruptions and to the photosynthesizing cyanobacteria covering the seafloor in mats, soaking up the rays of the sun shining through the pristine waters. The supercontinent of Rodinia had broken apart completely, providing an ever increasing distance between its constituent land masses that would ultimately become the Pacific Ocean but that in the early Paleozoic was what geologists refer to as a passive plate margin, a sheltered marine basin that was sheer pay-dirt for sediments seeking to secure a place for themselves in history. This basin, situated between two continents, supplied an ideal environment for rock layers to gather—and to stick—for the first time since the Supergroup hundreds of millions of years before. The 14 sedimentary strata overlying the Precambrian rocks span virtually the entire Paleozoic Era, nearly 300 million years.

The First Formations of the Paleozoic Era: Tonto Group

The Tonto Group includes the three bottommost strata of the postcard rocks: Tapeats Sandstone, Bright Angel Shale, and Muav Limestone. These three layers provide an example of what geologists refer to as a transgressive sequence, with a succession of sedimentary layers corresponding to the water's depth at the time of their formation. This series, from the dry sand that produced the sandstone of the Tapeats to the wet mud that created the Bright Angel Shale and on to the calcite that formed the Muav Limestone, grows increasingly thick to the west, the location of the Tonto Sea.

The Tonto was a shallow sea dotted with ancient weather-resistant crystalline rocks rising in islands, along with a few isolated blocks of Supergroup layers, stretched and tilted in low-lying faults that sheltered them from the erosion that had washed away the remaining sediments. Although the shores of the Tonto would remain barren of life for another 100 million years or so, its ocean floor hosted the evolution of numerous multicellular creatures of staggering variety and complexity. The rocks of the Tonto Group contain fossils of these organisms, including traces of soft-bodied sea animals, such as sponges, jellyfish, and mysterious crawling creatures that left burrows and trails. Trilobites, the first known creatures with eyes, abound in the Tonto Group. Even an imprint of *Anomalocaris*, a predator of the trilobite, may have been discovered, although this has yet to be confirmed.

The oldest layer of the Tonto Group, the Tapeats, appears as brown cliffs and ledges of sandstone, limestone, and shale. It's most dramatic aspect is the unconformity it marks, its horizontal sedimentary layers resting upon the craggy black metamorphic rocks that were laid down more than a billion years earlier. The Tapeats Formation is visible from most viewpoints along the rim. It is easily identifiable as a brown cliff that caps the steep vertical walls of the metamorphic basement rocks descending into the canyon. Along the Bright Angel Trail, the Tapeats first appears a few hundred yards below Indian Garden.

The Bright Angel Shale, the second formation of the Tonto Group, was laid down just offshore. It forms long, gentle slopes, its unmistakable greenish tinge testifying to the iron and other compounds reduced by waste products excreted by a multitude of creatures. Visible from all points on the South Rim, the Bright Angel Shale is easily identified both by its color and by its gradual gradients. Hikers walk through Bright Angel Shale near the base of Jacobs Ladder on the Bright Angel Trail.

The Muav Limestone, the third formation of the Tonto Group, accrued in deeper waters than the Bright Angel Shale, its layers composed of the remains of hard shells of some of the earliest creatures from the Paleozoic's first period, the Cambrian. It appears as gray cliffs that rise above the slopes of the Bright Angel Shale. It is best viewed along Bright Angel Trail, at the beginning of the switchbacks, just below Three Mile Rest House.

Earth's First Mass Extinction

An unconformity lasting throughout the Ordovician, Silurian, and the first part of the Devonian Periods follows the Tonto Group. This break in the rock record spans up to 150 million years in some stretches of Grand Canyon. These years represent a series of "firsts" in the history of the new and vulnerable biosphere—the first "first" of which was the first of Earth's five major extinctions.

Before the mass death, Earth was a watery wonderland. Coral reefs surrounded the Proto–North American continent of Laurentia. Fish—jawless but armored, with bony shielded heads, plate-like tails, and slits in the place of mouths—evolved in the sea while the first plants, tiny and leafless, invaded the land. But then came times of austerity, when the climate fluctuated wildly, only to grow steadily colder. The continent of Gondwanaland had migrated to the South Pole, becoming icebound; and three land masses—Laurentia, Baltica, and Avalonia—were converging. This shrank the ocean between them and, in the process, created vast mountain ranges. Glaciers spread, icing out the deep-water creatures while leaving the coral reefs high and dry. By the time that the global temperatures had finally warmed to their previous levels, the species that had initially depended on warmth had disappeared, while those that had evolved adaptations to the cold had lost their tolerance for heat. Hence, they too succumbed. By the Ordovician's end, 85 percent of Earth's flora and fauna had perished.

In the Silurian period that followed, new life forms would come to fill the vacated niches. Fish once again ruled the seas, and plants began to line freshwater marshes. Leafless, rootless, and lacking the protective waxy coating that would evolve in the future, these plants trapped the energy of sunlight by photosynthesizing along their stems. The first animal to climb out of the water may have been a scorpion-like arthropod. Because its lungs had not adapted to breathing air, it probably merely visited the shores for brief periods of time, long enough to molt before returning to the sea. The first known creatures to make their homes on land were most likely scavenger millipedes, followed by predatory centipedes and an extinct variety of arachnid.

If the Silurian Period bore witness to several firsts, the Devonian saw these first forms prosper, diversifying within the oceans and colonizing the continental interiors. Plants spread inland, developing roots and leaves. Wood evolved, allowing trees to become tall and to create the first shade the planet had ever known. The first four-limbed creatures—called tetrapods—evolved during the Devonian, sharing features with both fish and land animals. It is in this context, then, that the fourth formation, the Temple Butte, accumulated following the deposition of the Cambrian Period's Muav Limestone 150 million years earlier.

Temple Butte Formation

The Temple Butte formed when a shallow sea transgressed the region yet again, submerging the land to the west before dwindling into estuaries in the arid mudflats of the east. It is difficult to spot, despite its distinctive purplish hues and knobby texture, not only because of its inaccessibility and sparse, uneven distribution but also because of its similarity in composition both to the Muav Limestone below and the Redwall Limestone above. The thickest and most continuous Temple Butte outcroppings are visible at Grand Wash Cliffs in western Grand Canyon and even further west in the Basin and Range. Toward the east, one can best see the Temple Butte at the level of the Colorado River. Look for wedges slotted between the Muav below it and rocks of the Redwall above.

Earth's Second Mass Extinction

A 40- to 50-million-year unconformity follows the Temple Butte, a gap in time that corresponds to what is now known as the Devonian Extinction. It was a long, drawn-out, and complex disaster, beginning not with the failure of the biosphere but with its wild success. It was kicked off by the sudden appearance, for the first time in Earth's history, of dense, towering forests that sent oxygen levels soaring—and global temperatures tumbling. Glaciers and icebergs formed, robbing water from the seas, sucking the continents dry and killing off a swath of Devonian life. Trees up to 100 feet tall grew roots deep into the soil, introducing a new source of weathering that slowly turned Earth from a greenhouse into an icehouse. While life on land initially thrived, with insects, for example, growing huge, the oceans gradually choked, thanks to the carbon dioxide that the profusion of foliage had drawn out of the atmosphere. Meanwhile, the depletion of oxygen at the bottom of the sea suffocated numerous species living near the ocean's floor.

As with the Ordovician extinction that preceded it, however, the Devonian's loss was the next period's gain. Novel forms of life appeared, as revealed to us in the rocks of the Carboniferous Period to come. It was during this period that Grand Canyon's most prominent formation accrued—the Redwall Limestone—followed by its rarest, the Surprise Canyon Formation. The Carboniferous marked a peaceful time in Earth's geologic history, with a mild and stable climate that favored a wealth of life in the Redwall Sea—and the limestone layers that formed from their hard shells long after their soft parts had perished.

Redwall Limestone

As temperatures warmed and oceans transgressed the continent yet again, marine sediments commingled to produce Grand Canyon's Redwall Limestone. Redwall Limestone consists primarily of strong, interlocking crystals that make the rock resistant to erosion. With its steep stark cliffs, stained red by the sediments of the Supai above and blending with the Temple Butte and Muav formations below, the bold vertical lines of the Redwall Limestone stand out from most viewpoints on both North and South Rims. Despite its visibility from all viewpoints on the South Rim, the Redwall Limestone is largely inaccessible, due to its thick, steep, cliff-like vertical drop-offs. Fossil viewers longing for an up-close inspection of its layers will find the best access along the river, such as at Redwall Cavern.

Surprise Canyon Formation

From its curious way of plugging the sinkholes in the Redwall Limestone below to the manifold fossils within its layers, the Surprise Canyon Formation has surprised scientists ever since George Billingsley of the U.S. Geological Survey described this formation in the 1970s. Indeed, due to its infrequent and irregular outcroppings, the Surprise Canyon Formation is extremely difficult to spot from either the North or South Rim. Although the best views are from remote and rugged routes to the far west, one can also see Surprise Canyon sediments along the Hance and Grandview Trails.

Unconformity

Yet another unconformity spanning millions of years divides the limestone-rich Redwall and Surprise Canyon Formations from the sedimentary medley to come—a gap in time that prevails across the globe. The planet's land masses had been converging for some time, sliding ever closer, squeezing the seafloor beneath plate margins and raising the already epic Appalachians into towering giants. It was the movement of the magma on which the tectonic plates floated, then, that supplied the ongoing force that turned seafloor into continental crust, setting in motion a cycle of weathering, carbon dioxide depletion, oxygen elevation, and temperature drops that would culminate in the worldwide suppression of sediment by ice.

The upcoming amalgamation of the continents into one supercontinent, Pangaea, would add a new dimension to the interplay of water, weather, rock, and life. It was an experiment that would end not in cold but in warmth—and in the greatest extinction event our planet has ever known. In the meantime, however, Grand Canyon's four-tiered Supai Group tells the story behind this prehistoric merger, layer by layer. The first three formations accrued primarily during the latter half of the Carboniferous Period. An unconformity delays the fourth and final deposit, pushing it into Permian times.

The Supai Group

The Supai Group includes four formations: the Watahomigi, the Manakacha, the Wescogame, and the Esplanade. All four feature an assortment of deposits from saltwater limestone to terrestrial shale, siltstone, and sandstone. Due to the rapidly cycling climate, these were times of marine encroachments and retreats, when seas repeatedly flooded the region, only to recede and then return. During the tidal interstices, rivers drained gravel and mud from freshly formed mountain tops through floodplains and coastal dunes, leaving behind thin, alternating bands of sandstone cliffs and mudstone slopes. One can easily see the Supai Group from most vantage points on both the South and the North Rims. The Bright Angel Trail provides an especially up close and clear view.

The Permian Period

With Pangaea largely assembled by Permian times, the vast interior of this latest supercontinent became extremely dry. The swamps of the Carboniferous had dehydrated into fields of sand dunes lining seaside plains, the desert air oxidizing shales into bold reds as mountains to the east sloughed sediment into broad, slow-moving rivers.

Amphibians, needing water to reproduce, declined, while amniotes, with the ability to reproduce on land, prospered. Synapsids, an early version of amniotes that included reptiles, became the largest creatures of the Permian. From the synapsids, therapsids would evolve, their legs pulled under their bodies, the better to run with. Often described as mixing the features of reptiles and mammals, some therapsids may even have been warm-blooded and furry. Herbivores first appeared, with palates that allowed them to breathe and eat at the same time.

Hermit Formation

Despite a period of erosion that separates it from the Esplanade Sandstone, the slopes of the Hermit Formation blend seamlessly with the cliffs and ledges of the Supai layers. Derived from mud that sluggish streams carried down from the mountains rising to the east, the Hermit Formation features sandstone, siltstone, and shale, interspersed with ripple marks. Its most striking aspect, however, is neither its shape nor its composition, nor even the abundant fossils in its layers, but rather its colors. Bright reds from the reduction of iron by oxygen set off by soft yellow streaks of sandstone bleeding from above into the cracks on its surface distinguish this formation from all others. One can see the Hermit Formation from most points on both rims. Look for brick-red slopes scattered with trees, rubble, and brush and pierced with mud cracks filled by sandstone from the yellow cliffs of the Coconino Sandstone above.

The Coconino Sandstone

The Coconino Sandstone, consisting of fine grains of sand, sculpted by wind, compacted by weight and cemented by moisture from rare fogs and dews, was laid down over a span of 5 to 10 million years. In contrast to the larger, coarser grains of the Tapeats Sandstone, deposited hundreds of millions of years earlier by the Tonto Sea, the small rounded shape of the Coconino quartz and its luminescent color reveal its windblown origins. So, too, do the ripple marks and dramatic crisscross patterns on its surface, telling of huge dunes that alternately overlapped and were divided by deep ditches filled with water supporting multitudes of life.

Appearing as a thick bright band of yellow, the Coconino Sandstone is impossible to overlook. Its brilliant hues create a visual break between the distinctly red tones of the Hermit below and the pinkish browns of the Toroweap above. The Coconino Sandstone is visible for miles from virtually all points on the North and South Rims.

The Toroweap Formation

Although arid conditions prevailed and a period of erosion ensued, temperatures rose, polar ice melted, and the Kaibab Sea transgressed from the west. The persistently dry air evaporated the moisture from the sea, leaving gypsum and other compounds behind. The second layer from the top, the comparatively nondescript Toroweap Formation, is most easily identified in relation to the creamy Coconino below and the Kaibab above. It is easily spotted from all points on the North and South Rims.

Grand Canyon's Coconino Sandstone was deposited about 275 million years ago. This block of Coconino Sandstone is along the Trail of Time; directly above the block on the distant peninsula is the thick layer of Coconino Sandstone, which is visible from all viewpoints on both rims of Grand Canyon. (Randy Moore)

Kaibab Formation

For the next several million years, the Kaibab Sea continued to invade Pangaea, intruding ever further into its interior. When this limestone was deposited, the Grand Canyon region was the floor of a sea. It was in this environment that sediments collected for the next 5 million years—deposits that would form Grand Canyon's uppermost and last surviving layer. Three to four hundred feet of cream-colored cliffs make up the Kaibab Formation. The last layer to form yet the first we see, the bold limestone stripe that dominates the view from both rims provides a fitting end to nearly 300 million years of Paleozoic history.

Grand Canyon's End and Earth's Most Massive Extinction

Though the rock record vanishes for millions of years across most of the globe at this point in Earth's history, the scant layers that do remain tell the same sad story. It is a tale told twice before, but this time with a yet more menacing finish. In retrospect, it all seems so foreseeable. If the land masses hadn't sutured themselves into a giant patchwork quilt to form one sole supercontinent, the ratio of coastline to mainland would not have plunged—and Pangaea's interior might never have become so hot and so dry. If space and water had separated the chafing tectonic plates, life might have found a way to hide from the acid rain, solar dimming, global cooling, and eclipsed photosynthesis caused by the gas, dust, and ash let loose by the most destructive volcanic eruptions in Earth's history—the infamous Siberian Traps. And if the subterranean magma hadn't transited through all those carbon-rich layers on its journey to the surface, the carbon dioxide that the coal-rich sediments released might never have boomeranged the climate from hot to cold and back to hot again.

Then, the discharge from the melting coal might not have frayed Earth's ozone layer beyond repair. And if the ozone layer—that translucent pale blue shield way out in our planet's stratosphere—hadn't become so tattered and torn, the sun's searing rays might not have burned the distressed and light-starved plants so severely that they had no hope of recovery. Then, the heat wave that followed might not have lingered long enough to reach all the way down to the seemingly stable, deeply buried sheets of permafrost hidden within the margins of the ocean floor, where cold-loving microbes, impervious to earlier extinctions, busily converted carbon dioxide into methane. And if the warming effects of the methane hadn't escaped from the melting ice, raising temperatures to the hottest that the biosphere had ever known, a few more creatures might have had a fighting chance.

As it was, more than 90 percent of all species passed away. Those that did survive eked out lonely livings on a dry, barren, and weirdly quiet planet strewn with dead and rotting matter, a motley crowd of fungi and other decomposers as their only constant companions. Amphibians stuck it out but shrank in size from yards to inches. Synapsids, shedding their reptilian pasts in favor of warm-blooded lifestyles, went forth and multiplied but did not prosper, living in the shadow of some disturbing diapsids that reveled in the bleakness of the recently vacated planet. In fact, they would soon come to dominate it. Among these amazing creatures were those that would evolve into the dinosaurs.

Because the sediments of today's Grand Canyon end abruptly with the Kaibab Formation, no traces remain of the astonishing Mesozoic life forms whose fossils fill the younger rocks of neighboring regions across the Colorado Plateau. A different sort of relic would emerge from that era, beginning humbly, then growing with time, its origins bearing witness to the Mesozoic's cataclysmic end. It is the Colorado River and, along with it, its mile-deep legacy: Grand Canyon.

Further Reading

Beus, Stanley S., and Morales, Michael. 2003. *Grand Canyon Geology*, Second edition. New York: University Press.

Davidson, Gregg, et al. 2016. *The Grand Canyon: Monument to an Ancient Earth*. Grand Rapids, MI: Kregel Publications.

Sadler, Christa. 2005, 2006. *Life in Stone: Fossils of the Colorado Plateau*. Grand Canyon, AZ: Grand Canyon Association.

How the Colorado River Carved Grand Canyon

From the first people to peer over the rim thousands of years ago to its millions of annual visitors today, Grand Canyon has never ceased to perplex us, even as it inspires our wonderment and awe. Its vastness imparts an otherworldly quality that leaves us longing for answers to the overwhelming question: What fantastic force could have created a canyon that is ten miles wide and a full mile deep, while leaving its layered walls unbroken? The multitude of creation myths that have sought to explain Grand Canyon's origins attests to the human desire to understand how this magnificent site, one of the Seven Natural Wonders of the World, came into being.

Scientists have puzzled over this mystery ever since geologist John Strong Newberry accompanied the first U.S. Army expedition up the Colorado River in 1857–1858. Led by Joseph Christmas Ives, these men traveled by boat, then on foot, from the Gulf of California to Diamond Creek near River Mile 225 in western Grand Canyon. Newberry concluded that it was water—and only water—that carved Grand Canyon through a process he called Lake Overflow. So began a parade of competing hypotheses, from John Wesley Powell's theory of Antecedence to Alan Strahler's concept of Superimposition and on to the proposal of Stream Piracy conceived by one of the most original thinkers of them all, the 20th-century's late, great Edwin McKee.

Despite their differences, these scholars agreed that only a river could carve a canyon of such immense proportions. Yet the barrier that bedeviled them in their quest for clues resided in the river's very nature, for as its currents cut the canyon by eroding the rocks, so too did they wash away all evidence of the river's origins. Or so they thought. When the 21st century brought game-changing techniques for investigating the cooling history of a mineral called apatite, the ancient past revealed its secrets, exposing the hidden beginnings of both river and canyon.

Thanks to the latest research by Karl Karlstrom (2013) and many others, we now know that the river had forerunners that carved paleo-canyons of vastly different ages. Although the Colorado River would be the final common pathway, the driving energy that united these primordial gorges into a single canyon extending for 277 miles, the river did not always travel its present route, nor did it move in its current direction. And it did not act alone. Its waters emerged in response to a series of events that together molded the geography that gave the river its power. In fact, the following seven conditions would have to be met for the Colorado River to flow, to grow—and to cut Grand Canyon:

The Colorado River has been carving Grand Canyon for millions of years. This upstream view was taken near River Mile 33.3 from inside Redwall Cavern, a giant river-carved cave that is a popular rest-stop for river runners. (Randy Moore)

Uplift. First, the region would have to be higher than the level of the sea. Tremendous and continuous tectonic movement was needed to uplift this flat, low-lying landscape many thousands of feet without disturbing the sediments—a circumstance so rare that, when it did occur, it would be without prehistoric parallel.

Waterways. Even the highest of plateaus cannot spontaneously generate a canyon. The region required a waterway, since only a rapid flow could excavate and remove such a vast amount of sediment.

Weather Patterns. Rivers need water, and rivers that carve canyons need a lot of water. Yet epic dryness had plagued the area for millions of years. Where did the moisture come from? Movements beneath Earth's crust would have to alter the overlying topography so profoundly that glaciers on rising peaks would spill their ice melt onto the Colorado Plateau.

Elevated Headwaters. Even a large river draining an uplifted plateau could not have powered downward through so many thousands of feet of rock, the bottommost layers of which had been compacting for nearly 2 billion years. These waters needed a motor to keep their currents in a state of perpetual unidirectional motion. That motor was gravity. Hence, the fourth requirement was a steep downward gradient at the river's origin.

Postplateau Down-Warp. While a rise at the headwaters would kick-start a strong current, its muscle would inevitably weaken as it traversed the vast and flat landscape. To keep its waters rough and swift, the river needed an energy boost in the form of a geographical down-warp that followed the plateau's many miles of level terrain.

Outlet into the Sea. At this point in the story, the river would have consisted of a series of creeks and brooks draining hills, then pooling in valleys, each with its own current moving in the direction governed by its local topography. For these isolated rivulets and ponds to become one continuous river extending 1,450 miles, the lower segments would have to erode headward, flouting gravity by cutting uphill, to "capture" the streams directly above them. Otherwise, these separate canals would remain nothing but a series of ephemeral waterways lying end to end, yet segregated alternately by up-warps and lake basins. The only force powerful enough to defy gravity by pushing water upslope was a dramatic downslope—in the form of a precipitous fall at the river's mouth. The future river, in other words, needed an outlet into the sea.

The Hidden Seventh Factor. After uplift, a waterway, a wetter climate, elevated headwaters, a postplateau down-warp, and a final entry into the ocean, the Colorado River would reach the age of youthful maturity. Fully integrated, its raging rapids would unite Grand Canyon's disparate segments in 3 to 5 million years, a mere blink of Earth's eye in comparison to the nearly 2 billion years it took for the layers within the canyon to aggregate.

Yet a nagging question remains unanswered: Why did so many geologic events, each alone improbable, converge in the same place? The scientific answer—simple, self-evident, yet unfathomable from the vantage point of our human life spans—inspires a reverence all its own. It clarifies how this seemingly miraculous occurrence is not merely probable but ultimately inevitable. The final explanatory factor in the story of the river—and the canyon it carved—is deep time.

Between the time that Grand Canyon laid down its last layer and the emergence of the river, an entire geologic era—the Mesozoic—would pass and another, our era, would be well under way. The supercontinent of Pangaea, having grown ever drier and risen ever higher, would begin to break apart during the Mesozoic's first Period, the Triassic. Spurred on by a huge mantle plume forming ominously beneath its center, spewing great quantities of lava through an interconnected chain of volcanoes, Pangaea would gradually split across its middle. Seawater would slowly seep through the cracks, pushing the landmasses ever farther apart and ultimately producing the Atlantic Ocean.

Two more mass extinctions would take place, Earth's fourth and fifth. The fourth, at the end of the Triassic, would wipe out many of the fantastically varied life forms that had filled the niches that the Permian extinction laid bare. The dinosaurs, still small and meek, yet having somehow dodged the mass death that wiped out their worst enemies, the therapsids, would finally have their day. Diversifying, growing huge, and ruling Earth during both the Jurassic and Cretaceous Periods to come, they would disappear forever when a mountain-sized rock from outer space blasted into the Yucatan. This cataclysmic impact, creating tsunamis the likes of which had not been seen in over an eon, igniting vast explosions, and even vaporizing bedrock, would launch Earth into its fifth mass extinction, bringing the Mesozoic to an unforetold end.

As life in our Era, the Cenozoic, bounced back after the devastation that claimed the lives of the terrestrial dinosaurs and indeed all land animals weighing more than about 50 pounds, the continents would continue to drift apart until attaining the approximate configuration we see today. The geographic isolation that the continental separations engendered would stimulate an unprecedented diversity of new life, with each land mass evolving its own unique flora and fauna.

As the Colorado River was finishing the job of carving Grand Canyon, Lucy—the famous Australopith—would saunter upright across the African Savannah. Little more than a million years after that, our own genus, *Homo*, would come into existence. It was only a matter of time before modern humans would arrive at Grand Canyon, peer into its depths, and seek answers to the questions of what, how, and, above all, why?

In the following sections, we reply to these inquiries by describing the latest scientific evidence behind the birth, growth, and developmental milestones of the Colorado River. Beginning where Grand Canyon's layers end, we tell the story of how life—and rock—reemerged after the Permian Extinction. A broad sweep of geologic highlights provides the context for the accumulation of the dazzling rock formations that make up the Painted Desert we see today, as well as the forces that fostered the conditions that molded the river and its canyon. In a separate chapter, we examine the Colorado River as it is now, its flow strangled by dams, its volume drained by human consumption, and its distressed native fish gradually losing their struggle for survival.

The Mesozoic Era

As the mass death of the Permian Extinction spread across the planet, the formation of new rock halted. Thirty million years would pass before Earth would return to conditions that could support new life—and new sediment. The bloated supercontinent of Pangaea, dry and barren throughout much of the Permian, slowly rifted apart during the Mesozoic, releasing the hot gases that had puffed up its crust, thereby allowing the differentiating landmasses to sink to the level of the sea. Oceans transgressed the newly formed North American continent, first from the west and then from the east, depositing their sediments across the future Colorado Plateau throughout the Triassic, Jurassic, and much of the Cretaceous.

Triassic Period. Although the future Colorado Plateau was flooded for long stretches throughout the Triassic, the seas retreated during cooler spells—and when they did, an array of vegetation sprang forth along rivers and streams, providing food and shelter for a riotous assortment of land life alongside their shores. As the End-Triassic Extinction drew near, however, times got tough. Rivers and streams became choked with silica from volcanoes erupting to the south and west. Rather than sustaining the flora and fauna that had prospered along their banks, these waters now poisoned them. Enormous conifers were killed and mineralized, leaving behind the Petrified Forest as well as the radiant sediments of the Painted Desert, the manganese, iron, and other metals oxidizing and infusing the clay with brilliant shades of purples, oranges, pinks, and reds.

Jurassic Period. While we may never know how the early dinosaurs survived the End-Triassic Extinction when many larger, seemingly hardier creatures succumbed, we do know why they thrived: Many of their worst enemies had perished. And when the air cleared, Earth's just-right conditions allowed them to blossom. It was during this period that many dinosaur species on the Colorado Plateau—and across the globe—became huge.

Cretaceous Period. During the hot and humid Cretaceous Period, when Earth's oceans were advancing, shallow inland seas periodically flooded the Colorado Plateau, only to recede and then to return. The inhabitants of the Colorado Plateau continued to live their lives much as they had before, oblivious to the movements underfoot that would soon rumple Earth's surface to the southwest and northeast, raising mountains, shifting weather patterns, and altering the conditions to which they had so seamlessly adapted. It was in this context, then, that the region experienced the first pulses of uplift.

Uplift. With land masses drifting incessantly across the globe throughout the Mesozoic, it was only a matter of time before the western edge of North America would collide with the plate carrying the Pacific Ocean. And when it did, the relentless thrusting of the ocean crust beneath that of the continent set off a series of events, known as the Laramide Orogeny, which would lead to the Colorado Plateau's remarkable rise. As the region's final sea, the Mancos, was retreating to the east, the tectonic plates to the west struck head-on.

The oddly shallow angle at the site of the crash caused the ocean floor to slide under that of the continent many hundreds of miles farther inward than usual, penetrating its very interior. As the inserted slab of ocean floor sank, melting into the mantle, its buried tip pointed upward, forcing the ground above to fracture and rise, creating the Rocky Mountains to the northeast and the Mogollon Highlands near the site of the impact along the southwest coastline. Amid this upheaval, the ground that was sandwiched between the two mountain ranges rose smoothly and gently, its layers remaining placid, peaceful, and whole. So began the first inklings of the Colorado Plateau.

The First Paleo-Canyon. As the Mogollon Highlands to the southwest continued to rise, glaciers accumulated on their peaks. Streams formed that drained these snow packs, flowing in a northeasterly direction upon the Colorado Plateau. Their path blocked by the rising Rockies, the runoff pooled in gullies, creating ponds connected by streams that filled with sediment during dry spells and expanded into rivers when the climate grew wetter. By 70 million years ago—that is, less than 5 million years before the impact that brought the Mesozoic Era to its shocking end—these waters had begun to cut the first known paleo-canyon on the Colorado Plateau. Located at the Hurricane Fault in western Grand Canyon, this includes the area that John Strong Newberry saw in 1857–1858.

The Cenozoic Era

Earth's Fifth Mass Extinction. Soon after the Colorado Plateau began to elevate, an extraterrestrial object, 6 miles in diameter, struck the Yucatan. Life on the Colorado Plateau, thriving, prospering, and multiplying mere moments before impact, soon died off. Hit hardest were the largest, especially those who made their homes on land, due to the destruction of the food chain that had sustained them. Invertebrates residing in shallow water did not fare much better, succumbing to extreme and sudden temperature changes.

Climate Patterns and the Second Paleo-Canyon. As the Cenozoic Era progressed from its first period, the Paleogene, to its second, the Neogene, the rising Mogollon Highlands to the southwest grew increasingly snowy. When spring thaws loosened the grip of the glaciers that clung to their walls, ice melt cascaded down upon the Colorado Plateau. High, dry, and with little ground cover to protect them, the exposed sediments crumbled. By 25 million years ago, the runoff had begun to carve a second paleo-canyon, this time in eastern Grand Canyon. Recent evidence suggests that the watershed coursing through the region was far larger than previously believed, draining Nevada, Idaho, Montana, and Canada before emptying into the Atlantic Ocean. Its flow would soon ebb, however, when renewed unrest beneath Earth's surface pulled apart the great mountains to the southwest that supplied both its water and its power.

Although the tectonic plates on the western margin had been butting up against one another for quite some time, the direction of their movements gradually shifted. Instead of colliding as they had in the past, they began to rub, chafe, scrape, and eventually pull apart. In the process, they stretched, torqued, and thinned the continental crust. Rather than raising mountains as with the Laramide Orogeny that preceded it, this event, known as the Mid-Cenozoic Orogeny, would soon topple them.

Thus, as the Colorado Plateau continued to rise, the Mogollon Highlands began to collapse, and as these formidable mountains fell, temperatures rose. The snow and ice that hitherto blanketed their peaks evaporated. Groundwater soon followed. The northeast-flowing drainage, once massive, became what geologists call confused, meandering in its course, shrinking into isolated ponds, and at times disappearing entirely. The soil became so parched that a dune field called the Chuska Sand Sea emerged. Covering and killing the plants that would have provided moisture, this massive sand blanket deepened and spread until it buried the entire region.

Elevated Headwaters: The River Reverses Its Current. Even as the Mogollon Highlands proceeded to shrink, the Rocky Mountains continued to soar. Snow melt poured from their rising slopes along the path of least resistance, eventually traversing the Colorado Plateau and mingling with the preexisting northeast-flowing waterway. The older river, once rapid and raging when the Mogollon headwaters were high, had become sluggish and slow since the Highlands had caved. With the new southwest-flowing runoff from far up in the Rockies now the dominant force, the currents reversed. For the first time in the region's prehistory, a drainage system flowed across the Colorado Plateau from northeast to the southwest.

Postplateau Down-Warp: The River Begins to Integrate. Though the youthful river started strong, rushing, as it did, from high up in the mountains, it finished weak. Its power spent as it traversed a landscape that was as flat as it was dry, its once powerful rapids soon dwindled to a trickle, stopping and starting, then leaking into a chain of ephemeral ponds before petering out entirely. At this stage of its development, the existing river was nothing but an intermittent waterway, punctuated by lake basins, interrupted by dry ground and crisscrossed by intersecting streams. Even the gravitational force provided by its lofty headwaters was not enough to integrate its disparate segments into one continuous flow.

This changed when ongoing tectonic forces once again altered the landscape. The rifting of the continental crust that had previously collapsed the Mogollon Highlands had continued unabated. As the plates pulled ever farther apart, their overlying sediments became stretched so thin that they began to dip, eventually forming a regional depression that is now known as the Basin and Range Province. With the ground at a new low, the isolated drainage segments eroded headward, capturing streams above them or flowing into lakes that, due to their increasing tilt, had begun to spill into one another. Gradually, the various waterways—flowing downward from the Rockies across the now gently sloping Colorado Plateau and onto the Basin and Range—began to integrate.

Outlet to the Sea: The San Andreas Fault. Although the newly formed down-warp provided the gravitational impetus to stimulate the integration of the segmented waterways, the power of the emerging river continued to be limited by its landlocked state. Barring a decisive drop into the ocean, its currents lacked the strength to sculpt a giant canyon. It would take a dramatic event to empower the river to carve its legacy into the sediments of the high desert.

This event occurred 12 million years ago when an offshore section of Earth's crust fractured and then broke away entirely. In fact, the ocean crust moved so violently that it ripped off a chunk of the continent. With its sedimentary layers shredded beyond recognition, the mangled and dangling landmass sank into the sea, leaving only a sliver, Baja California, above water. This event marked the birth of the San Andreas Fault.

The Gulf of California had opened—and with it, a pathway down into the ocean. Within 6 million years, a fully integrated waterway would be flowing from its headwaters in the Rockies across the Colorado Plateau and into the Gulf of California. With its incredible elevation drop of nearly 2 vertical miles from head to mouth, the newly formed Colorado River would carve Grand Canyon in record time: an unheard-of 3 to 5 million years.

Further Reading

Belknap, Buzz, and Evans, Loie Belknap. 2017. *Grand Canyon River Guide.* Evergreen, CO: Westwater Books.

Hamblin, W. Kenneth, and Rigby, J. Keith. 1996. *Guidebook to the Colorado River, Part 1: Lee's Ferry to Phantom Ranch in Grand Canyon National Park.* Provo, UT: Brigham Young University.

Hamblin, W. Kenneth, and Rigby, J. Keith. 1969[1996]. *Guidebook to the Colorado River Part 2: Phantom Ranch in Grand Canyon National Park to Lake Mead, Arizona–Nevada*. Provo, UT: Brigham Young University.

Karlstrom, Karl E., et al. December 12, 2013. Formation of the Grand Canyon 5 to 6 million years ago through integration of older pelaeocanyons. *Nature Geoscience*. Accessed January 30, 2014, at http://geosci.uchicago.edu/~kite/doc/Karlstrom_et_al_2014.pdf.

Ranney, Wayne. 2012. *Carving Grand Canyon*. Grand Canyon, AZ: Grand Canyon Association.

Grand Canyon's First People

For visitors, Grand Canyon readily offers the most amazing views imaginable. The stunning panorama at Mather Point, for example, is a scant quarter mile from the main parking lot, and the scenery only gets better as one strolls along the rim-side promenade. From archaeologists, however, it withholds its secrets. This is no mere coincidence, for the same elements that created this enormous gorge—water, gravity, wind, and time—are precisely those that have deprived scientists of its artifacts. Most relics that might have revealed the lives of Grand Canyon's first people have been worn down, blown away, or buried under layers of sediment. Yet despite the hurdles that Grand Canyon's erosional geography poses, scholars have unearthed hints of human occupancy extending all the way back to the closing days of the Pleistocene, at least 13,000 years ago—and likely long before.

Also known as the Ice Age, these were the worst of times yet in some ways also the best. On the one hand, Earth's creatures faced a colder, more unforgiving world than the one we now know, with bitter cold, permafrost, and deep snow as the norm—especially at Grand Canyon. Yet it also bore witness to the first appearance of unique beings who spoke, painted, traveled, and traded; who killed animals more than 30 times their size with simple spear points chipped from stone; and who traversed an entire hemisphere to reach Grand Canyon. Who were these mysterious individuals? Were they like people today . . . or were they different? Perhaps most intriguingly of all, what compelled them to choose a landscape of exposed cliffs, high winds, and severe weather as their home?

When humans reached North America toward the end of the Pleistocene, they were at the height of their powers. Tall, strong, tough, and smart, they were ready to vanquish all manner of trouble. And trouble, in this frigid, new world of bursting ice sheets, flash floods, deadly avalanches, and slippery mountain ridges, was never far away. But even as they sought to keep their families safe from the harsh elements of the Ice Age, Grand Canyon's first people were threatened by a fate more frightening still: getting eaten by one of the giant carnivores known collectively as the North American megafauna. *Smilodon*, an oversized saber-toothed cat, with weirdly hinged jaws that opened into an uncanny gape, would gladly have sunk its fangs into the bodies of unsuspecting newcomers. Then there were the fast and ferocious short-faced bears—larger than grizzlies and swifter than horses—from whose paws one's only salvation was to run faster than the next person in the group. Packs of wolves cruised the countryside in search of vulnerable prey; humans, having only just arrived on the continent, would have made ideal victims. And never far away, as a reminder of their eventual fate, were scavenging vultures and long-legged hawks who, though not above carrying off small children, generally preferred to watch and wait, hovering overhead while other creatures did the dirty work of killing the wretched quarry.

But even as they evaded the clutches of the ferocious predators, humans, too, had to solve the daily quandary of finding food. While they learned by trial and error to track the novel game in this strange new world, they were also familiarizing themselves with edible plants, where to find them, and how to pound or cook them to render them digestible.

Grand Canyon's First Inhabitants

Following the megafauna, living in extended family groups and moving from rim to river in accordance with the seasons, the first people of Grand Canyon traveled light. Animal hides, bone tools, and wooden implements were their possessions—in addition, of course, to their weapons, which they crafted by pounding stone on stone. Yet despite their limited belongings and mobile lives, each group probably had its own culture of sorts and may even have traded with neighbors and passersby.

The large, exquisitely formed spear points that archaeologists have found in GCNP, called Clovis Points, tell us that the first people probably hunted the grandest of the Pleistocene megafauna. Their blades were large enough—and sharp enough—to have brought down mastodons, those magnificent mammals reminiscent of elephants, except that they were much larger, with 16-foot-long tusks that they used to dig up twigs and spruce needles from under piles of hardened snow. In GCNP, archaeologists have found the remains of a giant sloth, a colossal creature of the Pleistocene weighing 4,000 pounds. These herbivores had plenty of company: oversized beavers, pig-like peccaries, and woolly mammoths also inhabited the region. Dinner menus also frequently featured smaller game, such as rabbit and fox.

The World Grows Warmer

Just as Grand Canyon's first arrivals had learned to flourish in the thrilling, though perilous conditions of the Pleistocene, their environment began to change. Earth's tilt was increasing and would soon reach its greatest angle. Global temperatures would rise; desert habitats would inch northward; and the massive ice sheets that covered much of the continent would melt, releasing enormous quantities of water that created rivers and lakes in regions that had formerly been dry while disrupting the habitats of the creatures who had become adapted to the cold. The American Southwest warmed so dramatically that subarctic plant life now persisted only on the tops of higher mountains. Although some species adjusted to the cumulative and interactional changes in their life zones, others succumbed. Among the casualties were the North American giants, namely the megafauna—a loss that would forever alter the lives of the individuals who had made Grand Canyon their home.

There could be no doubt about it; the Holocene had arrived. It was an epoch that would see the greatest change in the shortest amount of time generated by the only species that had ever sought to set itself apart from the rest of the natural world: us.

The Archaic

Whereas artifacts from the Pleistocene are exceedingly rare, countless relics revealing the lives and times of the first people of the Holocene are strewn throughout GCNP. Indeed, these individuals left so many traces that scientists have given this span of time, namely the beginning of the Holocene, a name of its own: the Archaic. Unlike their Ice Age ancestors, this new generation had no need of blazing trails from rim to river; they simply traveled the paths of their predecessors. Rather than finding food, water, and lodging from scratch, they had learned from those who preceded them how to live in caves, camps, and rock shelters.

The Early Archaic Period

By early Archaic times, beginning about 9,000 years ago, the inhabitants of Grand Canyon could no longer be considered its first inhabitants. Indeed, their ancestors had been living in Grand Canyon for thousands of years, and, like us, they probably considered themselves quite modern. If anything, they may have longed for the good old days, when gargantuan game feasted on the foliage of lush forests and slurped ice-cold water from swollen streams. They may even have thought that there was nothing much new under the sun.

When it came to killing prey, however, the methods—and the mindsets—of Archaic hunters were changing. If the problem of the Pleistocene was to avoid being killed by the enormous beasts they hunted, the challenge of the Holocene was to cope with periodic food shortages caused by the highly changeable weather. Not only had the megafauna disappeared; the bison, deer, and bighorn sheep would periodically vanish during dry spells as well, leaving the Archaic people no option but to expand their diets to include rodents, reptiles, and insects. Attesting to this change are the numerous artifacts that archaeologists have found in GCNP. While the weaponry of Pleistocene hunters consisted of large spear points that were crafted by banging them with rocks, the armaments left by Archaic people were considerably smaller. These pressure-flaked, carefully crafted darts were affixed to lightweight shafts called atlatls that hunters would hurl from a distance.

Smaller game, combined with streamlined weaponry, transformed hunting from a dangerous activity requiring intense teamwork to a pursuit better accomplished alone. Rather than relying upon one another to bring down the kill—and to share the feast—Holocene hunters found themselves competing with one another to feed their families. Complicating matters further, new groups of people were flocking into Grand Canyon, most likely to avoid a drought that would build upon itself until eventually culminating in a full-on crisis.

But even as they fought for scarce resources, the people of the Early Archaic shared cultural innovations. It was near this time that they began to dig up roots and tubers and then to grind and mash them with stone utensils called metates. Though they continued to migrate seasonally from rim to river, their lives slowly became more settled, less mobile—and distinctly more territorial.

The Middle Archaic Period

Beginning about 6,500 years ago, the climate went from dry to downright arid. Drought descended upon the entire southwestern portion of the continent, parching the soil, eroding plant cover, releasing sand that, in turn, clogged rivers and streams—and setting in motion the dreaded dominoes of death as dehydration ascended the food chain. So quickly did this occur that it would have been felt over the span of one lifetime. There was no waiting it out either. The desiccation endured for 2,000 years.

If the early Archaic was characterized by new arrivals to Grand Canyon, the Middle Archaic saw a mass exodus as its inhabitants fled in search of sanctuary from thirst and hunger. Many died, but some found the oasis they were seeking along the tributaries of Glen Canyon. When others fled upstream from Lees Ferry, the population of Grand Canyon fell to a low not seen since the Pleistocene.

The Late Archaic Period

When the drought eased about 4,500 years ago, life bounced back, and people arrived from all directions. Dramatic changes had occurred in the interim, not only in hunting methods (gypsum points had replaced the pressure-flaked stone spearheads) but also, astonishingly, in symbolic forms of communication. Drawings on rock faces, known as pictographs, such as Shamans Gallery on the North Rim, as well as those in Stantons Cave, located 120 feet above the Colorado River at River Mile 32.0, conveyed meanings about which we can only wonder. No mere decorative art or child's play, these images often conveyed deeper concerns of a spiritual nature. Similarly, scientists have found figurines in remote caves in Grand Canyon. Constructed from twigs that had been split and folded to resemble deer or bighorn sheep, complete with horns and antlers and pierced by tiny twigs, these miniature sculptures were likely totems, offered to a Great Creator to confer success on a hunting expedition.

With the return of a more stable and sustainable climate, the standard of living at Grand Canyon during the Late Archaic was likely higher than elsewhere on the Colorado Plateau, where extreme dryness continued to dominate. People could once again hunt and gather a great diversity of food sources across the multiple life zones that spanned the vertical mile from the river to the rims. Bighorn sheep, deer, bear, bobcats, mountain lions, rock squirrels, mice, packrats, woodrats, eagles, hawks, waterfowl, chuckwalla, and small lizards were plentiful. Abundant pinyon nuts, herbs, wild fruit, edible cactus, greens, mesquite beans, agave (mescal), acorns, and walnuts would have provided the ingredients for delicious appetizers served with the main course. Medicinal plants and mineral pigments were also available in numerous side canyons.

No longer limited by the immediate exigencies of survival, Late Archaic people traded with passersby; created jewelry from serpentine; collected salt that had precipitated from the Tapeats Sandstone, peacefully resting upon the ancient metamorphic rocks since the days of the Cambrian; and even kept dogs, perhaps as companions, though more likely as hunting partners. But life was about to change yet again in ways that would have been unimaginable without the gift of hindsight. Farming, well under way for thousands of years in Central America and Mexico, had already spread to other regions of the Colorado Plateau. It was only a matter of time before it reached Grand Canyon's fertile soil. And once it did, life would never be the same.

Farming Cultures Settle in Grand Canyon

By this point in our story, about 3,000 years ago, 10,000 years and more had come and gone since Grand Canyon's first family stood upon one rim and gazed in wonderment at the other. The world had changed in the interim, in a manner that foreshadowed the struggles that we continue to face to this day. Humans had proliferated and prospered to such an extent that, in populous regions of the planet, they had created cities, built pyramids, calculated the distance to the moon, and composed epic sagas of war, adventure, and longing. But of all the achievements and discoveries that had sprung from the human mind, the one with the most lasting and far-reaching effects was the innovation of farming. Indeed, its ramifications were so life-altering that its arrival has been dubbed the Neolithic Revolution.

The farmers who were pouring into Grand Canyon brought much more than squash, corn, and beans. Arriving from all directions, they spoke a multitude of languages, enjoyed rich and vibrant traditions, worshipped in their own ways, and cherished their unique legends of creation. Each group claimed an area of Grand Canyon as its own—and then defended it. Gone were the days when one's home moved freely with the ripening fruits and the grazing game. Now a demarcated territory defined one's living space.

Possessions accrued, including baskets, sandals, and pottery, with designs that distinguished one culture from another. Weaponry had also changed, from the age-old atlatl of Archaic times to the iconic bow and arrow. An armament that, in the hands of a highly skilled warrior, could propel an arrow at 250 miles per hour, it would prove to be more accurate—and more deadly—than the legendary Colt Walker, employed by the U.S. Army to kill Indians in their mania for Manifest Destiny many centuries in the future.

The Basketmakers, so named for their tightly woven multipurpose baskets, would gradually develop into the Puebloan culture, which, in turn, branched into seven distinct subcultures that settled at Grand Canyon on their own separate segments of the river and the rims, of which one was the Zuni Tribe. Divided by multiple languages and customs, yet united by maize, they alternately traded and raided, aided and destroyed, and loved and hated one another for reasons that we now know little of but that we may imagine are not far different from our own.

Pit Houses and Kivas

Beginning about 1,500 years ago, the Puebloans constructed pit houses, which were erected below ground and into which they entered through a hole in the roof. Cool in summer and cozy in winter, the pit houses were the ultimate ecofriendly architecture. Over the centuries, larger circular buildings called "kivas" appeared, for rituals, ceremonies, and other gatherings. Storage rooms, bins, and fire pits were added by later generations, suggesting that village life initially blossomed.

Thousands of Puebloan ruins litter Grand Canyon. One such site can be found west of Kanab Canyon on the North Rim. Another, Bright Angel Pueblo, is at the bottom of Grand Canyon near Phantom Ranch. The Anasazi Bridge is located at River Mile 43.3, just upstream from President Harding Rapid. Constructed from driftwood, it spans a gap on the face of a cliff, affording the Puebloans a shortcut—albeit an exposed one—to the rim.

The most well preserved ruin—and the most accessible—is Tusayan Pueblo. Not only was it occupied for the briefest span of time, it was also among the last to be built. Constructed in 1185 on the South Rim, it originally featured a main building that was two stories high and divided into 15 rooms, housing perhaps 20 to 30 people at a time, as well as a plaza, fire pit, storage area, and kiva. Though far from any known source of water, it offered a glorious view of the San Francisco Peaks, which, according to the Hopis, housed spirits called *Katsinas* (or *Kachinas*) that brought blessings and, above all, rain. It also lay closer to the place of their mythical emergence, the *Sipapuni*, within the heart of Grand Canyon near the confluence of the Colorado and Little Colorado Rivers at River Mile 61.7. Perhaps to the ancient Puebloans, proximity to their spiritual origins was worth the extra effort of hauling water.

As luck would have it, however, the climate had changed yet again, becoming so dry that farming had become untenable. By 1225, virtually all of the many Puebloan farming communities had fled—except the people of Tusayan Pueblo. Then, by the year 1230, they too departed.

Although the Puebloans exited Grand Canyon, their culture never died—and neither did they. Rather, they traveled south, joining the Hopis, who accommodated the newcomers, absorbing the many and varied Puebloan groups into their society. Indeed, the Hopis had a genius for welcoming and learning from other cultures. Perhaps this is why they rightfully consider Grand Canyon their ancestral home.

Even as the Puebloans were staking out land, building pit houses, and tilling the soil, various groups of hunter-gatherers were also streaming into Grand Canyon. As with the farmers, they, too, arrived with different languages, customs, and traditions, claiming territories that they defended against their neighbors.

The Cerbat/Pai, a large confederation consisting of 11 separate clans, arrived in the 800s. They lived along the South Rim, moving from one water source to the next as they hunted deer and bighorn sheep while gathering mesquite, prickly pear, agave, and other plants. Living in makeshift shelters and roasting their meat over small pits, they used grinding slabs and handheld tools reminiscent of the people of Archaic times. Below the rims, their lifestyle was a bit more settled, as evidenced by larger campsites, more elaborate roasting pits, pots that they probably traded for deer hides or dried meat with the Puebloans, and, above all, pictographs.

One of the numerous members of the Cerbat/Pai confederation was the Havasupai Tribe. Settling in capacious Cataract Canyon, they stayed in touch with other Cerbat/Pai clans through runners who carried news up and down the South Rim from one band to the next. Closely related were the Hualapais, settling not far from the Havasupais along the western edge of the South Rim.

Meanwhile, the Southern Paiutes had sought safety along the North Rim from the devastating raids, initially by the Mountain Utes, Western Apaches, and Navajos and then by Spanish colonists in search of slaves. Though first and foremost hunter-gatherers, they moonlighted as farmers, having learned from the Puebloans how to plant corn and squash along springs and side canyons as a hedge against hunger. They etched petroglyphs and painted pictographs on rock faces to convey meanings to their family, friends, and allies. Like the Cerbat/Pai, they communicated with distant clans via runners, who also served as mnemonists, memorizing routes and various messages.

The Ongoing Story

The story of Grand Canyon's first people has yet to end, for the many tribes that made their homes there never entirely departed. Hunter-gatherers continued to live in Grand Canyon well into the 19th century and even into the 20th. The Havasupai Tribe has resided in Western Grand Canyon for more than 1,000 years. And the Hualapi, Yavapai, Southern Paiute, Pueblo, Zuni, Navajo, and Apache Nations also still call Grand Canyon home.

If there is a quality that unifies the great assortment of people from the Ice Age to our age—from those who have lived out their lives in Grand Canyon to those who have only briefly visited—it is their great and glorious diversity. While millions of visitors travel to Grand Canyon each year from every corner of the globe, our modern cosmopolitanism pales in comparison to the rare and radiant cultures that populated its rims and riverbanks for thousands upon thousands of years.

Further Reading

Christensen, Don, Dickey, Jerry, and Freers, Steven M. 2013. *Rock Art of the Grand Canyon Region*. San Diego: Sunbelt Publications.

Coder, Christopher M. 2006. *An Introduction to Grand Canyon Prehistory*. Grand Canyon, AZ: Grand Canyon Association.

Fagan, Brian. 2011. *The First North Americans: An Archaeological Journey*. London and New York: Thames & Hudson.

Page, Jake. 2003 [2004]. *In the Hands of the Great Spirit: The 20,000-Year History of American Indians*. New York: Free Press.

Walker, Steven L. 2009. *Indians of the American Southwest*. Scottsdale, AZ: Camelback Group.

Tribes That Call Grand Canyon Their Home

As visitors stroll down the walkway leading to the rim-side trail along Grand Canyon's South Rim, they come upon a large medallion of native stone inscribed, "TRIBES CALL THE CANYON HOME." Around the medallion's perimeter are the names of Grand Canyon's eight original tribes: Hopi, Zuni, Navajo, Apache, Yavapai, Hualapai, Havasupai, and Southern Paiute. Each of these names is oriented in the general direction of each of the tribes' current homelands. Dedicated in 2011, the medallion memorializes their joint commitment to keeping their cultures alive by nurturing the fragile but unbroken thread that ties the present moment to their remarkable past. The Landmark and Tribal Medallion, which was designed with input from the tribes, includes water symbols that depict the importance of rain and water in the desert, while the hand in the center reminds viewers of those who have come before.

While many aspects of their early cultures have been lost, a rich fund of knowledge about their customs and beliefs comes from a hitherto underappreciated source: the great oral historians, keeping traditions alive by passing them from person to person, century after century. Suppler than the written word yet truer than literary fiction, origin stories—the centerpiece of oral histories—breathe life into legends that would otherwise fade away. In so doing, these stories give sanctuary to the values, mores, and traditions that are the essence of a group's principles and ideals. Their word-of-mouth nature affords the narrator the flexibility to render the stories relevant to the present day while preserving the precious events that were never codified in print. Indeed, if not for the tribal elders faithfully transmitting these stories from one generation to the next, much of their history might have vanished forever.

Although the eight tribes that call Grand Canyon home are politically distinct, they have influenced one another in myriad ways over the course of their interlinked past. In addition to sharing the relatively recent trauma of forced boarding schools, suppression of religious practices, and theft of their motherlands, they have inherited a far longer legacy from a yet more distant past, of cherished lifeways that have been shaped, molded, and refined by the magnificent landscapes of their ancestral home. This, in turn, is reflected in the stories each tribe tells of its beginnings.

For the great agriculturalists, whose lives were devoted to planting, tending, and harvesting crops, their origination emerges from the soil, with their journey emanating from other worlds deeply buried beneath Earth's surface. In contrast, those who made their homes within the canyon or along the Colorado River and relied

Native American tribes who have called Grand Canyon home are honored by the Landmark and Tribal Medallion along the path from Grand Canyon Visitor Center to Mather Point. In this image, the Pollen Trail Dancers perform storytelling dances (traditionally featured at Navajo powwows) at a ceremony dedicating the medallion on June 15, 2011. (Grand Canyon Association Photo by William Brookins)

more on hunting and gathering connect their origins with cataclysmic events involving either floods or canyon walls that spasmodically open and close like broken elevator doors.

Although it is impossible to reproduce the vibrancy and nuance of stories told aloud to friends and family, this book would not be complete without honoring the oral histories recounted over the centuries by the elders of Grand Canyon's eight original tribes. Like all oral histories, there are many versions having their own unique emphases reflecting the lived experiences of various clan members within a given tribe. What follows are condensed versions of some of these tales of origination.

Below Earth's Surface: Hopi, Navajo, and Zuni Origins

We begin with the Hopis, renowned not only for their ingenuity at farming but also for their hospitality toward other tribes, especially those who grew corn. Although the ancient Navajos lived by hunting and gathering—and were not above raiding Hopi villages from time to time—they were also famously eclectic in their ability to absorb and adopt new ideas and methods from other cultures. The similarity of the Navajo origin story to that of the Hopis testifies to the age-old ties between them. The Zunis, like the Hopis, were among the earliest farmers of the region, growing primarily corn and supplementing their diet by hunting small game. Despite speaking unrelated languages and establishing settlements in separate areas, their origin stories are remarkably similar, suggesting enduring relationships that went beyond mere commerce.

Hopi Origins

The Hopis revere Grand Canyon as their spiritual home and the land of their origination. While Grand Canyon's salt mines have long been a destination for religious observances as well as a source of much needed salt, it is the Sipapuni, located deep within Grand Canyon, that the Hopis hold most sacred, for it marks the place of their first emergence from earlier worlds. A geologic dome consisting of mineral deposits created from a spring, the Sipapuni is located on the banks of the Little Colorado River, which is 4.5 miles upstream from its confluence with the Colorado River.

Hopi oral histories tell of a small, select group of morally upright, honest individuals who escape a decadent and corrupt society by climbing into this world through the domed Sipapuni after having first trekked through three former worlds. After surfacing and upon the advice of their guardian spirit, Masauwu, they venture north, south, east, and west until reaching the sea, whereupon they turn around, retrace their steps, and ultimately settle near the site of their first emergence. Two symbols, the square and the circle, represent this journey, the square symbolizing the four directions and the circle, their circuitous quest to find their way back home. The structure of their dwellings and ceremonial buildings, called kivas, features holes in the ceilings through which one enters and departs via a ladder, allowing the Hopis, with each exit and entry, to reenact their origin story, providing a reminder of their commitment to leading a life of decency and virtue.

The Hopis also have a creation story for Grand Canyon. When two gods, who are brothers, hurl a lightning bolt into the Earth, they accidentally sever the planet in two. To put it back together, they build the Canyon's walls from mud. Both events are recreated in the Hopi Snake Dance, the zigzag movement of the snake representing the brothers' lightning bolts and the quest for snakes, requiring dancers to head north, south, east, and west in remembrance of their original search for a proper homeland. Over the course of the dance, members of Hopi sacred societies reenact various other central events from their legends of emergence, as well as ceremonies concerned with creating community harmony by appeasing the spirits, called *Katsinas* (or *Kachinas*), who protect them.

Navajo Origins

Like the Hopis, the Navajo origin story involves passing through previous worlds before entering this one. Whereas for the Hopis there are three previous worlds, the Navajos count four. The first is dark, the second blue, and the third yellow. The fourth, submerged in a great flood, prompts the voyagers to travel to the present world, called the Glittering World, in honor of the dazzle of the sun, moon, and stars.

In another Navajo story, Grand Canyon originates when a humpback god, symbolized by the desert bighorn sheep, dragged his cane across a swath of Earth. Gradually, Grand Canyon filled with the multitude of life forms that accompanied the first humans on their journey upward though the worlds below earth.

The four mountains that encircle their ancestral lands—Hesperus and Blanca Peaks in Colorado, and Mount Taylor and the San Francisco Peaks in New Mexico and Arizona—are sacred to the Navajos, who believe that the Holy People—the Navajo gods—have placed the mountains in such a way as to delineate the boundaries of their ancestral lands. The mountains also remind the Earth People—that is, the Navajos themselves—to live their lives in balance with nature.

Zuni Origins

The Zunis locate their place of origin at Ribbon Falls on Bright Angel Creek, which flows into the Colorado River from the north rim wall of Grand Canyon. The Zuni origin story, though containing elements of both Hopi and Navajo accounts, is nevertheless distinct. Their origins find them in a dark and cramped place, deep within the center of Earth, called the Fourth World. Knowing of their suffering and wishing to alleviate it, the great Sun God sends his two sons to rescue them. They are aided in their journey by prayer sticks carved from the branches of a pine, a spruce, a silver spruce, and an aspen. After traveling through the Fourth, Third, and Second Worlds, upon their arrival in the First World and by the light of the red dawn, they behold themselves for the first time: Their fingers and toes webbed, their heads capped with horns, and their long tails wagging, they are horrified to find themselves drenched in slime with no mouths. It is then that they learn that they must travel further still before reaching their home and metamorphosing into humans. After a long and perilous trek, they reach the daylight world. They cry copious tears as they gaze upon the sun for the first time in their lives, causing sunflowers to rise from the soil. Their rescuers then bestow upon them mouths, allowing them to eat the gift of corn and teaching them to make delicious corncakes and other delicacies by pounding and grinding the kernels. In a final act of grace and mercy, their rescuers rid them of their horns, tails, and webbed digits, thereby conferring upon the newcomers the status of beautifully formed Zunis. Signs on the canyon walls near Ribbon Falls have been interpreted by Zuni cultural advisors as accounts of their departure from Grand Canyon (Hirst, 2016).

Canyons and Floods: Havasupai, Hualapai, Yavapai, Southern Paiute, and Apache Origins

Unlike the prior versions, the following five tales involve natural disasters such as floods and quaking canyon walls. For hunter-gatherers ever on the move, answers to the mystery of their beginnings lay in the waters and walls that sustained and sheltered them. Although each tribe has a unique perspective, all of them share the view that their origins are inextricably tied to the topography of Grand Canyon. While Havasupai origins begin with their entrance into Grand Canyon, the Hualapais and their closely associated neighbors, the Yavapais and the Apaches, believed themselves to have been sired either by the river or the canyon—or sometimes both. Southern Paiute mythology, in contrast, links the origins of Grand Canyon to the sorrow and the longing of a great chief for his dead wife.

Havasupai

There are many renditions of the Havasupai tale of arriving at Grand Canyon, some of which are told by a folk character, a trickster by the name of Coyote. In one of these tales, the walls of the canyon open and close, crushing those who wish to pass through, until two boys attack them with arrows, forcing them to remain open. The Havasupais enter the Canyon, where they establish their homeland. In a second version, a man throws a log between the oscillating walls, permanently propping them open, thereby allowing the Havasupais to enter. In a third account, a young mother teaches her sons to hunt by constructing arrows with feathers and points. As the sons wander in search of feathers for their arrows, they come upon the Canyon, where they establish a home for their people. As in the origin stories of some other tribes, animals are their first teachers. From the birds, they learned to speak and sing; from the mountain lions, how to hunt; from the beavers, how to build homes; and from the coyote, of course, how to engage in trickery.

Havasupai legends also tell of early relations with the Apaches. A benevolent god orders the Havasupais to live in harmony with the Apaches. Everything goes well until an ornery god persuades an Apache to kill a Havasupai hunter and abduct his wife, at which point the Havasupai drive the Apache from the canyon by force. This leads, in turn, to animosity between the two tribes, which they patch up eventually.

Hualapai Origins

The Colorado River plays the leading role in Hualapai mythology. Oral histories say that the first people were molded from the clay of the Colorado River. Ancient petroglyphs tell their own story of a great flood that submerges Earth, killing all its people, save for a select few. The petroglyphs are located at a site that remains sacred for the Hualapais, a stream gushing from a side canyon, called Salty Spring, or Ha'thi-el. Yet another tale passed down through the generations parallels the etchings in the rocks—and then builds upon it. A spiteful god, prone to violent attacks of rage, decides to destroy the world with a great flood. A benevolent god, determined to save his daughter from the inundation, hides her in a canoe. She becomes impregnated by the sun, bestowing upon her a daughter. She then gives birth to a second child, sired by a waterfall in Havasu Canyon. Her children repopulate the planet, giving rise to the first humans.

Still another variation recounts a conflict that started with children but quickly spread to their parents until the entire group became so embroiled that the clan dispersed, eventually becoming separate tribes. This may, in part, reflect their legendary animosity toward the Yavapai.

In addition to the creation of life, Hualapai oral histories emphasize the preservation of life through the healing power of shamanistic practices, including singing, shaking gourds, and sucking out diseases with tubes and herbs. While the

Hualapai or Walapai?

Visitors to Peach Spring, Arizona—the tribal capital of the Hualapai Indian Reservation—often encounter a spelling riddle. There, in town, the Walapai Gift Shop is near Hualapai Lodge. Is one of these names misspelled, or is there a difference between Walapai and Hualapai?

Hualapai has been spelled many different ways in various publications, treaties, and other documents over the past three centuries:

Jaguallapai—1776
Hah-wal-coes Whipple—1851
Walapai or Hualapai—1857, 1858
Walapais Bell—1869
Mataveke Paya Corbusier ("People of the North")—1873–1875
Wal-la-pais—1875
Walyepaid Kroeber—1907
Hawlapai—1908
Xawalapai ya ("Pine Tree Folk")—1908
Hwal'bay—1975
Walapai people—1975

Although the tribe uses the word "Hualapai," the tribe's official name—Hualapai or Walapai Tribe—was assigned in an executive order issued on January 4, 1883, by U.S. president Chester A. Arthur (1829–1886). This same order also created the 1,142-square-mile Hualapai Indian Reservation along 108 miles of the Grand Canyon and Colorado River.

shamans who successfully cured the sick received valuable buckskins, those whose patients did not survive risked execution. Clearly, it was a calling that required great confidence and courage.

Yavapai Origins

Historically, all Yavapai clans lived near the Colorado River and were in frequent contact with other neighboring tribes, including the Hualapais, with whom they often fought. Thus, it comes as no surprise that their origin story is similar to that of the Hualapais: A flood submerges the planet in water, killing all creatures except a little girl, who has been placed in a hollow cottonwood log by her parents. When she comes of age, she is impregnated by the water, and her progeny repopulate Earth. This explains why humans are made mostly of water.

Southern Paiute Origins

In one of the origin stories of the Grand Canyon's Southern Paiutes, a compassionate god, Taavotz, tries to ease the heartache of the great chief, Umbah, who has become deeply depressed since the passing of his beloved wife many moons earlier. To ease his suffering, Taavotz takes Umbah to the Spirit Land of the Dead to behold his wife one last time. They travel west, until reaching a vast mountain through which Taavotz cuts a long and deep gorge to reach their destination. As soon as Umbah sees his wife, he realizes that she is at peace. His sadness evaporates, and he again can feel joy. After returning to the world of the living, Taavotz covers their tracks by pouring the Colorado River into the enormous gorge that he had cut through the mountain, thereby burying the secret location of the Spirit Land—and honoring the privacy of the Dead who dwell there. To seal the deal, he imbues the river with the power to swallow anyone who attempts to ride its currents west.

The Southern Paiutes, who inhabited the North Rim of Grand Canyon National Park until the latter half of the 19th century, surely witnessed John Wesley Powell's famous expedition of 1869. One cannot help but wonder what they thought as they observed the rapids carry the small wooden boats ever further west toward the Spirit Land of the Dead.

Apache Origins

The Apaches ranged more widely than other tribes of Grand Canyon, hunting, gathering, raiding, and glorying in the stunning beauty of the natural world. Consequently, their creation story includes the entire universe. The tale begins with total darkness. Then, a disc emerges that serves as a conveyance of sorts for the Great Creator. He produces yellow for the dawn to the east, followed by a rainbow of colors for the setting sun to the west. To the surprise of the Great Creator, a little girl spontaneously appears. To keep her company, he creates a boy. He then creates the sun, the other stars in the firmament, clouds, lightning, and planet Earth. A tarantula is next, followed by a hummingbird. All the remaining plants and animals quickly follow. Then the Great Creator delegates authority to minor gods with specific territories over which they preside, including a god of the clouds and water, a god of the sky, a goddess of the crops, and a goddess of pollen who cares for the health of all creatures. The little girl presides over all other gods. The Great Creator then drifts skyward in a cloud of smoke until finally disappearing.

Origins at Grand Canyon

These stories of origination, although seemingly fanciful and full of whimsy, are one of the greatest gifts that oral historians of earlier times could have offered their descendants. Not only do they provide a glimpse, however limited and fleeting, of events of the past that otherwise would have faded away, but, perhaps more valuable still, they offer inklings of the inner lives of those who came before. It is perhaps this quality, more than any other, that makes these tales so precious and irreplaceable. They answer, over and over, the age-old question of where did we come from?—a question that remains a matter of great debate. Products of a long lost world, the origin stories of the eight tribes that call Grand Canyon home nevertheless respond to the unquenchable human longing to find meaning in our existence.

Today, the Landmark and Tribal Medallion remains a popular and culturally meaningful tribute to Grand Canyon's earliest tribes. However, because the federal government had arbitrarily divided the Paiutes into five distinct tribes, while artificially combining the culturally distinct Yavapai and Apache Nations into one, the current official count of Traditionally Associated Tribes is, in fact, 11: Havasupai Tribe, Hopi Tribe, Hualapai Tribe, Kaibab Band of Paiute Indians, Las Vegas Tribe of Paiute Indians, Moapa Band of Paiute Indians, Paiute Indian Tribe of Utah, San Juan Southern Paiute Tribe, Navajo Nation, Pueblo of Zuni, and Yavapai-Apache Nation (Hirst, 2016).

Further Reading

Bruchac, Joseph. 1992. Native American Animal Stories. In *Keepers of the Animals*, ed. Michael J. Caduto and Joseph Bruchac. Golden, CO: Fulcrum Publishing.

Champagne, Duane. 1994. *Native America: Portrait of the Peoples*. Detroit: Visible Ink Press.

Courlander, Harold. 1971. *The Fourth World of the Hopis*. New York: Crown.

Dennis, Yvonne, Hirschfelder, Arlene, and Flynn, Shannon Rothenberger. 2016. *Native American Almanac: More than 50,000 Years of Cultures and Histories of Indigenous Peoples*. Canton, MO: Visible Ink Press.

Hirst, Stephen, ed. 2016. *We Call the Canyon Home: American Indians of the Grand Canyon Region*. Grand Canyon, AZ: Grand Canyon Association.

Houlihan, Patrick, et al. 1987. *Harmony by Hand: Art of the Southwest Indians*. San Francisco: Chronicle Books.

Johnson, Sandy. 1994. *The Book of Elders: The Life Stories and Wisdom of the Great American Indians*. San Francisco: HarperCollins.

Locke, Raymond. 2005. *The Book of the Navajo*. Sixth edition. Los Angeles: Mankind Publishing.

Nabokov, Peter. 1999. *Native American Testimony: A Chronicle of Indian-White Relations from Prophecy to the Present, 1492–2000*. London, England: Viking Penguin.

Nequatewa, Edmund. 1994. *Truth of a Hopi: Stories Relating to the Origin, Myths and Clan Histories of the Hopi*. Flagstaff, AZ: Northland Publishing.

Nerburn, Ken. 2016. *Voices in the Stones: Life Lessons from the Native Way*. Novato, CA: New World Library.

Niethammer, Carolyn. 1977. *Daughters of the Earth: The Lives and Legends of American Indian Women*. New York: Simon & Schuster.

Shields, Kenneth, Jr. 2000. *The Grand Canyon: Native People and Early Visitors*. Chicago: Arcadia Publishing.

Thomas, David Hurst, et al. 1993. *The Native North Americans*. Atlanta: Turner Publishing.

Treuer, Anton. 2013. *Atlas of Indian Nations*. Washington, DC: National Geographic Society.

Waters, Frank. 1972 [1977]. *Book of the Hopi: The First Revelation of the Hopi's Historical and Religious Worldview of Life*. New York: Penguin.

Witt, Greg. 2010 [2012]. *Exploring Havasupai: A Guide to the Heart of Grand Canyon*. Birmingham, AL: Menasha Ridge Press.

Zolbrod, Paul. 1984. *Diné Bahane': The Navajo Creation Story*. Albuquerque: University of New Mexico Press.

John Wesley Powell

John Wesley "Wes" Powell (1834–1902), more than anyone else, introduced Grand Canyon and the Colorado River to the American public. Powell was born on March 24, 1834, in Mount Morris, New York. Powell's parents—Joseph and Mary, both dedicated to spreading the Methodist message—hoped that their son would join them in the ministry, but the young Powell had other plans.

In 1869, Civil War veteran John Wesley Powell became famous for leading the first intentional trip down the Colorado River through Grand Canyon. Powell turned his celebrity into a career as a Washington bureaucrat, but his influence waned when his science-based approach to water use in the Southwest challenged the political ideology of Manifest Destiny. (Library of Congress)

On April 14, 1861, after attending Illinois College (today's Wheaton College) and Oberlin College and working as a science/math teacher and superintendent in Illinois, the 27-year-old Powell joined the 20th Illinois Volunteers to fight in the Civil War. A year later, Powell—a staunch abolitionist—fought with General Ulysses S. Grant at the Battle of Shiloh (the bloodiest battle in American history up to that time); Powell was wounded, resulting in the amputation of his right forearm. Powell continued his Army career; while in trenches for 40 days during the siege of Vicksburg, Powell studied the area's rocks. Powell was discharged from the army in 1865 at the rank of brevet lieutenant colonel, but the title "major" accompanied Powell for the rest of his life.

In the fall of 1865, Powell returned to teaching geology at Illinois Wesleyan University. Powell decided to explore the last large area on the maps—an area the size of Texas—of the United

States that cartographers simply labeled "Unexplored." As Powell noted, "[T]he thought grew in my mind that the canyons [of the Colorado River] would be a book of revelations in the rock-leaved Bible of geology. The thought fructified, and I determined to read the book" (Powell, 2013). Powell first publicly announced his upcoming exploration of the Colorado River on December 18, 1867, at a meeting of the Illinois Natural History Society in Bloomington. Powell never met James White, who in 1867 claimed that he had floated through Grand Canyon on a makeshift raft.

On May 24, 1869, Powell began his 99-day expedition to explore the Colorado River and its canyons from Green River, Wyoming (elevation 6,075 feet); the expedition would end about 900 miles downstream at Rio Virgen, Arizona (elevation 700 feet). No one had intentionally gone through Grand Canyon, so there were no experts from whom Powell could get advice. The daring Powell, who knew his success would require stamina, preparation, and good luck, was accompanied by nine recruits, none of whom were scientists and virtually all of whom were underprepared for the challenge:

Oramel Howland, a hunter and printer (for *Rocky Mountain News*) who, at 36 years old, was the oldest member of the expedition. Unlike most of Powell's crewmembers, Oramel never fought in the Civil War.

Seneca Howland, a Union veteran of Gettysburg and the younger half-brother of Oramel. Seneca made topographical drawings of the course of the Colorado and Green Rivers. Seneca and Oramel were both aboard *No Name*, which sank two weeks into the expedition. Powell later blamed Oramel Howland for the sinking.

William H. "Bill" Dunn, a prospector, hunter, and trapper from Colorado. Dunn's job during the Powell expedition was to take barometric measurements.

John Colton "Jack" Sumner, a Union veteran, hunter, and prospector who was the expedition's head boatman. Sumner, who hoped the expedition would make him rich, rowed *Emma Dean*. Sumner (with George Bradley and William Hawkins) was critical to the expedition's success but resented Powell's personality and leadership style.

William Wesley Hawkins, a mountaineer, cook, and Civil War veteran. Hawkins and Andy Hall rowed *Kitty Clyde's Sister*. Hawkins and Sumner both condemned Powell after the trip.

Andy "The Character" Hall, a 21-year-old Scotsman, mule driver, and scout who was the youngest member of the group. Hall was a last-minute addition to the expedition.

Walter Henry Powell, the youngest brother of John Wesley Powell and a Civil War veteran who spent 10 months in a Confederate prisoner of war camp.

George Young Bradley, another Civil War veteran of Fredericksburg who joined the group for the excitement. Bradley rowed *Maid of the Cañon*, of which Walter Powell was a passenger. Ten days before the expedition started, John Wesley Powell worked with General Ulysses S. Grant to obtain Bradley's release from the army so that Bradley could join the expedition. Bradley was one of the few members of the expedition who had experience with boats.

Frank Goodman, an adventurous Englishman who was so anxious for an adventure that he offered to pay Powell for the opportunity to join the expedition.

Powell had four boats: *Maid of the Cañon*, *Emma Dean*, *Kitty Clyde's Sister* (named for a popular song of the time), and *No Name*, all of which were shipped to Green River, Wyoming, by Powell's patron, the Union Pacific Railroad. All of these boats were deep-keeled, heavy, Whitehall boats powered by one or two men pulling downriver while facing upstream. The boats, which were the fastest boats of their day, were designed for fast travel in the choppy water of lakes and shipping harbors, but the keel that kept them moving straight ahead made them poorly suited for maneuvering in Grand Canyon's shallow, rocky rapids. This forced Powell and his crew to portage tens of rapids. (In Tusayan, visitors can see a replica of one of Powell's boats behind the National Geographic Visitors Center.) Whitehall boats, which were poorly designed for shallow, rocky rapids, were seen, though rarely, in Grand Canyon up to the mid-20th century.

Naming the Grand Canyon

Just as the Colorado River has had many names, so too has Grand Canyon. Several of these names for Grand Canyon were bestowed by Native Americans. For example, the Navajos referred to it as *Bidahaztii* or *Bida haaazti tsekoh*, the Hopis called it *Suukotupqa* or *Sakwatupqa*, and Havasupais called it *Wikatata* ("Rough Rim") and *Chicamimi Hackataia* ("Large Roaring Canyon").

In 1776, Franciscan friar Francisco Tomás Garcés referred to Grand Canyon as Puerto de Bucareli, or Bucareli Pass (after a Spanish Viceroy). Garcés, like many people who came after, viewed Grand Canyon as a barrier more than as a scenic or geological wonder. In 1811, an atlas produced by Alexander von Humboldt showed the Grand Canyon as Puerto de Bucareli at the junction of the Rio Colorado and Rio Jaquesilla ("Little Colorado River"), but the name never became common. In 1857, Joseph Ives called Grand Canyon "Big Cañon."

The first published use of "Grand Cañon" was by Lorin Blodget in a book titled *Climatology of the United States, and of the Temperate Latitudes of the North American Continent*, which discussed the weather and geography of Grand Canyon. In 1868, General William J. Palmer included the name "Grand Canyon" (as a specific place) on a map that accompanied a railway survey, and in 1869, Samuel Bowles used the term "Grand Canyon" in print when he described it as "Grand Canyon of the Colorado."

Despite all of these earlier names, John Wesley Powell popularized and is usually credited with the first authoritative and consistent use of "Grand Canyon." Powell's "Grand Canyon" name was later made official by the United States Board on Geographic Names.

Powell's group—whose mission was more about survival than science—carried a 10-month supply of food, traps, guns, ammunition, clothing, tools to repair boats, and scientific instruments. Problems began almost immediately. On June 9, the 17th day of the expedition, *No Name*—carrying much of the crew's clothing and one-third of the expedition's supplies—sank at what Powell named Disaster Falls. Other disasters followed; a week later, a fire destroyed many of their clothes, blankets, cooking gear, and tents. On July 5, after the group had covered about 80 miles, Goodman—who had joined the expedition for the excitement it offered and was aboard the wrecked *No Name*—left the group after the first major rapid in northern Utah, noting "I've had more excitement than a man deserves in a lifetime. I'm leaving." By the time the group reached Lees Ferry, they were down to five weeks of food, and Powell placed the crew on reduced rations. Worsening their hunger were frayed nerves and a sense of impending danger. The group had no way of knowing what was around each bend in the river; they hoped for smooth water but knew that it could also be a waterfall that would kill them. They knew that no one could save them if more problems arose.

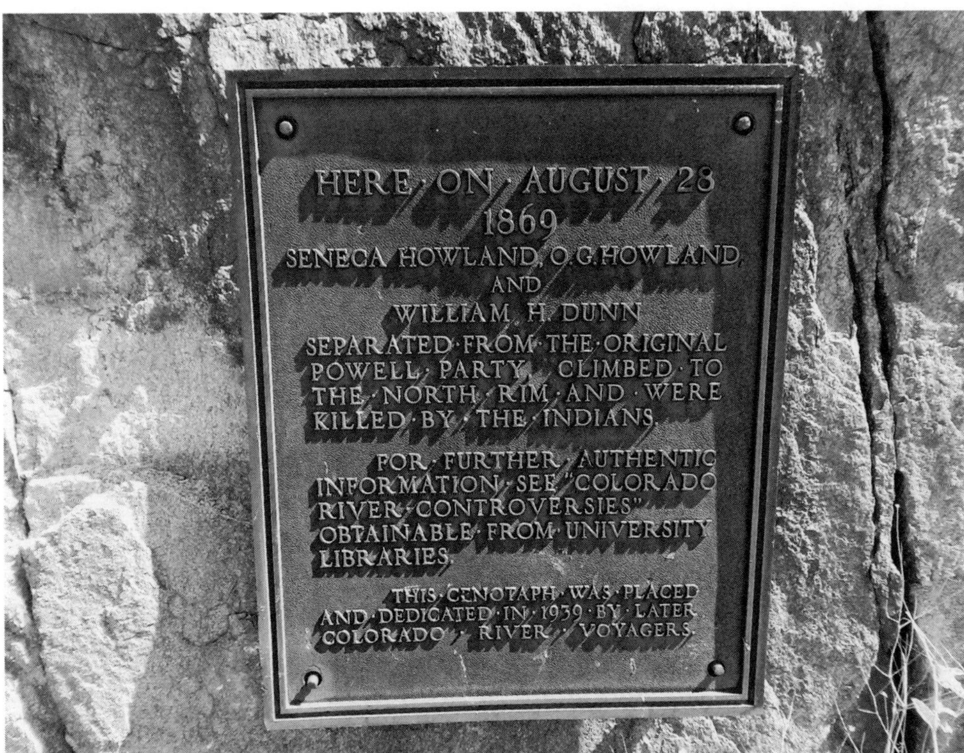

This plaque at Separation Canyon honors William Dunn, Seneca Howland, and O.G. Howland, the three crewmen who left John Wesley Powell's expedition through Grand Canyon here in 1869. The three men were never heard from again, and their fates remain an enduring mystery at Grand Canyon. Powell and his other crewmen (but not Dunn or the Howland brothers) are honored at the Powell Memorial at Powell Point. (Randy Moore)

After three months on the river, Powell faced an especially intimidating rapid at what is now called Separation Canyon. There was food for only five more days, and there was no way to portage the rapid. Moreover, the group faced, in Powell's famous words, "an unknown distance yet to run; an unknown river yet to explore. What falls there are, we know not; what rocks beset the channel, we know not; what walls rise above the river, we know not." On August 28, 1869, Dunn and the Howland brothers were convinced they would die if they continued and decided to leave, choosing the unknown dangers of a hike across the desert over the certain dangers of the upcoming rapids. They were never heard from again, prompting historians and others to speculate about their fates.

After Separation Canyon, only six of the original ten members of the expedition remained. What had been imagined as a thrilling, leisurely adventure was now a grim struggle to survive. However, two days after Separation Canyon (August 30, 1869), after covering almost 930 miles, Powell reached the mouth of the Virgin River (now under Lake Mead), and his journey was complete. There was no need to continue because the Colorado River had already been charted to this point by Joseph Ives in 1858. Crewmember George Bradley noted, apparently with relief, that "all our trials [are] over."

Powell's 1869 expedition was one of the last great explorations of land in the United States, and Powell's exploits (combined with Thomas Moran's stunning paintings of the Canyon) made him a national hero. Because Powell and his group had been cut off from all contact with civilization, many assumed that he had died, and a man named John Risdon—who was never associated with the trip—told newspapers that he personally witnessed the drowning of all of Powell's men. Powell's expedition was reenacted in the Disney film *Ten Who Dared* (1959), in which John Beal portrayed nerdish Powell, and 15-year-old Buzz Belknap stood in for the actor who portrayed Andy Hall. Top billing went to Brian Keith, who played Bill Dunn in his Disney debut. No photographs were taken during Powell's 1869 expedition.

In 1871–1872, Powell again started through Grand Canyon, this time—thanks to support from the Smithsonian Institution—better prepared with more food, better boats (*Cañonita*, *Nellie Powell*, and the second *Emma Dean*), and a different crew. At Lees Ferry, Powell gave one of his boats (*Nellie Powell*) to ferry operator John D. Lee, who had helped Powell and his crew. In October 1938, the charred remains of this boat were unearthed by rancher Leo Weaver near the ferry, identified by Eddie McKee, and given to NPS. This 20 × 60-inch remnant is the only remaining part of any of the boats used by Powell in Grand Canyon.

Whereas Powell's first trip through Grand Canyon was an expedition that focused on adventure and survival, his second trip was an exploration focused on science. This is why Powell chose the crew of his second expedition for their ability to document and explore the Canyon. Most were college-educated, sober, science-minded men; for example:

Almon Harris Thompson, Powell's brother-in-law (he married Powell's sister Ellen Louella, also known as Nellie, the namesake for the *Nellie Powell* boat) and second in command. Thompson, who had the notes from Powell's first trip through Grand Canyon, was the expedition's chief topographer. Thompson, who served as USGS's chief geographer from 1879 to 1906, is memorialized in GCNP at Thompson Point near River Mile 125.

Stephen V. Jones, a mathematician and surveyor.

Francis Marion Bishop, another Union veteran and assistant to Thompson (Bishop later became a professor).

William Derby Johnson Jr., a Mormon school teacher from Kanab hired as an assistant topographer

Andrew J. "Andy" Hattan, an acquaintance of Powell from the army. Hattan was the expedition's cook.

Frederick S. Dellenbaugh, a 17-year-old, self-taught artist who was Powell's artist and assistant topographer. Dellenbaugh is honored by Mount Dellenbaugh, a 6,750-foot-high peak on river right near River Mile 198.5.

John Karl "Jolly Jack" Hillers, a German immigrant (and last-minute replacement for Jack Sumner) whom Powell recruited in Salt Lake City. Hillers became the expedition's photographer after E. O. Beaman left the expedition (after a dispute with John Wesley Powell) and James Fennemore (Beaman's replacement) got sick. Hillers rowed *Emma Dean*, the expedition's boat named for John Wesley Powell's wife; Powell sat in an armchair lashed to this boat, from which he looked for danger ahead. Hillers, who became a photographer for the USGS in the American West, was one of the first people to photograph Grand Canyon. The first still extant photographs of Grand Canyon were taken by Timothy O'Sullivan, who photographed western Grand Canyon—aboard a boat appropriately named *Picture*—in 1871 during the War Department's expedition led by George M. Wheeler. O'Sullivan was a protégé of famed Civil War photographer Mathew Brady, one of the earliest photographers of American history.

Walter Clement "Clem" Powell, a 21-year-old assistant photographer and John Wesley Powell's orphaned cousin.

Before departing, Powell consulted with Mormon missionary Jacob Hamblin, who negotiated with local Native Americans to ensure a safe trip for Powell and his crew. Powell had no such trouble. Powell entered Grand Canyon on August 17, 1871, and terminated the expedition on September 8, 1872, at Kanab Creek (River Mile 144), telling his crew, "Well, boys, our voyage is done." (To which Hillers, the photographer, noted, "[E]verybody felt like praising God.") The mouth of Kanab Canyon, which is about 75 miles from Kanab, Utah, is among the few places having a well-defined route out of Grand Canyon. Interestingly, Powell used photographs from the 1871–1872 expedition to illustrate his 1869 expedition. Powell's second expedition produced what his first expedition could not: a published map and scientific papers. (Powell's first trip had produced an unpublished map that the second trip used, updated, and corrected.)

Powell published an abbreviated account of his Colorado expedition in 1869 as Appendix D of William Abraham Bell's *New Tracks in America* (1869). Powell's six-page report was titled "Major J. W. Powell's Report on His Explorations of the Rio Colorado in 1869." In this and future publications about his expeditions through Grand Canyon, Powell embellished and exaggerated the dangers of Grand Canyon. For example, in 1875, Powell published a report that blurred his first and second expeditions through Grand Canyon. Powell's stirring, suspenseful book, which opened the river corridor in Grand Canyon to further exploration, was originally titled *Exploration of the Colorado River of the West and Its Tributaries Explored in 1869, 1870, 1871, and 1872 Under the Direction of the Secretary of the Smithsonian Institution*. (In 1915, a National Academy of Science panel credited this book with creating national interest in Grand Canyon, adding that his book is "one of the best narratives of adventure anywhere to be found.") Powell never wrote about the second expedition separately or mentioned that crew. In 1895, Powell reissued his book as *The Exploration of the Colorado River and Its Canyons* and then as *The Canyons of the Colorado*, which he dedicated to his companions on the first expedition through Grand Canyon, including the three who left the group at Separation Canyon. In his book, Powell focused more on the adventure than on Grand Canyon. He also made several astute observations of the canyon's geology, including its unconformities, monoclines, and faults. However, as food dwindled and anxiety increased, Powell admitted that he missed much of the geology in Grand Canyon: "All about me are interesting geological records. The book is open and I can read as I run. All about me are grand views, too, for the clouds are playing again in the gorges. But somehow I think of the nine days' rations and the bad river, and the lesson of the rocks and the glory of the scene are but half seen" (Powell, 2013). Powell generally did not reference Grand Canyon's stratigraphy; the only rocks he referred to by name were those from the Carboniferous. Powell knew that Grand Canyon had been formed by erosion ("The carving of the Grand Canyon is the work of rains and rivers. Though storms are far apart and the heavens above are cloudless for most of the days of the year . . . an intermittent rill called to life by a shower can do much work in centuries of centuries."). However, there was nothing to indicate that Powell knew about John Strong Newberry's earlier geological studies in Grand Canyon.

In the early 1900s, Powell's crew began publishing their accounts of the voyages, beginning with Dellenbaugh's map-laden *The Romance of the Colorado River* (1902) and *A Canyon Voyage* (1908). These and other books revealed that Powell had exaggerated many events, that many of the events of the second voyage are described as happening during the first, and that Powell—who could not swim—was the only person to use a life jacket on the first trip (others crewmembers had them on Powell's second trip).

The imposing Powell Memorial at Powell Point on the South Rim lists only the names of those who completed Powell's journeys in Grand Canyon. The absence of the names of Dunn and the Howland brothers—who left Powell's first expedition at Separation Canyon—from the Powell Memorial is due to Dellenbaugh, a self-appointed consultant to the committee (which included Charles Walcott, H. C. Rizek, and chairman William Henry Holmes) that designed the monument. Although Dellenbaugh branded Dunn and the Howland brothers "deserters,"

John Wesley Powell and many of his crewmen are commemorated on the South Rim by the Powell Memorial at Powell Point. The imposing monument, which does not list the names of William Dunn or the Howland brothers (who left Powell's 1869 expedition at Separation Canyon), is the largest monument in Grand Canyon National Park. (M. Quinn/National Park Service)

Robert Stanton—who later led an expedition of his own down the Colorado River in Grand Canyon—claimed that the plaque's omission of Dunn and the Howland brothers "perpetuates a gross injustice." Indeed, Stanton had interviewed several of Powell's crew, many of whom contradicted Powell's stories, prompting Stanton to conclude that Powell had forced the three men to leave at Separation Canyon. In his *Colorado River Controversies* (1932), Stanton also condemned Powell's account of the journeys as "demonstrably false and, it would seem, deliberately misleading," "curiously inaccurate in many if not in all instances," and "guilty of suppression of the truth and unblushing exaggeration." Several members of Powell's expeditions are commemorated in GCNP and the surrounding area, including 6,990-foot-high Mount Dellenbaugh in western Grand Canyon; Dunn Butte, Hawkins Butte, Bradley Point, Sumner Butte, and The Howlands Butte near River Mile 84; and Hillers Butte, Jones Point, Hattan Butte, and Clement Powell Butte near River Mile 87.5. Walter Powell is honored by the Walter Powell Route, a short, steep trail 2 miles up the Little Colorado River from its confluence with the Colorado River.

In May 1875, *Scribner's Monthly* magazine paid Powell $500 for three articles and 12 engravings about his expeditions. These articles included Powell's first full written account of his 1869 expedition. Near the same time, Powell's *Exploration of the*

Colorado River of the West (his official report to Congress) was released and quickly sold out. Powell's book and articles made the Colorado River the essence of Grand Canyon, but otherwise included relatively little science. Powell's article titled "An Overland Trip to the Grand Cañon" in October 1875 also included Powell's first published map of Grand Canyon (measuring only 3.5 × 5 inches) and likened the river flowing through the uplifted Colorado Plateau to a log (i.e., the plateau) being raised into a buzz saw (i.e., the river).

Powell became a famous, influential Washington bureaucrat and the patron saint of Grand Canyon river-running. By 1876, however, his administrative duties required him to delegate scientific work to others. Powell became the second director of USGS (1881–1894, after Clarence King), the first director of the Smithsonian's Bureau of Ethnology (while with USGS from 1881–1902), and was one of the 33 founding members of the National Geographic Society. Powell created Illinois State University's first Museum of Anthropology, and, under Powell's leadership, the Smithsonian published an influential description of Native American languages. Powell, like John Muir, was a self-taught 19th-century Renaissance man, as well as an early and ardent advocate for Native America rights. Powell was succeeded at USGS by Charles D. Walcott. Powell and Walcott were among the last nationally prominent scientists who lacked college degrees.

Powell believed that science should benefit the public, and he knew from his expeditions that less than 5 percent of the American West was suitable for agriculture. Railroad companies, however, hoping to sell some of the 285,000 square miles of land given to them to build their railroads, lobbied for policies that promoted expansion in the West. Politicians and even most scientists also sided with the railroads, claiming that human habitation and agriculture would make arid regions such as the Great Plains and American West more humid and, in the process, "convert a desert into a farm or garden." Their campaign popularized the now debunked phrases, "Rain follows the plow" and "Rain follows the smoke." Congress passed the Desert Land Act, which enabled settlers to buy 640 acres of public land for $1.25 per acre. Although part of the acreage was to be irrigated within three years, the legislation included no provision for bringing water to the area. In fact, rich people and stock companies had already claimed the water and were selling it to other people.

Powell understood that the laws of nature would not bend to political ideology and that people would suffer if they expected settlement to change the weather in the American West. In 1878, Powell's remarkable science-based *Report on the Lands of the Arid Region of the United States* proposed an orderly and pragmatic development of the West that involved irrigation systems and state boundaries based on watershed areas to avoid conflicts between states. Powell argued that water, not land, was the limiting factor for agriculture in the West and that land should be allocated based on water availability rather than on uniform blocks of land as in the East. Powell's book included a map showing the average amounts of precipitation in the American West; that map made obvious a band of desert near the 100th meridian. Powell understood that the control of that land would belong to whoever controlled the water.

Powell's arguments were met with overwhelming opposition from land speculators and developers. Nevertheless, Powell continued to make his science-based claims. In a speech delivered in 1893 at the International Irrigation Congress, Powell prophetically noted that government officials were creating a major problem: "[Y]ou are piling up a heritage of conflict and litigation over water rights, for there is not sufficient water to supply these lands" (Powell, 1893). He also argued that the government should not sell land unless it was irrigable.

Powell's repeated emphasis on the lack of water in the arid West infuriated politicians and businesspeople. Powell was condemned and quickly lost power; by 1892, the USGS budget was cut by nearly half, and Powell was left powerless. But Powell was right; today, the Colorado River is the most litigated river in North America. Nevertheless, his ideas were largely ignored until the 1930s Dust Bowl, when storms caused immeasurable suffering and damage to the agriculture and ecosystems of the American and Canadian prairies.

Powell died in the family cottage in Brooklin, Maine, on September 23, 1902, at age 68 of a cerebral hemorrhage. He was buried in Arlington National Cemetery beneath a bronze portrait and a chiseled inscription reading, "Soldier. Explorer. Scientist." When informed of Powell's death, colleagues noted that "the greatest of scientific men is gone" and "our brightest exemplar of human knowledge is no more." Powell's wife Emma lived on a government pension of $30 per month (raised to $50 per month in 1903) until her death in 1924, at which time she was buried beside her husband.

For many decades, Powell's advocacy of the intelligent use of water and land in the American West (i.e., via local instead of federal regulations) was seldom recognized. However, in 1953, Wallace Stegner's *Beyond the Hundredth Meridian* resurrected Powell's work while urging people to include the environment when reviewing history.

Powell Plateau

Powell Plateau is a seldom visited plateau that juts from the North Rim at the eastern end of North Bass Trail. The plateau, which was named in 1882 by Clarence Dutton, is visible in the west from Grand Canyon Village and is surrounded by formations named for Grand Canyon surveyors (e.g., Dutton Point, Ives Point, Walthenberg Canyon, Newberry Point). Powell Plateau honors John Wesley Powell, who named tens of places in Grand Canyon (as well as Grand Canyon itself) and hundreds more in the American West but named none for himself. Powell is also honored by Lake Powell, which formed behind Glen Canyon Dam after it was completed in 1963.

John Wesley Powell remains a towering figure in the history of Grand Canyon. His name commemorates Lake Powell, Powell Plateau, and many other sites in Grand Canyon and elsewhere, and in 1969 the U.S. Postal Service issued a 6¢ stamp honoring Powell's 1869 expedition through Grand Canyon. The Powell Museum in Page, Arizona, honors Powell, and the John Wesley Powell Award is the highest award given by USGS to a person outside the federal government. The Powell Memorial at the South Rim is Grand Canyon's largest monument.

Further Reading

Berger, Todd R., ed. 2008. *Reflections of Grand Canyon Historians: Ideas, Arguments, and First-Person Accounts*. Grand Canyon, AZ: Grand Canyon Association.

Dolnick, Edward. 2001. *Down the Great Unknown: John Wesley Powell's Discovery and Tragedy through the Grand Canyon*. New York: Harper Perennial.

DuBuys, William E. 2001. *Seeing Things Whole: The Essential John Wesley Powell*. Washington, DC: Island Press.

Ghiglieri, Michael P. 2010. *First through Grand Canyon: The Secret Journals and Letters of the 1869 Crew Who Explored the Green and Colorado Rivers*. Revised edition. Flagstaff, AZ: Puma Press.

Lago, Don. 2017. Andy Hall's promotion. *Boatman's Quarterly Review*, 30 (3), 22–26.

Lago, Don. 2018. *The Powell Expedition: New Discoveries about John Wesley Powell's 1869 River Journey*. Reno: University of Nevada Press.

Marston, Otis Reed "Dock." 2014. In *From Powell to Power: A Recounting of the First One Hundred River Runners through the Grand Canyon*, ed. Tom Martin. Flagstaff, AZ: Vishnu Temple Press.

Powell, John Wesley. 1878. *Report on the Lands of the Arid Region of the United States, with a More Detailed Account of the Lands of Utah*. Washington, DC: Government Printing Office.

Powell, John Wesley. 1893. *Official Report of the International Irrigation Congress, Held at Los Angeles, California, October, 1893*. Los Angeles: Los Angeles Chamber of Commerce.

Powell, John Wesley. 2013. *The Exploration of the Colorado and Its Canyons*. New York: Simon & Brown.

Stegner, Wallace. 1953. *Beyond the Hundredth Meridian*. Boston: Houghton Mifflin.

Whose Canyon Is It?

Native Americans have lived and visited Grand Canyon for millennia. Today, tribes such as the Navajo, Hualapai, and Havasupai continue to live in and along Grand Canyon. Grand Canyon has sustained and nurtured these and other tribes for centuries, and they, in return, have revered and protected the canyon. Native Americans have left few visible marks on Grand Canyon.

By the 1800s, territorial disputes were increasingly common in the Southwest. In 1848, the Treaty of Guadalupe Hidalgo ended the Mexican-American War. That treaty gave the United States much of northwest Mexico, which includes Grand Canyon. (The Gadsden Purchase in 1853 completed the modern boundary with Mexico.) By the mid-1800s, settlers in the American West were fighting Native Americans for resources and land ownership. These wars, including the Navajo War (1858–1864) and Hualapai War (1865–1870), were often fueled by Manifest Destiny, an ideology claiming that the United States was destined to expand from coast to coast in North America. In light of this ideology, Native Americans such as the Navajo and Hopi were merely obstacles and were therefore condemned as savages, heathens, and barriers to American expansion. (The same tribes would later be patronized as models of environmentalism, peaceful sources of wisdom, and New Age mystics.) Tribes such as the Navajo and Hualapai fought the U.S. military, while others surrendered.

In 1868, the U.S. government signed a treaty with Navajo leaders that created the Navajo Nation in northeastern Arizona and parts of Mexico. The Navajo Nation, which was expanded to include parts of eastern Grand Canyon in 1884, 1900, and 1930, today is the largest Indian reservation in the United States: it includes more than 300,000 people spread across 27,000 square miles (this area is larger than 10 of the United States). The Colorado River forms 76 miles of the western border of the Navajo Nation, which includes attractions such as Canyon de Chelly National Monument and Monument Valley, a scenic backdrop made famous in many Western movies.

The Navajo secured a formal treaty, which acknowledges that all parties participating in the treaty are nations. With several other tribes of Native Americans at Grand Canyon, however, the U.S. federal government typically used presidential proclamations to create reservations, thereby making the reservations dependent on the U.S. government. For example:

In 1880, President Rutherford B. Hayes established the Havasupai Indian Reservation around Supai village and nearby cultivated land. The Havasupai reservation was originally 5 miles wide and 12 miles long, but—after the U.S. Army surveyed the area—the reservation was reduced to 518.6 acres so that the remaining land could be exploited by prospectors. In 1975, the GCNP Enlargement Act added more than 200,000 acres to the reservation and gave the tribe access to land atop the South Rim. Today, the Havasupai Reservation borders part of the south side of Grand Canyon.

In 1882, President Chester Alan Arthur created what became the 2.4-million-acre Hopi Reservation, which was surrounded by the Navajo Reservation. The reservation's headquarters are at Kykotsmovi, on the Second Mesa. The vague wording of Arthur's Executive Order began decades of disputes with the Navajo about their reservations' borders and land use.

In 1883, President Arthur established the Hualapai Reservation in the tribe's original homeland along the south side of Grand Canyon. In 1947, this reservation was doubled to nearly a million acres when the Atchison, Topeka and Santa Fe Railroad exchanged some of its land for transportation rights through the reservation. The reservation includes 108 miles of the Colorado River and canyon frontage. The Hualapai Tribe's headquarters are in Peach Springs, Arizona.

Today, the Hopi live east of Grand Canyon, the Havasupai live inside Grand Canyon, the Navajo live along GCNP's eastern edge, the Hualapai border GCNP along the south, and the Southern Paiute live north of the Colorado River on the Arizona Strip. The Zuni Indians, who were once associated with Grand Canyon and now live in New Mexico, also consider Grand Canyon to be their place of origin. Approximately one-fourth of Arizona is occupied by the state's 27 tribes of Native Americans.

Preserving Grand Canyon

Even before the tourism boom at Grand Canyon began in the early 1900s, some farsighted politicians began trying to protect and preserve the area. In 1882, Indiana senator Benjamin Harrison (1833–1901) sponsored legislation to create GCNP, despite the fact that he had never visited Grand Canyon. That legislation, which would have made GCNP the second national park (after Yellowstone in 1872), failed because of strong opposition from miners, loggers, ranchers, and others, as did Harrison's reintroductions of the legislation in 1883 and 1886. On February 20, 1893, then president Harrison—invoking the Forest Reserve Act passed by Congress only two years earlier (which gave presidents the authority to establish forest reserves on land in the public domain)—made Grand Canyon a U.S. Forest Reserve. (Harrison created 16 other such reserves while president.) The creation by Congress of the U.S. Forest Service (as part of the Department of Agriculture) in 1905 resulted in forest reserves (including Grand Canyon Forest Reserve) being renamed National Forests two years later. Because the U.S. Forest Service was primarily concerned with forestry rather than with tourism, there were no entrance stations to Grand Canyon Forest Reserve. Their stations were in places best suited for fighting fires and managing the forest, not in the growing village at Grand Canyon's South Rim.

Grand Canyon was given additional federal protection in 1906 when President Theodore Roosevelt created Grand Canyon Game Preserve and even higher status two years later when Roosevelt—wielding power given him by the 1906 American Antiquities Act—made it Grand Canyon National Monument on January 11, 1908. Grand Canyon Forest Reserve and Grand Canyon National Monument were managed by the U.S. Forest Service from 1905 to 1919. Most of the area's roads, hotels, and other facilities were created by individuals and from informal partnerships with the Atchison, Topeka and Santa Fe Railway and the Fred Harvey Company.

In 1910 and 1911, there were more failed attempts to create GCNP. However, on April 4, 1917, Senator Henry Ashurst introduced Senate Bill 390 to create GCNP. That bill was approved on May 16, 1918, after which World War I and disputes about park boundaries delayed the legislation. On October 18, 1918, the bill passed the House of Representatives' Committee on Public Lands and was sent to President Woodrow Wilson. Wilson signed the legislation on February 26, 1919, thereby creating Grand Canyon National Park as America's 15th national park. This ended logging in the area that included GCNP, but it did not mean that all of the land belonged to GCNP because, beginning in 1862, a series of Homestead Acts had allowed settlers in the American West to claim up to 160 acres (0.25 square mile) of land. Similarly, mining laws during the same period allowed anyone to claim public land for mining. Although anyone could have claimed up to 160 acres of land at Grand Canyon before 1908 (e.g., as people did for hundreds of mines), only two people—John Hance and Pete Berry—filed patents for homesteads on land that was later made part of GCNP.

Although Native Americans' centuries-old claims to land in the Grand Canyon area were ignored, the existing "inholdings" and "private entries" of 391.48 acres of private land in GCNP were protected when Woodrow Wilson created GCNP in 1919, as they had been when Roosevelt created Grand Canyon National Monument in 1908. However, park officials soon realized that this private land within GCNP contradicted the mission of the NPS "to conserve the scenery and the natural and historic objects and the wild life therein and to provide for the enjoyment of the same in such manner . . . as will leave them unimpaired for the enjoyment of future generations." For example, William Bass—who built the first rim-to-rim trail in Grand Canyon—could have developed a commercial corridor from one side of Grand Canyon to the other if he'd had the money to do so.

Of all the privately owned land in GCNP, mining claims were the most problematic for NPS because they were so extensive; indeed, entrepreneurs such as Ralph Cameron filed hundreds of mining claims spanning large tracts of key land in Grand Canyon (e.g., Indian Garden) and along the South Rim (e.g., at Bright Angel Trailhead), thereby enabling him to impose tolls on visitors wanting to access popular parts of Grand Canyon. Cameron also operated the Cameron Hotel atop a mining claim at the South Rim from 1903 to 1907. These claims had no relation to mining, but—thanks to Cameron's influence and political power—remained in effect for decades.

To accomplish the NPS mission, NPS's first director (Stephen Mather) began trying to reclaim GCNP's private landholdings. As Mather soon learned, this was a difficult chore, for there was usually little public support for governmental agencies to take land away from pioneers. Most citizens supported pioneers being able to use unprotected federal lands for business activities such as mining, grazing, and forestry. Stephen Mather spent much of his own money to eliminate private holdings in national parks and convinced several companies to donate their land and facilities to GCNP. For example:

In 1926, Mather convinced the Santa Fe Land Development Company to buy William Bass's many claims and associated properties (e.g., roads, camps) for $25,000 to ensure that they would not be developed by private companies. When the successor of the Santa Fe Land Development Company ended its tourism-related businesses in GCNP in 1954, these properties were sold to the Fred Harvey Company, who later gave them to GCNP.

In the 1890s, tourism entrepreneur Sanford Rowe claimed several blocks of land south of today's Grand Canyon Village. This land remained in private control until the NPS bought the land in 1956. Five years later, GCNP tore down all of the buildings there.

In 1906, Dan Hogan and Charlie Babbitt patented a 20.6-acre claim for Orphan Mine 13 miles east of Grand Canyon Village. Hogan opened a trading post at the site in 1936 and sold the land in 1947 to Madeleine Jacobs, who leased the land to a mining company that extracted large amounts of uranium from the site. Mining ended in 1969, and the land (now contaminated with uranium and closed to the public) was returned to GCNP in 1987.

In the early 1890s, Peter Berry and other investors bought land at Grand View Point to access their Last Chance Mine at Horseshoe Mesa. Berry and his partners sold this land to newspaper publisher William Randolph Hearst in 1913. In 1939, NPS—for the first and only time—used a condemnation (i.e., eminent domain) lawsuit to obtain the land from the Hearst estate for $85,000. The two so-called Hearst Tanks, which are ponds that often fill with water during the rainy season, are 0.9 mile southeast of Grand View Point. Aside from the trailhead and some signage, there are no visual signs of Grandview Point's colorful past.

When Pete Berry's homestead at Grand View Point was returned to the park in 1939, only one private homestead—that is, the one patented by John Hance southeast of Grand View Point in 1884—remained in GCNP. Hance sold his land to James Thurber who, in 1906, sold it to Martin Buggeln, the former owner of Bright Angel Hotel. When NPS refused to help Buggeln launch a tourism business at the South Rim, Buggeln refused the NPS request for a road right-of-way through his property when it built a road from Grand Canyon Village to Desert View, thereby forcing the NPS to detour the road around Buggeln's property. In 1948, NPS bought Buggeln's land from Buggeln's widow, Eva Moss Buggeln, and later rebuilt the road to Desert View in its current location.

Today, there is only one site within the original boundaries of GCNP that remains under private control: the 325 acres patented in 1901 by John Hance for his asbestos mines on the north side of the Colorado River just below Hance Rapid. NPS did not know of this claim until 1930, but since then has done relatively little to obtain the land. (The land is hard to access and most covers steep slopes.) In the 1950s, William Randolph Hearst bought the land in exchange for paying delinquent taxes. Today, this land—near River Mile 78.7—continues to be owned by the Hearst Estate and is the only privately owned land within the original boundaries of GCNP. Visitation is not allowed at this site.

Conservation and Exploitation at Grand Canyon

As Grand Canyon became increasingly popular, many people began viewing the area as an exploitable source of riches. Early hunters and gatherers were replaced by explorers and prospectors, who came to Grand Canyon to find gold, copper, and other valuable ores. By the early 1900s, people at Grand Canyon began claiming land for businesses. Many of these claims involved mining, and many of these mines—such as those by Pete Berry, Bill Bass, and Dan Hogan—were profitable, working mines. Later, many of these early settlers and prospectors—realizing that they could make more money by catering to tourists than by digging in their mines—transformed themselves into tour operators. In the process, the canyon itself became a lab and museum as the public's image of Grand Canyon was passed from John Wesley Powell, Robert Stanton, and other explorers and pioneers, to businesses and businesspeople, such as John Verkamp, Emery and Ellsworth Kolb, and the Fred Harvey Company, and finally to the artists, cameras, postcards, and websites that promote GCNP today. Throughout it all, Grand Canyon has endured as a unique natural wonder and World Heritage Site.

Grand Canyon continues to lure and inspire millions of visitors from all over the world. Its geologic formations remain as spectacular and impressive as when they were first described, and its vistas remain among the most beautiful in the world. However, despite the plea of Teddy Roosevelt to "do nothing to mar [Grand Canyon's] grandeur, sublimity and loveliness" and the protections afforded to Grand Canyon after it was made a national park, the balance between preserving and exploiting Grand Canyon has become increasingly tenuous. In many regards, the area remains unchanged; its vistas and geological formations are nearly identical to those encountered by Native Americans and other pioneers in Grand Canyon.

However, Grand Canyon—which *has* been changed by people who have ignored Roosevelt's advice—is not as it was when first visited by the canyon's Native peoples and early explorers. One of the places where these changes are most obvious is along the Colorado River. To best appreciate this, consider the 1869 voyage of John Wesley Powell and his crew through Grand Canyon. As Powell's expedition proceeded down the Colorado River, they were awed by a pristine canyon showcasing native and endemic plants and animals, all sustained by a powerful, untamed river whose water was warm and laden with silt. The speed of the river on which they floated varied and was changed by local rains.

Now consider what today's river runners encounter as they retrace Powell's journey through Grand Canyon. Although the rock formations have changed relatively little, the Colorado River has been tamed; its flow, instead of being wild and variable, is now relatively steady and determined not by nature but instead by releases from Glen Canyon Dam to meet the energy demands of the area's residents. In 1869, the water on which Powell and his crew floated through Grand Canyon in August was silt-laden and near 80°F. Today, however, the Colorado River's water comes from the bottom of Lake Powell and is near 49°F year-round. Because 90 percent of the river's silt is now deposited behind Glen Canyon Dam instead of in and beyond Grand Canyon, water flowing through Grand Canyon is unnaturally clear and virtually silt-free (except after a storm, when silt from the Colorado River's tributaries stains the water). Many of the large beaches in Grand Canyon described by Powell and other explorers are gone. At the core of these changes is the conflict between the creation of GCNP and the economic benefits of Glen Canyon Dam.

These environmental changes in Grand Canyon have also changed life in Grand Canyon. Most species of native fish, such as the endangered humpback chub (*Gila cypha*) and razorback sucker (*Xyrauchen texanus*), are now gone, having been replaced by exotic species such as trout. Similarly, many species of native plants in Grand Canyon, such as sentry milk-vetch (*Astragalus cremnophylax*), have been replaced by more than 180 species of exotic and invasive plants such as Russian olive (*Elaeagnus angustifolia*), Sahara mustard (*Brassica tournefortii*), tamarisk (salt cedar; *Tamarix ramosissima*), camelthorn (*Alhagi maurorum*), and ravenna (*Saccharum ravennae*). GCNP rangers and volunteers have pulled up hundreds of thousands of these invasive plants, but their efforts have had few long-lasting effects; indeed, invasive species' ability to grow fast, reproduce vegetatively, and produce large numbers of seeds (many of which can remain dormant for years) make them virtually impossible to eradicate. This is why several side-canyons in GCNP have vast populations of exotic and invasive plants that continue to spread throughout the park. GCNP officials have given up on trying to eradicate many exotic and invasive species in many parts of the park and instead try to "manage" their populations. Although the first attempts to eradicate of these plants began three decades ago, in 2018 there were more than 160 non-native species of plants in GCNP. Invasive plants such as salt cedar are periodically removed from parts of the river corridor, but they will never be eliminated from GCNP.

Although GCNP is no longer the "natural experience" it once was, more than 6 million tourists continue to visit GCNP each year to experience Grand Canyon. Regardless of when, where, why, or how these tourists see Grand Canyon, the canyon is less a place merely to see than it is a cultural, visual, and even spiritual place to experience. Early visitors saw what they were told to see about Grand Canyon through pioneers' ideas and images, including John Wesley Powell's exaggerated descriptions of the canyon's dangerous rapids, Clarence Dutton's descriptions of immense panoramas, and Thomas Moran's beautiful paintings of the canyon's overlooks. For most early visitors, Grand Canyon was beautiful and overwhelming, and for others it was so vast as to inspire religious awe. As Dutton noted about

Grand Canyon in 1882, "Dimension means nothing to the senses, and all we are left with is a troubled sense of immensity" (Dutton, 1892).

For these and other reasons, Grand Canyon in 1979 was made a United Nations Educational, Scientific and Cultural Organization (UNESCO) World Heritage Center, a designation that protects the canyon and recognizes its collective value to humanity worldwide. Since it became a national park, GCNP has attracted more than 200 million visitors.

Further Reading

Arizona State University. 2017. Who owns the Grand Canyon? Retrieved July 9, 2017, at http://grandcanyonhistory.clas.asu.edu/history_whoownsgrandcanyon.html

Dutton, Clarence E. 1882. *Tertiary History of the Grand Canyon*. Washington, DC: U.S. Government Printing Office. (Reprinted 2001 by the University of Arizona Press, Tucson, Arizona, with an introduction by Wallace Stegner and a new foreword by Stephen J. Pyne.)

Hirst, Stephen. 2007. *I Am the Grand Canyon: The Story of the Havasupai People*. Third edition. Flagstaff, AZ: Grand Canyon Association.

Neumann, Mark. 1999. *On the Rim: Looking for the Grand Canyon*. Minneapolis: University of Minnesota Press.

Pyne, Stephen J. 1998. *How the Canyon Became Grand*. New York: Penguin Books.

Schwartz, Douglas. 1989. *On the Edge of Splendor: Exploring Grand Canyon's Human Past*. Santa Fe, NM: School of American Research Press.

Shields, Kenneth, Jr. 2000. *The Grand Canyon: Native People and Early Visitors*. Charleston, SC: Arcadia Publishing.

Thybony, Scott. 2007. *The Incredible Grand Canyon: Cliffhangers and Curiosities from America's Greatest Canyon*. Grand Canyon, AZ: Grand Canyon Association.

Whose Water Is It?

After John Wesley Powell (1834–1902) scouted the Colorado River through Grand Canyon in 1869, one thing was obvious to him: There was not enough water to irrigate all the land that needed to be irrigated. Although Powell was right, most other people—especially businesspeople and politicians—disagreed. Railroads lobbied for policies that promoted expansion in the Southwest because they wanted to sell some of the more than 280,000 square miles of land they had been given in exchange for building railroads in the West. Politicians and others claimed that homesteading in the Southwest (and the farming that would accompany it) would somehow make the area wetter and more suitable for agriculture and, in the process, transform much of the region into a large farm. All of these people and even some scientists invoked Manifest Destiny, which claimed that it was the divine right and duty of the United States to spread from the Atlantic to the Pacific. In the Southwest, this belief was accompanied by the now discredited slogan, "Rain follows the plow." In response to these promises, more and more people began moving to the Southwest.

Until his death in 1902, Powell continued to point out problems with governmental policies regarding water in the American Southwest. For example, at an 1893 conference on irrigation, Powell prophetically noted that government officials were setting themselves up for trouble and litigation, for there was not enough water to supply the land. Despite Powell's fame and stature as a Grand Canyon explorer, as well as the evidence supporting his claims, politicians pushed forward. The results have been as Powell predicted: In most years, the Colorado River does not reach the Gulf of California, its natural destination. Bridges in Mexico that once crossed the Colorado River now cross a dry riverbed. The Colorado River's delta, once a lush wetland covering more than 2,000 square miles and nourished by the river's water and silt, is now gone.

How Did These Problems Begin?

The Colorado River, which flows through or along seven Western states, is the lifeblood of the American Southwest. The Colorado River supplies drinking water for 10 percent of Americans, and the daily lives of more than 30 million people depend directly on water from the Colorado River. Indeed, most of the water that people drink in Las Vegas and Los Angeles, use to flush toilets in Phoenix, and bathe with in San Diego comes from the Colorado River. Water from the Colorado River is used to irrigate more than 80 percent of the nation's winter vegetable crops.

Providing enough water for this many people and this much agriculture has required the construction of 10 major dams and 80 major diversions of the Colorado River. All of these dams and diversions, including those related to Grand Canyon, are regulated by a piecemeal group of federal and state laws, contracts, court decisions, regulatory guidelines, and international treaties collectively called the Law of the River. The foundation of the Law of the River is the Colorado River Compact. The flow of the Colorado River through Grand Canyon is a focal point of that Compact.

In the early 1900s, California's increasingly rapid use of water from the Colorado River began to concern several states in the Southwest. These states were worried that California would make legal claims to the large amounts of water in the Colorado River that they were using and, in doing so, deprive other states of much needed water. By 1911, even President William Howard Taft was interested in the problems that had arisen regarding the control of the Colorado River. In response to these concerns, seven states—Colorado, New Mexico, Utah, Wyoming, Arizona, California, and Nevada—formed the League of the Southwest in 1917 to ensure the orderly and equitable development of the Colorado River and the division of the Colorado River's water. The following year, as the League did its work, Congress passed the Kincaid Act, which ordered the Secretary of the Interior to study and report on the use and possible diversions of the Colorado River's water. While this study was being done, all of the seven states—each still worried that most of the Colorado River's water would be stored elsewhere—began their own negotiations about the distribution and use of water from the Colorado River.

Underlying these states' negotiations was a political question: Can groups of states form agreements among themselves that exclude other states? This question is addressed by the so-called Compact Clause, which is contained in Article 1, Section 10, Clause 3 of the U.S. Constitution. In the early 1920s, this clause, which stipulates Congressional approval of any foreign or interstate agreements that "encroach upon the supremacy of the United States," had been used to settle a variety of questions related to issues such as boundaries and criminal jurisdictions but had never been applied to interstate rivers. On August 19, 1921, Congress—realizing the growing importance of the Colorado River to the Southwest—consented to the seven states' negotiations about the distribution of water of the Colorado River. After 27 meetings (chaired by then Secretary of Commerce Herbert Hoover), delegates from the states—meeting on November 24, 1922, in Santa Fe, New Mexico—produced an agreement called the Colorado River Compact. All of the seven states except Arizona ratified the Compact the next year (i.e., in 1923), and Congress approved the Compact in Section 13(a) of the Boulder Canyon Project Act in 1928. This legislation was signed into law the following year by President Herbert Hoover, thereby making the Colorado River Compact official. The Colorado River Compact also authorized the construction of Hoover Dam and related irrigation facilities in Arizona, California, and Nevada.

Arizona's first governor, George Wylie Paul Hunt (1859–1934), opposed the Colorado River Compact because he believed it gave California too much water. In the mid-1930s, Benjamin B. Moeur (1869–1937)—the fourth governor of Arizona—even called out the National Guard and commissioned a two-boat "navy" to protect what he called "Arizona's water" from being rerouted to Los Angeles. (When one of those boats, a dilapidated steamboat, became stranded, workers from California rescued the boat's crew.) Arizona finally approved the Compact in 1944, and the dispute about the allocation of water from the Colorado River was settled by the U.S. Supreme Court in 1964 in *Arizona v. California*. The 16 hours of oral arguments in this case set a modern record for the Supreme Court.

The Colorado River Compact, which was the first application of the Constitution's Compact Clause to interstate rivers, is only about 2,000 words long. However, it remains the defining document for the use of Colorado River's water, and its impact on Grand Canyon has been tremendous. The Compact, which was implemented by Floyd Dominy of the Bureau of Reclamation, allocates water from the Colorado River equally between two groups of states called the Upper Basin and Lower Basin. Members of these groups, each of which divides their allocation of water among themselves, are as follows:

Upper Basin. Colorado, New Mexico, Utah, and Wyoming (plus a small part of Arizona that is tributary to the Colorado River). Most of the water in the Colorado River comes from snow melt in the Upper Basin.

Lower Basin. Arizona, California, and Nevada (plus small parts of New Mexico and Utah that are tributary to the Colorado River). Most of the water in the Colorado River is used in the Lower Basin. Precipitation in the Lower Basin often has important local impacts but seldom affects water levels in Lake Powell or Lake Mead.

The Colorado River Compact provided assurances for both groups of states. To ensure that the Upper Basin states do not hoard the river's water, the Upper Basin must provide an aggregate flow of 75 million acre-feet of water to the Lower Basin over any 10-year period. (One acre-foot of water covers an acre with water one-foot deep; this is approximately 325,000 gallons of water, which is the amount of water used in a year by a typical family of four.) This allocation, which was based on precipitation patterns over several years before the Compact's signing in 1922, meant that states in the Upper Basin no longer had to worry that faster growing states of the Lower Basin would use more water and thereby win rights to that water. However, as John Wesley Powell predicted, it also meant that drought and shortages of water in Lake Powell—such as those that have occurred since 2000—create important legal, scientific, and public policy problems.

There were several caveats. For example, during periods of water surplus, the Lower Basin can get an additional 1.1 million acre-feet of water, and The Mexican Water Treaty (signed on February 3, 1944) allocates 1.5 million acre-feet of the Colorado River's water to Mexico. According to Minute 319 passed on November 20, 2012, by the International Boundary and Water Commission of the United States and Mexico, the annual flow to Mexico will increase by 40,000 acre-feet when the elevation of Lake Mead exceeds 1,145 feet and will diminish by 50,000 acre-feet if the surface elevation of Lake Mead is between 1,075 and

1,050 feet, by 70,000 acre-feet if the surface elevation is between 1,050 and 1,025 feet, and by 125,000 acre-feet if the surface elevation drops below 1,025 feet. The Compact's allocations of water mean that California's annual allocation of water from the Colorado River would drop Lake Mead by 44 feet, Arizona's annual allocation would drop Lake Mead by 28 feet, and Mexico's annual allocation would drop Lake Mead by 15 feet. The annual allocation to Nevada, the least populated of these states, would drop Lake Mead by only about 30 inches.

The Colorado River Compact and Grand Canyon

The point at which the Colorado River's water is allocated to the Upper and Lower Basins is Compact Point (also called Lee Ferry), which is 1.6 miles downstream from Lees Ferry (just past the Paria Riffle). Lees Ferry, which includes a pillar-shaped USGS gaging station on river left, is also where most river trips through Grand Canyon begin.

Although the Colorado River Compact allocates prescribed amounts of the Colorado River's water to different states, the flow of the river at Lees Ferry fluctuates widely from year to year. Hoover Dam, which was completed in 1935 just past the western end of Grand Canyon, controls only the lower parts of the Colorado River, thus leaving states in the Upper Basin with no way to store water to counteract low-flow years. To help the Upper Basin states and to moderate the Colorado River's flow, Dominy fought hard to build more dams on the Colorado River in order to store water and generate hydroelectric power. One of these dams is Glen Canyon Dam, which is just 15.5 miles upstream from Lees Ferry; this dam is the physical manifestation of the Colorado River Compact.

Dams in Grand Canyon

The first dams *in* Grand Canyon were proposed in 1913 but none were built then. When GCNP was created in 1919, most of its land was protected as in other national parks and monuments. However, its water—that is, the water in the Colorado River—was not. Although the National Park Service Act of 1916 prohibited water development projects (e.g., dams) in national monuments and parks, Arizona Congressman Carl Hayden (a sponsor of the bill that created GCNP) inserted special language into the GCNP legislation that enabled the Secretary of the Interior to authorize such projects in GCNP at his/her discretion.

By 1923, a variety of proposals sought to exploit the water and power of the Colorado River through Grand Canyon. Because all of these projects required accurate maps, Congress authorized USGS geologist Claude Hale Birdseye to lead a two-month, 19-day survey of Grand Canyon to determine the best places to build dams along the Colorado River that would enhance California agriculture and produce hydroelectric power. Birdseye's survey was the first (and only) instrument-based survey of the Colorado River in Grand Canyon. Birdseye's group of 12 men (which included Emery Kolb as head boatman) used four 800-pound boats stocked with two tons of equipment, including the first radio used on a run down the Colorado

River through Grand Canyon. (This radio was a public relations coup; radio station KHJ [Kindness-Happiness-Joy] in Los Angeles aired special broadcasts to and about Birdseye's expedition.) The Birdseye Expedition, which was later featured on the cover of *Scientific American*, was also the basis for the first *Belknap Grand Canyon River Guide*. The *Belknap Guide*, which was issued in 1969 to commemorate the centennial of John Wesley Powell's expedition, became a standard resource for river runners in Grand Canyon.

Birdseye identified 29 promising sites for dams between Lees Ferry and Black Canyon. At the end of Birdseye's expedition, USGS issued press releases that made sensational headlines throughout the United States. Birdseye's expedition was the last great exploration of the American West; subsequent expeditions used staged disappearances, movies, cute animals, and other gimmicks to get the public's attention. However, none of the dams suggested by Birdseye were ever built.

In 1932, President Herbert Hoover invoked the American Antiquities Act of 1906 to create the 426-square-mile Grand Canyon National Monument (that included Toroweap) to recognize the scientific value of western Grand Canyon. The earlier Grand Canyon National Monument became part of GCNP. However, the legislation that created this monument included no provisions to block dams and other reclamation projects. Grand Canyon National Monument remained a separate NPS unit until 1975, when it too became part of GCNP.

Around 1930, newspaper magnate William Randolph Hearst (1863–1951), who owned property at Grand View Point and elsewhere in Grand Canyon, threatened to develop his property by operating a train powered by electricity provided by a dam. Hearst threatened to build a dam on the Colorado River, but he never did.

In 1946, interest was revived in building dams in GCNP. That year, the U.S. Bureau of Reclamation's *The Colorado River, A Natural Menace Becomes a National Resource* proposed high dams at Bridge Canyon (named for a small natural bridge near River Mile 237.5, which was then 78 miles below the western boundary of GCNP) and Marble Canyon (River Mile 39.5, initially called the Redwall site, which was above the eastern boundary of GCNP). These dams would have profoundly changed GCNP. For example, Marble Canyon Dam would have flooded 53 miles of the Colorado River, and Bridge Canyon Dam would have backed up water for 93 miles, thereby flooding Lava Falls and the lower end of Havasu Canyon. Talus and scars from the test holes made in the canyon's walls are still visible at the construction sites of these dams.

Plans for these dams were initially supported by Sierra Club executive director David Brower (1912–2000), who later (in the 1960s) led the Sierra Club's *opposition* to dams in Grand Canyon. The dams were also supported by the Federal Power Commission and the Hualapai tribe, the latter of whom were to be paid $1 million per year for their flooded land. (Bridge Canyon divides the Hualapai Reservation on the south side of the Colorado River from GCNP on the north side of the river.) However, the Bureau of Reclamation later recommended building five major dams above Lees Ferry; these dams included Glen Canyon Dam and Echo Park Dam. Arizona, which wanted to get the Colorado River's water to Phoenix and Tucson, strongly supported construction of Glen Canyon Dam; California, however, opposed

In the 1940s, the U.S. Bureau of Reclamation began building Marble Canyon Dam, a 310-foot-high hydroelectric dam proposed at River Mile 39.5 that would have flooded more than 50 miles of Grand Canyon. Public opposition to the dam, led by the Sierra Club, later helped stop construction of this and other dams in Grand Canyon. This exploratory tunnel on river left at the dam's proposed site is a monument to that halted work. (Randy Moore)

the dam because it could eliminate surplus water from the Colorado River that California had grown accustomed to using.

To block the construction of Echo Park Dam in Colorado (which would have flooded much of scenic Dinosaur National Monument), environmentalists eventually consented to construction of Glen Canyon Dam just upstream from Lees Ferry. This dam was authorized in 1956 by the Colorado River Storage Project Act, a part of the Law of the River. The dam's site was chosen for the height of the canyon's walls and the narrowness of the canyon there. The Colorado River Storage Project Act also approved several irrigation projects and other dams for "regulating the flow of the Colorado River, storing water for beneficial consumptive use, providing for reclamation of arid and semi-arid lands, providing flood control, and generating hydro-power."

Glen Canyon had been proposed as a national monument in 1935–1941 and 1946, and some people considered its many deep, colorful sandstone canyons to be more beautiful than Grand Canyon. The fight over Glen Canyon Dam—which flooded Glen Canyon—was one of the first major battles of the conservation movement in the United States. David Brower and politician Barry Goldwater both later admitted that their support for Glen Canyon Dam was their biggest mistake.

After it was authorized in 1956, construction of Glen Canyon Dam began in 1957, and from 1961 to 1964, workers poured concrete 24 hours per day. It took 400,000 24-ton loads of concrete to build Glen Canyon Dam, whose primary purpose was to store water and sediment in the arid Southwest. The dam's incidental purposes were to generate electricity and opportunities for recreation. Glen Canyon, which had neither the fame nor historical cache of Grand Canyon, was flooded, and Grand Canyon was not. In the process, Glen Canyon became known as "The Place That No One Knew."

The completion of Glen Canyon Dam in 1963 stabilized the flow of the Colorado River and produced a longer river running season. For many years, the dam was operated without consideration of the downstream environment. Because the dam trapped sediment and drastically reduced the sediment load of the Colorado River, beaches in Grand Canyon that were home to many native species began to erode. Four native species of fish soon disappeared, and a variety of invasive species changed the riparian ecosystems. The closure of the gates of Glen Canyon Dam in 1963 removed any pretense that the Colorado River is a wild river through Grand Canyon or that Grand Canyon is a natural environment.

The dramatic and extensive environmental changes in Grand Canyon produced by Glen Canyon Dam finally demanded action. In 1992, concerns about the environmental impact of Glen Canyon Dam led to passage of the Grand Canyon Protection Act, which was sponsored by Arizona senators John McCain and Dennis DeConcini to help shield Colorado River resources from degradation caused by Glen Canyon Dam. The legislation required that the dam be operated "in such a manner as to protect, mitigate adverse impacts to, and improve the values for which Grand Canyon National Park and Glen Canyon National Recreation Area were established." The act directed the Secretary of the Interior to protect the Colorado River while trying to integrate Glen Canyon Dam with the natural environment. Today, the Grand Canyon Monitoring and Research Center is responsible for scientific (e.g., natural history, recreational, ecological, and cultural) research along the Colorado River in Grand Canyon. Despite these efforts, however, Grand Canyon remains profoundly changed by Glen Canyon Dam.

In 1963, the U.S. Bureau of Reclamation again proposed building dams in Grand Canyon. Soon thereafter, the proposed sites of these dams even appeared on roadmaps. However, these plans were made moot in 1968, when President Lyndon B. Johnson signed Public Law 90–537, which banned hydroelectric dams in any part of Grand Canyon without Congressional approval. Arizona senator Carl Hayden, in his 47th and final year as senator, had earlier cosponsored legislation authorizing Glen Canyon Dam but declared that the ban on dams in Grand Canyon was "the high water mark" of his career as a U.S. senator. The following year, just before leaving office, President Johnson created Marble Canyon National Monument, thereby bringing the proposed site of Marble Canyon Dam under the jurisdiction of NPS. In 1975, the Grand Canyon National Park Enlargement Act made this monument a part of GCNP and created today's boundaries of GCNP.

The power that was to be generated by dams in Grand Canyon was later provided by the Navajo Generating Station, whose groundbreaking ceremony on

April 22, 1970, coincided with the original Earth Day. This coal-powered plant produced twice as much energy as was predicted for the dam. In 2017, officials announced that the Navajo Generating Station would close at the end of 2019.

Today, the Colorado River Compact continues to affect Grand Canyon and regulate the allocation of water from the Colorado River. As Powell predicted in 1893, the allocation remains contentious. Unbeknownst to negotiators of the Colorado River Compact in 1922, the average flow of the Colorado River on which the Compact is based (approximately 16.4 million acre-feet) overestimated the river's flow because it was based on decades of data from one of the wettest periods in the past 800 years. Because today's flows of the Colorado River are significantly less than those on which the Colorado River Compact was based, there is now a "structural deficit" of more than a million acre-feet of water in the Colorado River Compact's allocation.

Compounding the problem is the ongoing drought in the American Southwest. In early 2018, the surface of Lake Mead was at an elevation of 1,083 feet. This means that if the surface drops another 13 feet, water to Arizona could be cut. Colorado River Indian Tribes has some of the oldest rights to water from Lake Mead, but federal law does not allow these tribes to share or ship water. However, Congress may change that if the drought continues.

All of the water in the Colorado River—and then some—is promised to someone, and none of the river's water reaches the river's delta. Colorado, where precipitation and snow melt account for more than two-thirds of the river's water, is considering a variety of projects to further divert water from the river. One such project would take water from the river and pump it over the Rocky Mountains to Colorado's densely populated Front Range. In most areas, farmers' legal rights to the river's water trump those of others.

In 2017, the environmental group American Rivers named the Lower Colorado River America's most endangered river, as it had done four other times since 2003.

Further Reading

Boyer, Diane E., and Webb, Robert H. 2007. *Damming Grand Canyon: The 1923 USGS Colorado River Expedition*. Logan: Utah State University Press.

Kuhbander, Alexis. 2017. Parched by drought, Lake Mead water levels continue to fall. *Williams Grand Canyon News*, June 21, 2017.

Stegner, Wallace Earle. 1982. *Beyond the Hundredth Meridian: John Wesley Powell and the Second Opening of the West*. Lincoln: University of Nebraska Press.

The Colorado River's Historic Sites

Every year, millions of visitors visit the North and South Rims of Grand Canyon to see the canyon's stunning panoramas and learn about how the canyon was discovered, explored, and preserved. These visits to the canyon's rims are marked by ubiquitous questions such as, "What's that down there?" and "How far away do you think that is?" that are accompanied by the names of sites such as Desert View Watchtower, Hermits Rest, and Grand Canyon Village. Although views from the rims are from different overlooks and feature different geological formations, the views are all similar in that they all peer down into the canyon. Relatively few of these views include the Colorado River, which defines Grand Canyon. Indeed, many people who visit Grand Canyon leave the park without ever seeing the river.

Although the North and South Rims attract more than 98 percent of the visitors to GCNP, some visitors venture to the bottom of Grand Canyon along a variety of trails to get a different view and perspective of GCNP. At the bottom of Grand Canyon along the Colorado River, the experience, sounds, lifestyle, schedule, and views are dramatically different; most of the rims' iconic sites are invisible, all of the views are up toward the rim, and conversations among visitors are peppered with names of sites such as Lees Ferry, Phantom Ranch, Hance Rapid, and Lava Falls Rapid. The experience at the bottom of Grand Canyon has little in common with that at the top.

Following are the iconic landmarks along the Colorado River through Grand Canyon. Although the Colorado River's many exciting rapids comprise part of this history, the history, culture, and experience at the bottom of Grand Canyon is far more than just running the river's rapids. The history of people along the Colorado River is marked by relics, graves, inscriptions, and memorials of countless stories of exploration, development, and tragedy that are integral parts of GCNP.

Mile 0–Mile 50

Lees Ferry is a relatively calm area along the river that designates River Mile 0 in GCNP. The ferry is named for John Doyle Lee (1812–1877), a Mormon who began operating the ferry in the early 1870s. The ferry stopped operating in 1928. Most river runners begin their trips here. Downstream travel from Lees Ferry requires a permit.

Navajo Bridge (also known as Lees Ferry Bridge and Marble Canyon Bridge) at River Mile 4.2 is about 470 feet above the river. The 834-foot-long upstream bridge, which is now a pedestrian footpath, cost $385,000 and opened in

George Wheeler

In 1871, the U.S. War Department sent Lt. George Montague Wheeler (1842–1905) to survey land west of the hundredth meridian. Wheeler, a graduate of West Point, and his crew left Needles, California, and went 200 miles up the Colorado River to the area near Grand Wash Cliffs. Guided by Mohave Indians, the Wheeler Survey took three boats up the river to Diamond Creek. With Wheeler was Timothy H. O'Sullivan, who is credited with taking the first photographs of Grand Canyon. After Grand Canyon, Wheeler continued his survey of present-day California, Nevada, Utah, Arizona, and New Mexico

The 63-year-old Wheeler died in New York City on May 3, 1905. He is buried in the U.S. Military Academy Post Cemetery in West Point, New York.

1929. This bridge's opening was attended by more than 6,500 people in 1,217 cars. A similar $15 million bridge located 330 feet downstream opened in 1995. Near Navajo Bridge is the first appearance of Coconino Sandstone at river level. Navajo Bridge is the last downstream car crossing until Hoover Dam, which is 342 miles away.

Frank Brown Inscription at River Mile 12.1 was carved by Peter Hansbrough about 15 feet above river level on river left. On July 10, 1889, Frank Mason Brown—president of the Denver, Colorado Canyon and Pacific Railroad—drowned near this point when the boat he was in capsized. Brown, who was hoping to build a railroad through Grand Canyon, was not wearing a life jacket. Five days later, Hansbrough also drowned. Brown's railroad was never built.

Norm Nevills Inscription "N.N. No 1 7-12-47" from Nevills's 1947 trip through Grand Canyon is on river left at River Mile 12.4.

USGS Inscription at River Mile 16.1 consists of a small rock pick (part of the USGS logo at the time) and "USGS" etched beneath an overhang on river right. The inscription, which is pecked in Supai rock, was made on August 4, 1923, during the USGS expedition led by Claude Birdseye.

Driftwood Logs at River Mile 18.7 atop the Supai boulder in midstream were probably deposited in 1958 by two floods that exceeded 100,000 cubic feet per second (cfs). This area, which is called Boulder Narrows, is 1,800 feet below the rim; the boulder here (Supai Group) is the largest boulder you'll see in the river. Just downstream from this site, in August 1986, a monoplane buzzed friends who were enjoying a rafting trip through Grand Canyon. The plane then crashed, killing the pilot and his son.

Lone Cedar Beach at River Mile 23.5 is where the Robert Stanton Expedition through Grand Canyon stopped for lunch on July 13, 1889. About 80 yards upslope from the river on river left is a lone one-seed juniper tree in which prospector Harry McDonald (part of Stanton's crew) carved "H McD 1889." This is the only known tree carving in GCNP.

Harry McDonald was a prospector and boatman with Frank Brown's expedition in 1889 to plan a railroad through Grand Canyon. On July 10, the boat carrying Brown and McDonald flipped near River Mile 12.1; Brown drowned, but McDonald swam safely to shore. Three days later, during a lunch break, McDonald carved "H.McD." in the base of a juniper tree at Lone Cedar Beach (River Mile 23.5). This is the only known wood-carved inscription in Grand Canyon National Park. (Randy Moore)

Marble Canyon Dam was proposed to be built between River Miles 30 and 40 in the 1920s. The project identified two possible sites—one at River Mile 32.2 and another at 39.4—for a dam that would have impounded water for the generation of hydroelectric power while dramatically diminishing the flow of the Colorado River through much of GCNP. Work began at both sites in the 1940s but was later abandoned. Today, the worksites are marked by drill holes and drifts of excavated soil and rock. The upper site was located just above Redwall Cavern (see below) and was often called the Redwall Dam site.

South Canyon on river right at River Mile 31.5 is where, in 1889 after the drowning of Brown, Henry Richards, and Hansbrough, Robert Stanton and his remaining men—Harry McDonald, George Washington Gibson, Frederick Nims, and John Hislop—hiked out of Grand Canyon. Stanton named the point opposite this canyon Point Retreat.

Stantons Cave at River Mile 32.0 is approximately 120 feet above the river on river right. After the disastrous first expedition to map a railroad route through Grand Canyon in 1889 that killed three people, Robert Brewster Stanton stored cameras, tobacco, and other supplies in this dry, 140-foot-deep cave before hiking out of Grand Canyon. In 1890, Stanton retrieved the supplies and continued down the river. In 1934, Bus Hatch found several split-twig figurines in this cave, which were later estimated to be 4,000 years old. When Robert Euler excavated the cave in 1963, he found 165 more of the figurines. Today, a 2,000-pound steel barrier keeps looters out of the cave but does not impede a summer colony of Townsend's big-eared bats (*Corynorhinus townsendii*) that live in the cave.

Vaseys Paradise on river right near River Mile 32.2 features a verdant spring and lush area that includes ferns, native poison ivy (*Toxicodendron rydbergii*), and the dime-sized, federally endangered Kanab ambersnail (*Oxyloma haydeni kanabensis*). The waterfall here, which flows at an average rate of 10 gallons of water per second (at 58–66°F), emerges from a cave in the Redwall Limestone created by North Rim rain and snow melt. (There are more than 2.5 miles of caves above the river here.) The Toroweap, Coconino, and Redwall formations are prominent here. Powell named this place for botanist George W. Vasey (1822–1893), who had helped him explore the Rocky Mountains in 1868.

Redwall Cavern at River Mile 33.3 on river left is a giant natural amphitheater carved from Redwall Limestone at river level. Virtually all river runners stop here to enjoy the cavern, view, and broad, sandy beach. Camping and fires are not allowed at Redwall Cavern. During 1957's high water (96,000 cfs), this cavern was flooded.

Nautiloid Canyon at River Mile 35.0 on river left is an excellent place to see fossils of nautiloids (*Rayonnoceras* spp., which are relatives of octopus and squid) in bare limestone. Unlike modern species, which have a coiled shell, the shells of nautiloids at Nautiloid Canyon are cylindrical and taper to a point at one end. Some of the best of these fossils, which were discovered and named in 1967 by geologist William "Bill" Breed, are about 150 feet up the drainage. Nautiloids were common in the Paleozoic, but only six species of *Nautilus* live today in the Pacific. Caves typical of Redwall Limestone are abundant around Nautiloid Canyon.

Bridge of Sighs on river right at River Mile 36.0 is one of the few bridges in GCNP visible from the river. This 15 × 30-foot arch (about 200 feet above the river) in Redwall Limestone was named in 1911 by the Kolb brothers (Ellsworth and Emery), who believed the area resembled a dungeon because it was "gloomy and prison-like." The Kolbs named the formation after Ponte dei Sospiri ("Bridge of Sighs"), an enclosed bridge in Venice, Italy, across which convicted prisoners walked to a dungeon.

Remains of the boat *Grand Canyon* at River Mile 41.6 are what's left of the boat in which famed river runner Bert Loper died while rowing through Grand Canyon in 1949. In November 1951, Loper's widow Rachel visited this site for a final look at the boat that she had helped Bert build. Loper's remains were found downstream by a hiker in 1978 on river left near Cardenas Creek.

Anasazi Bridge at River Mile 43.3 is high in a nook at the base of the Redwall, 200 feet above the north side of the river. The site is closed to all visitation.

Peter Hansbrough Inscription ("PMH 1889") was carved here by Robert Stanton's boatman Langdon Gibson near President Harding Rapid at River Mile 44.0 on river left to mark the grave of Peter Hansbrough. Five days after carving the Frank Brown Inscription (see earlier in this list) at River Mile 12.1 to memorialize drowning victim Frank Brown, Hansbrough also drowned; his remains were found the next year and identified by his shoes and clothing. Given Hansbrough's fate, most of the remaining men were reluctant to etch Hansbrough's inscription, fearing that they, like Hansbrough, would be the next to die. Point Hansbrough at River Mile 44 is named for Hansbrough.

Grave of David Quigley is just upstream from the Peter Hansbrough Inscription. The body of Quigley, a 15-year-old Boy Scout who drowned more than 60 miles upriver in Glen Canyon on June 26, 1956, was found near here by river runners two weeks later and buried here on July 10, 1956.

Grave of Willie Taylor is just behind the Willie Taylor Camp on river left at River Mile 44.8. Wilson "Willie" Taylor was on a river trip in 1956 with eight other people (in three boats) when he had a heart attack at Redwall Cavern (River Mile 33). He died a few hours later after landing downstream and was buried above the high-water mark. Taylor's grave is marked with a plaque installed by Taylor's friend, Dock Marston, who helped bury Taylor. The plaque reads, "Willie Wilson Beigle Taylor December 27, 1896–June 6, 1956 He loved the mighty rock and water structures of the Grand Canyon and all the creatures that lived there. Grand Canyoneer 1949 1950 1951 1954." Taylor was the 100th person through Grand Canyon after John Wesley Powell.

Mile 50–Mile 100

Nankoweap Granaries at River Mile 52.3 (high on the northern cliff wall on river right) are eroded caves in Redwall Limestone more than 600 feet above the river. The granaries were used in 900–1150 CE by Anasazi Indians to store beans, squash, corn, and other food. Hikers can reach these granaries, which were featured on the fourth America the Beautiful Quarters Program in 2010, from Upper Nankoweap Camp 0.8 miles downstream. Just past the Nankoweap Granaries, the 8 miles beginning near River Mile 58 are the windiest of Grand Canyon.

Little Colorado River enters at River Mile 61.7 from river left from limestone springs a few miles upstream. This marks the end of the Marble Canyon segment of Grand Canyon. (Marble Canyon was the section of Grand Canyon that John Wesley Powell feared most and that Robert Brewster Stanton described as "the most beautiful sight I had ever looked upon.") In the dry season, the spring-fed water is milky blue because of dissolved travertine and limestone in the water; during the wet season, the water is muddy because of runoff. The confluence of the Colorado and Little Colorado Rivers also marks the end of the Navajo Reservation and the beginning of GCNP on river left. The Little Colorado River provides the principal drainage of the Painted Desert region.

Site of the 1956 Midair TWA–United Airlines crash is near River Mile 63. On June 30, 1956, United Airlines flight 718 and TWA flight 2 collided, after which the United DC-7 crashed into the top of the Redwall Limestone above the camp here. The TWA Super Constellation crashed onto the Tonto Platform about a mile downstream from Crash Canyon Camp on river right at River Mile 62.9. All 128 people aboard the planes died. There is a memorial to the victims of the crash at Desert View Watchtower.

Desert View Watchtower is visible straight downriver on the South Rim near River Mile 64.0. This is one of the few rim-side buildings that are easily visible from the river.

Hopi Salt Mines on river left at River Mile 63.5 to 65 are sacred to Hopi and are closed to all visitors. To Hopis, Grand Canyon is known as *Öng'tupqua*, which means "Salt Canyon." To get to the sheets and white stalactites of salt near here, Hopi males would descend Salt Trail Canyon to the Little Colorado River and then go approximately 7 miles to the Colorado River. Along the way, the males would leave prayer offerings at selected sites. (Hopi females, as a cultural rule, are advised not to enter the depths of Grand Canyon; Hirst, 2016). Although famed Grand Canyon hiker Harvey Butchart rediscovered the route to the mines from Salt Trail Canyon in 1956, subsequent theft and vandalism at the mines around 1980 prompted the Hopis to suspend trips there.

Unkar Delta and Puebloan Dwellings at River Mile 72.7 on river right are one of the largest known concentrations of ruins on the river. You can access the ruins on a 0.8-mile trail that includes brochures at the trailhead. The site was excavated in 1967–1968 by Douglas Schwarz, who uncovered homes, ceremonial kivas, and crop terraces that were used by up to 11 families at a time for three centuries between 850 and 1180 CE. These families seasonally migrated up Unkar Creek to Walhalla Plateau on the North Rim. A small Puebloan granary is just above the river on river right a mile downstream (River Mile 73.8).

The Great Unconformity, which was named by John Wesley Powell, is at eye level on river right in Blacktail Canyon at River Mile 120.6.

Some of John Hance's 58 Mines, which are now closed, are just upstream from Asbestos Canyon at River Mile 78.4. Nearby, Hance Camp at the end of New Hance Trail is where you'll often see backpackers at the river. Hance, who died in Flagstaff's hospital for indigents, was one of the first people to be buried in what became Grand Canyon's Pioneer Cemetery. Visitation is banned at Hance's asbestos mine.

Kaibab (Black) Suspension Bridge at River Mile 88.1 was completed in 1928, spans 440 feet, and is 75 feet above the river. This bridge replaced a flimsy bridge near here from 1920 to 1928, which Brighty of the Grand Canyon was the first to cross. This bridge, in turn, had replaced David Rust's cable car that operated near here from 1907 to 1920. Near the Kaibab Suspension Bridge is a USGS gaging station that was installed in 1922.

Grave of Rees B. Griffiths is at River Mile 88.2 on river right between the Bright Angel Pueblo and Phantom Ranch. Griffiths died in 1922 while working as a foreman during construction of the Kaibab Trail.

Bright Angel (Silver) Suspension Bridge at River Mile 88.3 carries pedestrians and a waterline from Roaring Springs to the South Rim.

Mile 100–Mile 150

Ross Wheeler, a small, heavy, steel boat at River Mile 108.5 abandoned by prospector and erstwhile newspaperman Charles Silver Russell (and two other passengers) in 1915, is on river left near the foot of the South Bass Trail. Russell—backed by $5,000, four miles of film, and promotional contracts with the Atchison, Topeka and Santa Fe Railway (Santa Fe Railway) and Denver & Rio Grande Railroad—believed he could make a better film about Grand Canyon river running than had the Kolb Brothers, but his film was never made. The *Ross Wheeler* is all that's left of Russell's expensive dream.

Remnants of a cable crossing built by William Bass at River Mile 108.7 to haul livestock, people, and ore are along both banks. The 300-foot-long cable, which was 50 feet above the water to avoid spring floods of the river, was used from 1906 to 1925. NPS cut the cable in 1968 to protect low-flying planes and helicopters.

Another Cable Crossing at River Mile 111.4 above Hakatai Rapid was installed around 1920 by Bass for his asbestos mine, which is about 0.7 miles up Hakatai Canyon on river right. In 1926, all of Bass's properties were purchased by the Santa Fe Railway and later transferred to the NPS. All of Bass's mines are now closed to visitors.

Monument Fold is a small monoclonal fold (and associated minor faulting) in Tapeats Sandstone at River Mile 116. The fold is visible at river level as a 20-foot upward displacement on river right. Nearby is Explorers Monument, which honors Grand Canyon explorers Joseph Ives, George Wheeler, Edward Beale, Almon Thompson, and John Newberry. The plateau here between the river and North Rim is Powell Plateau.

Elves Chasm at River Mile 117.2 is a popular side canyon with waterfalls, pools, and a clear flowing stream. A short walk across Royal Arch Creek here leads to Inscription Cave, an overhang under which are written (in charcoal) now barely visible historic names, including Nathan Galloway and Clyde Eddy. (The "Nathan Galloway 1897" inscription is one of the oldest known river runner inscriptions in Grand Canyon.) While at Elves Chasm on the morning of February 20, 1890, an inspired Robert Stanton described the view near Elves Chasm:

> The long stretch of river . . . is wonderfully beautiful, grand, and picturesque. The regular form of the inner granite gorge, capped with the '"beaded" [Tapeats] sandstone. The beautiful stretch of smooth but swift water. The rich green coloring of the grass, bushes, and the flowers among the granite breaks, and the shiny sparkling black granite. The marble cliffs on left coming close to the river in great frowning buttresses. On the right, the marble cliffs running almost parallel with river; broken into smaller buttresses of every imaginable shape, capped with several benches of the upper red sandstone, all a flaming red.

Camping is not allowed within 0.25 mile of Elves Chasm.

Upper Blacktail Canyon on river right at River Mile 120.6 offers excellent eye-level views of The Great Unconformity. Here, about 130 feet into the canyon, the 550-million-year-old Tapeats Sandstone (Grand Canyon's oldest Paleozoic formation) sits atop the tilted, 1.6-billion-year-old Precambrian rocks called Vishnu Schist. The Grand Canyon Supergroup of rocks is missing in The Great Unconformity. Blacktail Canyon, which was named for blacktail deer (now called mule deer) in the area, is where the first black schist is visible near the beach.

Deer Creek Falls dominates this popular stopping point for river runners at River Mile 135.9. Here, a 160-foot-high waterfall terminates Deer Creek, a perennial, spring-fed stream. A half-mile hike up Deer Creek leads hikers to the Narrows and Throne Room.

Anasazi Granaries at River Mile 137.4 are on river left at the base of Tapeats Sandstone. One granary is authentic, but the other was built by Boy Scouts.

Kanab (Paiute for "willow") Creek on river right at River Mile 144.0 is where John Wesley Powell ended his second journey through Grand Canyon in 1872. Kanab Creek's total dissolved chemicals, including metals such as chromium and copper, are the highest in GCNP. A report of gold in Kanab Canyon in 1871 by prospector George Riley triggered a 4-month "gold rush" here by more than 400 prospectors the following year. Camping is not allowed at Kanab Creek.

Deer Creek Falls is a popular place for river runners to stop and enjoy the 160-foot-high waterfall that terminates Deer Creek. In this photo, river runner Jennifer Sprague admires the spectacular waterfall. (Randy Moore)

Mile 150–200

Jesse "Shorty" Burton flipped his boat here at River Mile 150.4. Burton was trapped beneath the boat and drowned. A pie plate plaque honoring Burton (and his baking skills) is on river right.

Hualapai Tribal Lands begin at River Mile 165 and stretch along river left for the next 109 miles.

Cable at a USGS Gaging Station at River Mile 166.8 was hit by a Bell 206 helicopter in 1983, killing Charlie Houser.

Whitmore Helipad at River Mile 187.4 on river left is where some river runners end their rafting trips.

Mile 200–Mile 250

Pumpkin Spring on river left at River Mile 213.3 is an orange, pumpkin-shaped travertine dome made of calcium carbonate jutting out of the canyon's wall. The water in the "pumpkin" is warm, but it contains high concentrations of arsenic, chromium, zinc, nickel, lead, and other metals. NPS describes this water as "the only truly dangerous water in Grand Canyon."

Diamond Peak on river left dominates the view downriver from River Mile 222.5. The 3,512-foot-high diamond-shaped peak, which is approximately the same height as Navajo Bridge at River Mile 4.5, is 403 feet higher than where most river runners start their trips at Lees Ferry.

Diamond Creek Takeout on river left at River Mile 225.9 is the first place since Lees Ferry where a road reaches the Colorado River; this is also the access point used by Joseph Ives to reach the river with U.S. Army troops in 1858. Tourists have been visiting Diamond Creek since at least 1883, and Grand Canyon's first commercial hotel—The Farlee—was operated by Julius and Cecilia Farlee about a mile up Peach Springs Canyon at the confluence of Diamond Creek from 1883 to 1889. The Farlees took tourists by wagon from a railroad stop in Peach Springs, Arizona, to their hotel. Diamond Creek, which enters Grand Canyon from river left and designates the start of the Hualapai River Runners rafting trips down Grand Canyon, is famous for flash floods, one of which washed vehicles into the river. Visitors who come to the takeout point drive 21 miles across Hualapai land to Route 66. The first tourists to drive to the Colorado River (L. Wing and O. K. Parker) did so along this creek bed in 1914 in a 22-horsepower Metz car.

Farlee Hotel

In 1883, Julius H. and Cecelia Farlee and a colleague named James Young opened the rustic, eight-room Farlee Hotel (also known as the Diamond Creek Hotel) about a mile from the Colorado River where Diamond Creek intersects Peach Springs Wash. This frame hotel, which was 100 miles west of where John Hance had settled the previous year, was the first hotel at the bottom of Grand Canyon. The Farlees brought tourists from Peach Springs, Arizona, to their hotel in stagecoaches. The hotel averaged 100 guests per year, but business later declined when tourists began choosing hotels at the more scenic sites to the east.

The Farlees and Young, who advertised in the March 25, 1883, issue of *Mojave County Miner*, were Grand Canyon's first outfitters. By 1899, the hotel was closed, and by 1901, all of the lumber had been taken away for other uses.

Bridge Canyon Dam was proposed to be built at River Mile 237.7. On river left are remnants of equipment and a table, and on river right is a 4×5-foot wooden door built in 1939 to protect a small cave used to store dynamite. Construction of this 736-foot-high dam was stopped by the GCNP Enlargement Act signed in 1975 by President Gerald Ford.

Bridge City Camp site at River Mile 238.7 includes cable, pipes, and steel that were to be used to build the Bridge Canyon Dam.

Separation Canyon at River Mile 239.8 refers to the canyons that enter from the north and south. This was once a formidable rapid but is now buried under Lake Mead. Here, on river right about 70 feet above the water, you'll find a plaque (dedicated in 1939) that reads as follows: "Here on August 28, 1869, Seneca Howland, OG Howland and William H. Dunn separated from the original Powell Party. Climbed to the North Rim and were killed by the Indians." Many raft trips through Grand Canyon transfer their passengers to speedboats here. Personally owned recreational watercraft coming from Lake Mead are not allowed upstream past this point.

Mile 250–Mile 277

Hualapai Glass-Bottomed Skywalk is visible on river left at River Mile 265.5 opposite Bat Cave on river right. (Visitation is not allowed at Bat Cave.) The Skywalk, which is 1.5 miles up a side canyon, is on the Hualapai Indian Reservation (i.e., outside GCNP). Guano was found in Bat Cave in the 1930s, and mining began in the 1940s. In 1950, U.S. Guano Corporation built a 9,010-foot-long tramway across the river to the cave 2,897 feet below. The fertilizer was sold in stores in one- and three-pound packages for 69¢ and $1.89, respectfully. However, the amount of guano had been overestimated, and prices of fertilizer soon dropped. This ended mining at the site. By the mid-1960s and after the tramway was featured in the movie *Edge of Eternity* (1959), the site's three cables across the river were soon gone. In 1995, NPS considered taking the nearby towers down, but Mohave County objected, so today the towers stand as reminders of the area's history.

Further Reading

Belknap, Buzz, et al. 2017. *Grand Canyon River Guide*. Evergreen, CO: Westwater Books.

Berger, Todd R. 2007. *It Happened at Grand Canyon*. Guilford, CT: TwoDot Publishers.

Hirst, Stephen, ed. 2016. *We Call the Canyon Home: American Indians of the Grand Canyon Region*. Grand Canyon, AZ: Grand Canyon Association.

Grand Canyon and Modern Christian Fundamentalism

Before scientists began studying Grand Canyon, Native people told stories that explained the canyon's beautiful colors, stunning sediments, and mysterious geology and origin. Some of the oldest of these stories came from the Havasupai. According to one of their origin stories, the canyon was carved by a flood of the Colorado River that happened when the god Ho-ko-ma-ta unleashed a vast rainstorm. Another presumably more benevolent deity, Toc-ho-pa, saved his daughter Pu-keh-eh by putting her in a hollow pine log. After the floodwaters carved Grand Canyon (Chic-a-mimi) and receded, Pu-keh-eh came out of the log and repopulated Earth.

By the 1700s and early 1800s, most geologists were Christians who interpreted their discoveries through their faith. They believed that understanding geology and other aspects of nature would lead to a deeper understanding of God. By the early 1800s—that is, decades before Charles Darwin conceived of evolution by natural selection and John Wesley Powell boated through Grand Canyon—these geologists had rejected claims that Noah's Flood had covered Earth and sculpted Earth's topography, because there was no evidence for such a flood. They concluded that the story of Earth's history was written in its rocks and that the rocks don't lie.

By the 1850s, geologists knew that Earth was vastly old. Claims of a "young Earth"—that is, an Earth approximately 6,000 years old, as famously proposed by Irish prelate James Ussher and "scriptural geologists"—were rejected outright because of overwhelming evidence to the contrary. The Noachian Flood, a foundation of scriptural geology and the biblical literalism on which it rested, was assumed to have been confined to the mid-East, where its effects had been erased by time.

However, George McCready Price (1870–1963)—a Seventh-Day Adventist and self-taught geologist—believed otherwise. Price's view of the world was determined by Seventh-Day Adventism cofounder and prophetess Ellen White (1827–1915), who claimed to have had "visions" in which God took her back in time to witness creation. White reported that creation lasted a week and included a worldwide flood that sculpted Earth's surface and buried organisms that became fossils. When White claimed that anything contradicting her "vision" was a "compromise," Price set out to invalidate an old Earth by invoking Flood Geology to discredit the so-called geological column (on display at Grand Canyon and elsewhere) and thereby make the six-day creation week acceptable. Price's 723-page *The New Geology* (1923) claimed that the well established sorting of fossils in particular sediments of the geological column "is clearly a scientific blunder . . . all such schemes of an alleged historical order which have been hitherto proposed are now seen to be wholly unscientific" (Price, 1923). Although Price was cited at the Scopes Trial by prosecutor William Jennings Bryan, Bryan did not endorse Price's claims, and Price's ideas were generally ignored by mainstream Christians.

Ignored, that is, until they were repackaged and developed in 1961 in *The Genesis Flood*, the book by John C. Whitcomb and Henry M. Morris that began the modern creationist movement. Whitcomb and Morris claimed there was a violent, worldwide flood produced by rain and the eruption of underground seas. To Whitcomb and Morris, there was a clear choice regarding their claims: "Either the Biblical record of the Flood is false and must be rejected or else the system of historical geology which has seemed to discredit it is wrong and must be changed" (Whitcomb and Morris, 1961). Whitcomb and Morris based their claims on "letting the Bible speak for itself and then trying to understand the geological data in light of its teachings." Whitcomb and Morris used the Bible to determine Earth's geologic history and then looked for data supporting their prescribed conclusions. Young-Earth organizations such as the Institute for Creation Research (ICR), which reject the long-standing Christian belief that the geological record tells Earth's story, require their students, faculty, and employees to sign a statement of faith to fundamental religious doctrine and biblical literalism.

Young-Earth creationism, a religious ideology that rejects modern geology and other sciences in favor of biblical literalism, is a rapidly growing influence at Grand Canyon National Park. In 2017, the Christian fundamentalist business Canyon Ministries told its more than 1,500 customers that Noah's Flood formed Grand Canyon only a few thousand years ago. This photo shows one of Canyon Ministries' "A Different View" vans at Powell Point. (Randy Moore)

As their name suggests, Young-Earth creationists (YECs) believe that Earth is only about 6,000 years old and that Grand Canyon formed within days after Noah's Flood. For example, Answers in Genesis (AiG)—a YEC organization based in Kentucky—claims in its publications and its popular Creation Museum that because "the Bible says that Earth is only about 6,000 years old, Grand Canyon could not have formed millions of years ago" (Answers in Genesis, 1996). To make such claims, YEC rejects a variety of modern sciences, including radiometric dating, and alleges that conspiring scientists suppress data inconsistent with their conclusions. Modern scientists reject claims by YEC that Earth (including Grand Canyon) is only about 6,000 years old. Nevertheless, biblical literalism and YEC are resilient; several tenets of YEC—including that humans lived with dinosaurs—are popular with Americans.

Young-Earth Creationism and Grand Canyon

Grand Canyon is an icon of Young-Earth creationism. Indeed, it is cited prominently in their books, videos, school curricula, radio and television programs, churches, conferences, and museums. YECs use a literal and selective interpretation of the Bible to claim that Grand Canyon records events before, during, and after Noah's Flood: The deepest rocks of Grand Canyon (e.g., Vishnu Schist, Zoroaster Granite) were deposited during Creation Week and were therefore present before the Flood. For example, AiG claims that Grand Canyon's bottom layers of granites and metamorphic rocks were formed by Day 3 of Creation Week. Similarly, in *Grand Canyon: Monument to Catastrophe* (1995), YEC Steven Austin argues:

> The crystalline-basement rocks exposed deep within the Canyon (schist, granite, and gneiss) represent some of earth's oldest rocks, probably from early in Creation Week. Tilted, deeply buried strata (the "Grand Canyon Supergroup") show evidence of catastrophic-marine sedimentation and tectonic associated with the formation of an ocean basin midway through Creation Week, and may include ocean deposits from the post-Creation, but pre-Flood world (Austin, 1995).
>
> The spectacular flat-lying rock layers of Grand Canyon were rapidly and continuously deposited during the early part of Noah's flood, about 4,300 years ago. After regional uplift, the horizontal layers of rock were cut by floodwater while the sediments were still soft. An earthen dam then ruptured, releasing a barrage of water that carved the Canyon itself.
>
> Fossils in the rocks of Grand Canyon represent animals that died in the Flood. Animals did not die before Adam and Eve sinned; as claimed by Tom Vail of Canyon Ministries, "[H]uman and animal physical death, disease, suffering, and other natural evils (earthquakes, tsunamis, tornados, etc.) entered into this world subsequent to, and as a direct result of God's curse in response to Adam's sin" (Vail, 2003). According to AiG, fossils in the sediments of Canyon should remind us of God's judgment on sin and of the salvation He offers.

YECs reject decades of ongoing research confirming that Earth is 4.5 billion years old, as well as consistent and independently verifiable data from radiometric dating, fossils, and geological mapping that document Grand Canyon's history. In *Grand Canyon in Arizona* (2010), AiG claims that "Since the Bible says that the earth is only about 6,000 years old, Grand Canyon could not have formed millions of years ago."

Advocates of YEC, who base some of their conclusions on never before seen events and unreproducible chemical reactions, also reject basic tenets of several other sciences, including biology, geology, cosmology, and other disciplines whose conclusions contradict their interpretations of the Bible. For example, AiG's Steve Austin—who claims that there was no illness, pain, suffering, or death before Adam and Eve—claims that mainstream geologists are baffled because the facts indicate that Grand Canyon was formed by a flood. In his *Grand Canyon: Monument to Catastrophe* (1994), Austin claims that "the creationist perspective, unlike the evolutionist's, is fully rational" and that environmental degradation has been caused by sin, "especially [by] liberal churches which have led the way in compromising Scripture with evolutionism and other secular ideas" (Austin, 1994). Austin claims that mainstream scientists who disagree with his religion-based claims are willingly ignorant and deliberately choose to reject evidence, as predicted in the Bible for the "last days."

ICR claims that Grand Canyon's strata were deposited by Noah's flood, uplifted later in the flood, and then eroded when the floodwaters retreated to ocean basins, all within the one-year timeline given in Genesis. In his "Grand Canyon: Is It Really 'Exhibit A' for Evolution and the Old Earth?" ICR's John Morris claims that "many old-earth advocates are revising their views. Now many geologists who study the Canyon are talking about catastrophic deposition (during the Biblical flood)" (Morris, 1994). Morris also believes that there is no evolutionary sequence of fossils in Grand Canyon, that the rim of the canyon is hardly touched by erosion, and that Grand Canyon is best understood as being the result of a great flood.

Scientists' Responses to YEC Claims about Grand Canyon

For decades, scientists using testable evidence rather than religious claims have understood that Earth is billions of years old. GCNP's conclusion about Earth's age is displayed at the Trail of Time, where a trailside marker and interpretive panel near Maricopa Point documents Earth's age as 4.56 billion years. An old Earth is also part of a prominent, permanent exhibit by GCNP at the Grand Canyon Visitors Center: "The rugged walls of Grand Canyon record billions of years of our planet's history, shelter plants and animals, and have sustained people for thousands of years." Scientists have rejected YEC's claims that Noah's Flood created Grand Canyon, sculpted Earth's topography, and deposited fossil-bearing rocks all in a year. Here are some of the reasons why:

Sand doesn't have the strength to form vertical walls; indeed, dry sand can support slopes less than 40°, and wet sand (e.g., after a flood) can support slopes of only about 20°. If Grand Canyon was cut rapidly through soft, water-soaked mud within a few days after the flood, how could a mile-high stack of saturated sediments remain upright without slumping into a growing canyon?

Fine sediments settle slowly and would have been resuspended by turbulent floodwaters. However, fine sediments occur throughout the geologic column at Grand Canyon. How could fine sediment be located anywhere in the canyon walls except at the surface?

Most fossils in Grand Canyon are of extinct species. If a flood buried all of the creatures in Grand Canyon's walls just a few thousand years ago, why aren't any modern animals—*not even one*—entombed there?

How could fossils be sorted in a violent flood in the chronological sequence documented for more than 200 years by paleontologists? For example, why are there no fossils—*not one*—of fish, reptiles, clams, snails, or mammals in the Grand Canyon Supergroup?

Cliff-forming sandstones (e.g., Tapeats Sandstone, Coconino Sandstone) and shale (e.g., Bright Angel Shale) are deposited mechanically, whereas Grand Canyon's limestone (e.g., Muav, Redwall, Kaibab, and Toroweap) are made of dead marine organisms (i.e., skeletons made of calcium carbonate, $CaCO_3$). How could these different processes occur if all the sediments were in turbulent floodwaters and deposited atop them at the same time?

Scientists accept radiometric dating, which consistently indicates that the oldest rocks on Earth (and the moon) are ancient. There is no evidence that the basic laws of physics that underlie radiometric dating change during floods or when rocks are wet.

How could delicate fossils of raindrops and animal tracks such as those found in Grand Canyon (e.g., along Hermit Trail) have survived a flood strong enough to remodel Earth's surface? How could animals still have been alive to leave footprints in the Coconino Sandstone when layers were being deposited and buried during a flood?

YECs claim that The Great Unconformity—that is, the missing sediments between the 1.7-billion-year-old Vishnu Schist and the 525-million-year-old Tapeats Sandstone—is the result of erosion, the onset of Noah's Flood, and "a sober reminder that God judges sin" (Matthews and Neville, 2017). YECs are silent about how Coconino Sandstone and other layers form thousands of feet *above* The Great Unconformity, as they are about the many other unconformities *above* The Great Unconformity that would have required long, repeated periods of exposure in the middle of a worldwide flood.

Book Controversy

The claims of YEC about Grand Canyon made national news in 2003 when YEC Tom Vail—a river guide in Grand Canyon—published the controversial *Grand Canyon: A Different View*, an edited, 104-page compilation of essays by several of "the leading creation science writers, researchers, and theologians of our time" (e.g., Steve Austin, Duane Gish, Ken Ham, John Morris, and Kurt Wise). All of the articles in Vail's beautifully illustrated book were "peer reviewed to ensure a consistent and Biblical perspective" of biblical literalism (Vail, 2003). Earlier in his career, Vail had presented scientific information about Grand Canyon, but he rejected science after becoming a Christian because he claimed that believing in the Bible requires Grand Canyon to be only a few thousand years old. In 1996, Vail formed Canyon Ministries, which began offering "Christ-centered trips through Grand Canyon."

Vail argues that his book is a "one of a kind exploration of the Grand Canyon" that refutes "the evolutionary view of the Canyon's origins" by claiming that "the vast majority of sediments everywhere were deposited as a result of a global flood that occurred after, and ultimately as a result of, the initial sin that took place in the Garden of Eden about 6,000 years ago" (Vail, 2003). Vail believes that Grand Canyon was carved "in a matter of just days, shortly after the global flood of Noah's day about 4,300 years ago" (Vail, 2003). Vail claims that his conclusions are accurate because "you can trust God's word" and that anyone who disagrees with his conclusions is undermining Christianity and denying God's involvement in nature. Vail believes that Grand Canyon is a testimony to God because it "presents evidence of God's judgment of the world . . . The Canyon gives us a glimpse of the effects and scale of a catastrophic global flood, the biblical Flood of Noah's day . . . [This evidence] brings us to the feet of our Lord Jesus Christ" (Vail, 2003).

Vail's geological claims—for example, about Noah's flood, Grand Canyon forming 4,300 years ago, and the canyon's sediments being deposited in a few days—contradict mainstream science practiced and endorsed by geologists at GCNP and elsewhere. Nevertheless, *Grand Canyon: A Different View* was unanimously approved for sale in Grand Canyon National Park bookstores in 2003. Soon thereafter, many people—including GCNP superintendent Joe Alston and the presidents of seven major scientific societies (the American Paleontological Society, American Geophysical Union, National Association of Geoscience Teachers, Association of American State Geologists, Society of Vertebrate Paleontology, American Geological Institute, and Geological Society of America)—objected because they did not want visitors to believe that Vail's religion-based claims were endorsed by GCNP or have anything to do with science. The resulting controversy, which made headlines and national news programs, was reignited in 2008 with the publication of Vail's *Your Guide to the Grand Canyon: A Different Perspective* (Vail et al., 2008), which repeated many of Vail's claims about Grand Canyon.

Soon after Vail's book appeared in Grand Canyon bookstores, David Shaver—chief of the Geologic Resources Division of the NPS—condemned Vail's book as inaccurate, unprofessional, unscientific, and inconsistent with the mission of NPS. Another GCNP geologist noted that selling Vail's book in GCNP was equivalent to Yellowstone National Park selling a book entitled "Geysers of Old Faithful: Nostrils of Satan" (Paleoblog, 2007). Although scientists and others criticized GCNP officials for promoting "faith-based parks," Alston and all the others who objected to Vail's book being sold at GCNP were overruled by NPS. Although the NPS told Congress and others that the book's sale at GCNP was under review at the national level by several offices, no such review took place. This prompted an official of Public Employees for Environmental Responsibility to note that NPS was suspending its belief in geology so as to not offend anyone's religious faith. Later, when GCNP bookstores moved Vail's book from the natural science section to the inspirational section, AiG—erroneously claiming that the First Amendment entitles them to have their books sold by the government or nonprofit organizations—denounced the removal of Vail's book as censorship and an attack on free speech.

In 2005, NPS approved a policy (Interpretation and Education Director's Order #6) stating that the "history of the Earth must be based on the best scientific evidence available, as found in scholarly sources that have stood the test of scientific peer review and criticism . . . Interpretive and educational programs must refrain from appearing to endorse religious beliefs explaining natural processes." Nevertheless, in 2017, Vail's book remained for sale at several bookstores at and around Grand Canyon, including at the National Geographic Visitors Center. Canyon Ministries, which claims that Earth is only a few thousand years old, tells its clients that Grand Canyon is "a memorial to the judgment of the Flood" (Matthews and Neville, 2017).

YEC at Grand Canyon continues to grow. In 2017, Canyon Ministries escorted more than 200 people through Grand Canyon (up from 14 in 1999) and provided Bible-based tours for more than 1,500 people at the South Rim. Its popularity and outreach are predicted to more than double by 2020. Canyon Ministries also offers numerous "Christ centered" train rides, sunset tours, and other activities at Grand Canyon in which people "learn how the global effects of the Biblical Flood of Noah's day, not millions of years of erosion, carved Grand Canyon." The director of rim tour operations for Canyon Ministries is Nate Loper, a descendant of famed river-runner Bert Loper.

In 2017, the controversy about the promotion of YEC at GCNP again made front-page headlines. YEC promoter Andrew Snelling, the director of research at AiG and an advisor to Canyon Ministries, asked GCNP for permission to collect 30 pounds of rocks from the Inner Gorge so he could explain "geological phenomena and other endeavors from a Biblical perspective." When his application was rejected because it lacked scientific reasoning and was out of date and because Snelling had no track record as a scientist, Snelling—represented by Alliance Defending Freedom—filed a 22-page lawsuit against GCNP, NPS, and the Department of the Interior, claiming that he was the victim of religious discrimination and "an abuse of government power" (Santos, 2017). Soon thereafter, GCNP relented and approved Snelling's request, at which time Snelling withdrew his lawsuit (*Snelling v. United States Department of the Interior*).

Virtually all geologists accept that Grand Canyon is vastly old. In 2016, a group of these geologists—many of whom are Christians—published *The Grand Canyon, Monument to an Ancient Earth: Can Noah's Flood Explain Grand Canyon?*, a beautifully illustrated book that refutes the claims of YEC and shows that "Grand Canyon provides overwhelming evidence that the Earth is old" (Hill, et al., 2016). The book's conclusion is emphatic:

> When the data are considered in their totality and allowed to take us wherever they lead—without foreknowledge of the answer or a predetermined outcome—we are invariably led to a history of the canyon that extends back millions of years. A recent age for the canyon can only be imagined by deciding on such an answer in advance, carefully selecting bits of data that can be construed to fit the preconceived model, and ignoring data that do not fit. Herein lies the difference between science and flood geology—science goes where the data lead, flood geology does not. By deciding in advance what the answer to the question will be, flood geology does not study nature to discover what processes have been at work or what events may have transpired (or, for that matter, what God actually did). Rather, flood geology starts with an answer and studies nature only to find those ways that fit with the prescribed model. In this respect, flood geology is the antithesis—the very opposite—of science . . . The message of flood geology is that what is observed in nature today cannot be used to inform us of what may have happened in the past, that fundamental laws of physics and chemistry cannot be assumed to be well understood, and—critically—that nature cannot be trusted to tell its own story. In this regard, flood geology is not only unscientific, it is unbiblical. The first chapter of Romans states that the Creator's divine nature is manifest in His physical creation—in nature. If nature cannot be trusted to tell a truthful story, what does this say about flood geologists' conception of God? (Hill, et al., 2016).

Further Reading

Answers in Genesis. 1996. Grand Canyon: Monument to the worldwide flood. *Creation*, 18 (2), 28.

Answers in Genesis. 2010. *Grand Canyon in Arizona*. Retrieved February 2, 2014 at http://www.answersingenesis.org/store/product/grand-canyon-arizona/?sku=90-4-029

Austin, S.A. 1992. Excessively old "ages" for Grand Canyon lava flows. *Acts & Facts*, 21 (2), 1–4.

Austin, Steve. 2003. *Geologic Evidences for Very Rapid Strata-Deposition in the Grand Canyon*. DVD. Petersburg, KY: Answers in Genesis.

Austin, Steven A., ed. 1995. *Grand Canyon: Monument to Catastrophe*. Santee, CA: Institute for Creation Research.

Hill, Carol, Davidson, G., Helbe, T., and Ranney, W., eds. 2016. *The Grand Canyon, Monument to an Ancient Earth: Can Noah's Flood Explain Grand Canyon?* Grand Rapids, MI: Kregel Publications.

Matthews, Mike, and Neville, Chris. 2017. Grand Canyon: Nine days below the rim. *Answers: Building a Biblical Worldview*, 12 (4), 50–67.

Morris, John D. 1994. Grand Canyon: Is it really "Exhibit A" for evolution and the old earth? *Acts & Facts*, 23, 6.

Morris, John D. 2001. How long does it take for a canyon to form? *Acts & Facts*, 30, 12.

Morris, John D. 2014. Lots of Grand Canyons. *Acts & Facts*, 43 (6).

Morris, John D. 2004. Revisiting the Grand Canyon. *Acts & Facts*, 33, 5.

Paleoblog. 2007. Grand Canyon and the nostrils of Satan. Retrieved July 1, 2017, at http://palaeoblog.blogspot.com/2007/01/grand-canyon-nostrils-of-satan.html

Price, George McCready. 1923. *The New Geology*. Mountain View, CA: Pacific Press.
Santos, Fernanda. 2017. A creationist wants rocks to study. The Grand Canyon says no. Retrieved May 29, 2017, at https://www.nytimes.com/2017/05/29/us/grand-canyon-study.html? r=0
Snelling, Andrew A. 2005. Radioisotope dating of rocks in the Grand Canyon. *Creation*, 27 (3), 44–49
Vail, Tom, ed. 2003. *Grand Canyon: A Different View*. Green Forest, AR: Master Books.
Vail, Tom, et al. 2008. *Your Guide to the Grand Canyon: A Different Perspective*. Green Forest, AR: Master Books.
Whitcomb, John C., Jr., and Henry Morris. 1961. *The Genesis Flood: The Biblical Record and Its Scientific Implications*. Philadelphia: Presbyterian and Reformed Publishing.

The Future of Grand Canyon

Since before it became a national park in 1919, there has been a tension between exploiting and protecting Grand Canyon. Generous policies and few environmental restrictions enabled Grand Canyon's pioneers such as William Bass, John Hance, and Ralph Cameron to claim land anywhere in the canyon and along the canyon's rim in an unregulated, first-claimed/first-owned free-for-all. As these and other entrepreneurs pursued personal wealth with little regard for earlier Native American inhabitants, hundreds of private mining claims began to litter the rim and inner canyon. When prospectors realized that they could make more money from tourists than from mining, these parcels of land—especially those at today's Grand Canyon Village—became especially valuable.

Although GCNP managed to regain virtually all of the park's privately owned land, the ongoing growth in tourism produced more and more commercial opportunities at Grand Canyon. Today's concessionaires at GCNP—both on the rim and along the river—are regulated by GCNP, which tries to produce an enjoyable experience for the more than 6 million tourists who visit Grand Canyon each year. However, meeting the demands of these tourists is challenging and has changed Grand Canyon.

Although the mission of NPS is to preserve "unimpaired the natural and cultural resources" at Grand Canyon, many parts of GCNP bear little resemblance to those seen by the Native Americans, John Wesley Powell, and other Grand Canyon pioneers. For example, Glen Canyon Dam drastically altered the Colorado River and wildlife in Grand Canyon. Moreover, for many years, GCNP intentionally introduced invasive species in the park, and Glen Canyon has made today's Colorado River in Grand Canyon far calmer, clearer, and colder than during pre-dam days. The temperature and flow of the river, which were once dictated by nature, are now determined by the power needs of the area. At the South Rim, parts of the canyon's rim described as "Rehabilitated" and "Ecologically Sensitive" areas remain contaminated with radioactivity and unsafe for visitation.

To its credit, GCNP has done much to protect the park and return it to a more natural condition. Along the river, GCNP now regulates the sizes and departures of groups leaving Lees Ferry and bans visitation at several fragile sites. Although motorized boats are still abundant, river runners must carry out all of their trash and solid human waste, and campfires have been replaced by cook stoves and fire pans. Some of the invasive species (e.g., burros) are gone.

At the rim, GCNP has replaced the Kentucky bluegrass lawn (installed in the late 1960s) around El Tovar Hotel and other rim-side lodges with native vegetation, and now uses shuttle buses to move thousands of tourists from site to site.

The rims are now cleaner and safer than in decades past. In recognition of its efforts, GCNP in 2014 was designated a Globally Important Bird Area, a global network of places having outstanding value for bird conservation.

The park also has plans to address other problems. For example, much of the air pollution that diminishes views at GCNP comes from elsewhere (e.g., forest fires, sulfates from urban emissions from Southern California and Arizona smelters), and 1999's Regional Haze Rule aims to reduce this pollution to natural background levels by 2064. Crime rates are down in GCNP, thus explaining why Safewise, a home-security system provider, named GCNP one of the top 13 vacation destinations in the United States in 2017.

Despite these improvements, however, many problems persist, largely driven by the tension between conservation and tourism discussed at the beginning of this book. The ongoing popularity of Grand Canyon has produced massive crowds that often clog parking lots, overlooks, and entrance stations. These crowds, combined with beleaguered budgets, often force administrators to manage GCNP not like a protected resource but instead like an amusement park. And as is true for many amusement parks, visitors to GCNP often cannot find places to park, after which they stand in long lines to take crowded buses to crowded overlooks. In some parts of the park, the sounds of nature are polluted by the nearly constant drone of airplanes and helicopters. For decades, we have ignored Teddy Roosevelt's admonition that we "not have a building of any kind, not a summer cottage, a hotel or anything else, to mar the wonderful grandeur, the sublimity, the great loveliness and beauty of the Canyon." Roosevelt, who did more than any other president to make conservation a national priority, would be disappointed by many of the things that have happened to and at Grand Canyon.

Threats to Grand Canyon continue. For example, developers hope to build a dude ranch, spa, hotels, and more than 2,200 new homes in Tusayan. This project would remove much of the area's groundwater, threaten in-canyon seeps and springs, and increase Tusayan's population by 1,000 percent. In 2017, the Tusayan Town Council unanimously approved raising the height of commercial buildings from 35 to 65 feet, which would change the landscape and increase the light pollution at Grand Canyon. However, in November 2017, residents of Tusayan rejected the proposal by a vote of 71–60. (Only about half of Tusayan's registered voters voted.) Other developers want to build a 420-acre, industrial-scale resort whose 1.6-mile gondola would take up to 10,000 people per day to the confluence of the Little Colorado and Colorado Rivers. The developers claim that this project—called the Escalade—would bring much needed jobs, provide infrastructure, and save the Navajo culture. Critics respond that the development would destroy one of the most picturesque parts of GCNP that is sacred to the Navajos, Hopis, Zunis, Havasupais, and other tribes of Native Americans. Tribal leaders have so far balked at the developers' demands that they pay no taxes to the Navajo Nation, that they return only 8–18 percent of their revenues to the Navajo Nation, and that they must get $65 million from the Navajo Nation to help pay for infrastructure. In 2017, the Navajos' Naabik'iyati' Committee (which vets proposed legislation for the larger Navajo Nation Council) rejected the development. However, developers have promised to continue their fight for the project's approval.

Escalade

Escalade is a proposed $1 billion entertainment complex including a tramway, art studios, RV park, tourist information center, rim trails, public safety office, gas station, mall, boutique hotels, restaurants, trailer park, cultural center, and associated structures on the western edge of the Navajo Nation, approximately 100 miles by road from Flagstaff. The project is spearheaded by Confluence Partners of Scottsdale, Arizona, so named for the complex's proposed location at the confluence of the Little Colorado and Colorado Rivers. The site would include 47,000 square feet of restaurants and retail space on the rim.

Escalade is the largest tourism venture ever proposed by the Navajo Nation. A 1.6-mile-long tramway would carry as many as 10,000 people per day 3,200 feet below the rim to a river walk about 300 feet above the confluence, where visitors could visit gift shops, restaurants, and other attractions. The complex is touted by investors and supporters as a way to bring in 3,500 jobs and $5 billion (over the next 10 years), as well as infrastructure (e.g., water lines, electric services, roads, sewer treatment plants), to the tribe. However, many Native Americans oppose the development because of concerns about its impact on the environment and damage to the confluence. (For many tribes of Native Americans, the spirits of their ancestors reside at the bottom of Grand Canyon.)

Uranium Mining at Grand Canyon

Uranium has been mined from Grand Canyon and the surrounding area since the 1950s. The first such mine, Orphan Mine near Powell Point, is closed and is still being "rehabilitated," and some people worry that more such pollution from uranium mines threatens today's canyon. These fears intensified in 2010, when USGS reported radioactive dust and uranium-contaminated water at several mines near Grand Canyon. In 2012, Secretary of the Interior Ken Salazar responded to these findings by announcing a 20-year ban on new uranium claims (i.e., on "mineral withdrawal") on more than a million acres adjacent to Grand Canyon. This ban, which was meant to give scientists time to understand how water flows around and through Grand Canyon, was cheered by environmentalists, but it applied only to new mining claims; the area's 3,156 existing inactive "zombie mines" (2,215 of which originated since 2003) were not affected by the ban. In early 2018, environmentalists worried that the 20-year ban would be shortened, or that the acreage covered by the ban would be reduced, or both. Much is at stake; the million acres around Grand Canyon hold about 79 million pounds of uranium ore which, if extracted and enriched, could satisfy the domestic market for uranium for 1.5 years.

One of the existing mines that worries Native American leaders is Canyon Mine, which is six miles south of Tusayan in the Kaibab National Forest. The mine, which is owned by Energy Fuels, Inc., was approved by the U.S. Forest Service and opened on September 26, 1986. However, a drop in uranium prices in 1992 prompted Energy Fuels to inactivate the mine. Around 2006, however, a global nuclear renaissance raised the price of uranium, and the mine was reactivated, and Energy Fuels discovered more high-grade uranium and copper that could be economically extracted from the mine. In 2017, this mine was the leading producer of uranium and the only conventional uranium-mining company in the United States.

When the 1,475-foot-deep shaft of Canyon Mine flooded in 2017, water that was pumped from the shaft into an evaporation pond contained 130 parts per billion uranium; this is more than quadruple the federal standard (30 parts per billion) for uranium in drinking water. Arsenic concentrations were 30 times the federal limit. At some sites, concentrations of uranium exceeded background levels by more than tenfold, and animals' tissues had uranium concentrations that exceeded standards for drinking water by more than 70-fold. Elsewhere around Grand Canyon, 15 mines and five wells near uranium mines have concentrations of dissolved uranium that also far exceed standards for drinking water. Native American leaders worry that these mines and wells that provide their water are permanently polluted.

Canyon Mine is in the shadow of Red Butte, a culturally sacred site to the Havasupai Tribe; tribal members often visit the site for ceremonies and to gather medicinal plants. The mine overlies groundwater that supplies Havasu Spring and Havasu Creek. Ores extracted from Canyon Mine also must be transported in dozens of 30-ton-capacity trucks each day through Flagstaff, Cameron, Tuba City, and other communities (including along 180 miles of the Navajo Nation) to Energy Fuels' White Mesa Uranium Mill. (This mill is the only fully licensed and operating conventional uranium mill in the United States.) Between 1987 and 2017, at least two accidents spilled uranium ore in the Navajo Nation.

Environmental groups such as Grand Canyon Trust have protested the resumption of mining at Canyon Mine because its pollutants could affect the water used by more than 30 million people (including thousands of Havasupai, Hopi, and Navajo) and 4 million acres of cropland. In 2015, Havasupai councilwoman Carletta Tilousi warned that Canyon Mile was "going to contaminate our groundwater . . . and when this contamination reaches [us], there will be no more Havasupai."

Politicians such as Congressman Raul Grijalva, Native American leaders, and others were disappointed when President Barack Obama refused to create the Greater Grand Canyon Heritage National Monument, which would have protected (and permanently banned new uranium mines on) 1.7 million acres of tribal homeland around Grand Canyon. Many tribal leaders and others worry that radioactive pollution continues to leach from thousands of active and inactive mines in and around the South Rim.

Generating Power at Grand Canyon

Grand Canyon is also being affected by environmental concerns and governmental regulations about generating power. The coal-powered Mohave Generating Station closed in 2005, and, in 2017, the Navajo Generating Station (located just outside of Page, Arizona) announced that it would close by the end of 2019. The Navajo plant, which opened in 1969 as one of the West's largest and most polluting coal-powered plants, had invested more than $400 million in 1997–1999 to meet demands by the Environmental Protection Agency that it reduce its sulfur dioxide (SO_2) emissions by more than 90 percent and its nitrogen dioxide (NO_x)

emissions by 40 percent. Closure of the Navajo Generating Station will improve air quality and visibility at Grand Canyon but will also eliminate 1,000 mining and plant jobs (plus more than 3,000 indirect jobs). Because these jobs (and those at northern Arizona's Kayenta Mine, which provides 8 million tons of coal for the generating plant every year) account for 50–70 percent of Navajo Nation's jobs and 40 percent of its budget, the closing of the power plant will dramatically hurt the Navajo Nation's economy at Grand Canyon.

Grand Canyon and Climate Change

Grand Canyon is also threatened by climate change. For example, the flow of the Colorado River between 2000 and 2014, when compared to the 20th-century average, decreased by 19 percent. Approximately 15–50 percent of this decrease has been attributed to climate change. Not surprisingly, the plants and animals at Grand Canyon—which includes five of North America's seven "life zones"—are also threatened by development and climate change. For example, the Mexican spotted owl (*Strix occidentalis lucida*)—one of the largest owls in the Americas—has been on the Endangered Species List since 1993. This owl, of which only 2,000 remain in the United States, depends on old-growth forests, which are now being lost to fires, logging, and development. Similarly, the southwestern willow flycatcher (*Empidonax traillii extimus*) is also increasingly rare along the Colorado River because of the degradation of the riparian habitat in GCNP. These birds thrive in dense, low-lying wetlands, which are disappearing in GCNP. Endemic fish such as humpback chub (*Gila cypha*), razorback sucker (*Xyrauchen texanus*), and bluehead sucker (*Catostomus discobolus*) are rare in GCNP, while others, including Colorado pikeminnow (*Ptychocheilus lucius*), roundtail chub (*Gila robusta*), and bonytail chub (*Gila elegans*), are gone.

Climate change is also predicted to increase the abundance of drought-resistant plants at Grand Canyon. Diminished river flow, less frequent flooding (caused by Glen Canyon Dam), and increased temperatures (which may result from climate change) are predicted to increase the abundance of drought-tolerant plants such as tamarisk and desert grasses, which will replace native plants in GCNP.

The Future of Grand Canyon

Grand Canyon's history and its peoples' stories are diverse. For example, the earliest people at Grand Canyon considered the canyon to be home, and later explorers, prospectors, and entrepreneurs considered it a challenging adventure and a potential source of wealth and fame. Countless artists came to Grand Canyon seeking not just picturesque views but inspiration. People have found all of these things and much more at Grand Canyon.

Today's visitors to Grand Canyon, like those of earlier centuries, experience the canyon anew. These experiences are often unique and personal, and it becomes easy to understand the reverence many Native Americans and others have for Grand Canyon. Grand Canyon is an experience that far surpasses what visitors have read in books or what they've seen on television shows, websites, postcards, and brochures. It is also exceedingly hard to fully appreciate; as J. B. Priestley noted, "It is useless to try to describe the Grand Canyon. Those who have not seen it will not believe any possible description, and those who have seen it know that it cannot be painted in either pigments or words." Early Grand Canyon concessionaire Fred Harvey had a simpler conclusion: "No one can describe it to you." The claims by Priestley and Harvey underlie Grand Canyon's diverse impacts on visitors: People see and feel different things at Grand Canyon, and everyone interprets their experience differently. For some, the experience is humbling, whereas for others it is beautiful, serene, and spiritual. For example, Native Americans consider Grand Canyon to be alive and sacred. To these people, Grand Canyon is their ancestral home—where their ancestors originated and where their spirits will return to Earth. No one leaves Grand Canyon untouched by its immensity, beauty, history, and spectacle.

Grand Canyon faces several ongoing changes and challenges, yet it remains one of the most popular, iconic landscapes on Earth. The area's intricate, colorful, and visually overwhelming panoramas continue to make Grand Canyon the canyon to which all others are compared. At Mather Point, where most visitors get their first glimpse of Grand Canyon, the canyon's impact on people is often stunning: Most people stare quietly—and often reverently—at the canyon, searching for words to describe what they see. These people soon realize that seeing Grand Canyon is accompanied by a feeling for it, and they will often spend much of their remaining time at Grand Canyon (and after leaving the Canyon) trying to understand and express this feeling. For many people, the feeling lingers for a lifetime.

Regardless of their motivation for coming to Grand Canyon, everyone is awestruck by Grand Canyon's immensity, harshness, and beauty. As American poet Carl Sandburg noted in 1928, "Each man sees himself in the Grand Canyon—each one makes his own Canyon before he comes, and each one brings and carries away his own Canyon." Everyone, from longtime residents to vacationers who visit Grand Canyon for only a few hours, is inspired, humbled, or overwhelmed by something at Grand Canyon. These feelings are why, on September 3, 1964, President Lyndon B. Johnson signed into law the U.S. Wilderness Act, which required government agencies to survey all federal lands (including GCNP) for possible designation and protection as wilderness. At that signing ceremony, Johnson warned his and future generations of the importance of protecting places such as Grand Canyon:

> If future generations are to remember us with gratitude rather than contempt, we must leave them something more than the miracles of technology. We must leave them a glimpse of the world as it was in the beginning, not just after we got through with it. (Martin, 2012)

Further Reading

Harvey, Fred. 1905 [2017]. *El Tovar: A New Hotel at the Grand Canyon of Arizona*. London: Forgotten Books.

Martin, Thomas C. 2012. *Big Water, Little Boats: Moulty Fulmer and the First Grand Canyon Dory on the Last of the Wild Colorado River.* Flagstaff, AZ: Vishnu Temple Press.

Part II
Topical Entries

A

AIRPLANES AND HELICOPTERS

The first recorded flight over Grand Canyon occurred in February 1919, with Army Lieutenants R. O. Searle and E. D. Jones aboard. Thereafter, commercial flights often deviated from their designated routes to show passengers the vast canyon. In 1933, the first regularly scheduled flight to Grand Canyon brought 107 passengers to the canyon; by 1966, planes brought 9,842 passengers to Grand Canyon. That same year, another 11,239 passengers came to GCNP aboard the Atchison, Topeka and Santa Fe Railway, 4,614 came aboard buses, and 73,034 came in cars.

In 1950, Arizona Helicopter Services flew the first commercial helicopter tour from the South Rim. That same year, NPS used helicopters for the first time for a search-and-rescue mission in GCNP. Today, helicopters are abundant at Grand Canyon. They are also the most important means of emergency rescues in the park.

The first casualties from air flight over Grand Canyon occurred in 1956, when two commercial jets from Los Angeles—one 5 miles off its planned route and the other 25 miles off—collided in uncongested airspace over Chuar Butte in eastern Grand Canyon, killing all 128 people aboard. This accident, which more than doubled the number of people killed in Grand Canyon before 1956, was also the worst peacetime civilian airline disaster in commercial aviation history (to that time) and remains the worst of several disasters at Grand Canyon. This crash led to the development of anticollision radar and the formation of the Federal Aviation Administration (FAA) in 1958 but did not end airline fatalities at Grand Canyon. For example, a midair collision in 1986 of a Grand Canyon Airlines Twin Otter plane and a Helitack Bell 206 JetRanger helicopter near Crystal Rapid killed all 25 people aboard and focused national attention on air tourism. This was the 14th air crash in the past decade in Grand Canyon (for comparison, only 16 river runners died in Grand Canyon from 1889 to 1990).

Before 1987, planes and helicopters could fly virtually anywhere in Grand Canyon. In December 1983, however, a helicopter returning to Las Vegas along the Colorado River hit a USGS cableway in Grand Canyon near River Mile 65, killing one of the helicopter's passengers. (Between 1966 and 1986, 10 crashes in Grand Canyon had killed 41 people.) These and other accidents led to the passage in 1988 of Special Federal Aviation Regulation 50-2, which established flight-free zones below 14,500 feet in 44 percent of GCNP. Within a few years, more accidents in and around Grand Canyon prompted Congress to pass the National Parks Overflights Act (Public Law 100-91, spearheaded by Arizona Senator John McCain), which created more flight-free zones over Grand Canyon, banned flights below the rim, and imposed other flight restrictions at Grand Canyon and several other

Jo Mora

Joseph Jacinto "Jo" Mora (1876–1947) was a cartoonist, painter, and sculptor who in 1931 was commissioned by Scenic Airways to produce a whimsical color map of Grand Canyon. When the deal fell through, Mora—who had lived with the Hopis and Navajos—produced a black-and-white map himself. Mora's detailed maps, which incorrectly placed *sipapu* (a sacred Hopi spring near the confluence of the Little Colorado and Colorado Rivers) at Phantom Ranch, combined Mora's humor with his cartooning skills and knowledge of Grand Canyon's history and are now valued at more than $3,000 each. Mora died on October 10, 1947, in Monterey, California.

Mora was often recognized as a "Renaissance Man of the West." A year after he died, a colorized version of Mora's original map (colored by Mora's friend Ferdinand Burgdorff) was released. In 1968, 20 years after his death, Mora's art was featured on the cover of the Byrds' influential *Sweetheart of the Rodeo* album.

national parks. With the exception of films taken during takeoffs and landings at the Havasupai and Hualapai Reservations, the last aerial video below the rim of Grand Canyon was filmed in 1984. Although helicopters may land at the Havasupai and Hualapai Reservations, all other flights at GCNP must remain at least 1,500 feet above the rim.

In 2000, the FAA limited scenic flights in GCNP to about 90,000 per year, reduced noise, and expanded the no-fly zone in GCNP. (The FAA later reauthorized this many flights in GCNP, but only 55,185 commercial tours were flown.) Soon thereafter, the FAA approved 1,721 additional flights in GCNP, providing the planes and helicopters were using quiet-aircraft technology. As an incentive for the installation of this technology, GCNP reduced the per-flight fee paid by such planes and helicopters by 20 percent, thereby saving the industry $250,000 per year. Critics, however, noted that the quiet technology was defined by noise per seat, thereby allowing aircraft to be noisier as long as they could hold more passengers.

In 2017, the air tourism industry around GCNP generated more than $2 million.

See also: Desert View; Pioneer Cemetery and Other Gravesites

Further Reading

Berger, Todd R., ed. 2008. *Reflections of Grand Canyon Historians: Ideas, Arguments, and First-Person Accounts*. Grand Canyon, AZ: Grand Canyon Association.

AKIN, LOUIS

Louis Benton "Mapli" Akin (1863–1913) was raised in Portland, Oregon, after which he moved to New York City to study art. In 1903, the Atchison, Topeka and Santa Fe Railway commissioned Akin to produce paintings of Hopi Indians and their culture. This suited Akin, who believed that moving to Arizona would remedy his tuberculosis. Akin went to Arizona and lived with the Hopis at Oraibi, where he rented a room for 75¢ per week. Many of the 25 or so paintings that Akin painted there were made into postcards and posters that the Santa Fe Railway used to promote the Hopis and Grand Canyon. The Hopi accepted Akin into their society; they named him Mapli, which means "sleeveless," for his favorite shirts. Akin lived in Arizona for the rest of his life.

Akin then moved into the South Rim's El Tovar Hotel, at which time his paintings began focusing on romantic landscapes. In 1906, Akin produced his most commercially successful painting, titled *El Tovar Hotel, Grand Canyon*, which included the hotel with Hopi and other Native American figures in the foreground. This painting, which was commissioned by the Santa Fe Railway and printed on posters, postcards, and advertisements, featured El Tovar Hotel and Hopi House, which was shown much closer to the rim than it is. The painting also deleted Verkamp's store, which was a competitor of Santa Fe Railway.

Today, one of Akin's most viewed paintings is his 6-by-9-foot *Evening—Grand Canyon*. That painting, which Akin almost destroyed after painting it in 1907, was purchased by GCNP after Akin's death at an auction of the artist's unsold paintings. Ironically, the painting now hangs in Verkamp's Visitor Center, the building that Akin had deleted from his most commercially successful painting.

Akin, who was the first artist to live at Grand Canyon for an extended period, moved to Flagstaff, where he lived in a Babbitt Brothers Trading Company store. Akin, who suffered from depression and financial problems, died there of pneumonia on January 2, 1913, while working on paintings depicting Indian life and culture for the Southwest Indian Room of New York's American Museum of Natural History. He is buried in Flagstaff's Citizen's Cemetery. Whereas virtually all other graves in Citizen's Cemetery face east to west, Akin's grave faces north to south so that, as per his wishes, he faces San Francisco Peaks, his great inspiration.

Further Reading

Kinsey, Joni L. 1998. *The Majesty of the Grand Canyon: 150 Years in Art*. Petaluma, CA: Pomegranate Communications.

ANTIQUITIES ACT OF 1906

The Antiquities Act of 1906 arose from concerns of Iowa Congressman John Lacey regarding the plundering of prehistoric Native American artifacts, cliff dwellings, missions, and other ruins (collectively called antiquities) on federal lands in the American West. The legislation, which was signed into law by President Theodore Roosevelt on June 8, 1906, was the first federal law to protect historic, prehistoric, natural, and cultural resources in the United States. Specifically, the Antiquities Act empowered the president of the United States—without Congressional approval—to set aside for protection "historic landmarks, historic and prehistoric structures, and other objects of historic and scientific interest that are situated upon the lands owned or controlled by the Government of the United States" (American Antiquities Act of 1906). This meant that the president could, by proclamation, create national monuments from land in the public domain to protect important natural, cultural, or scientific features. The law also allowed the president—with nearly unfettered discretion—to reserve and accept private lands to stop the excavation or destruction of antiquities. The Antiquities Act enabled faster action than was possible by Congress when creating a national park or national monument. The first use of the Antiquities Act to protect a large area occurred on September 24, 1906, when President Theodore Roosevelt created Devils Tower National Monument. This 600-foot-high guidepost and religious site in Wyoming was the nation's first national monument.

In 1908, President Roosevelt again invoked the Antiquities Act when he issued Presidential Proclamation 794 to establish Grand Canyon National Monument, the precursor of GCNP. This ended opportunities for individuals to make new claims (i.e., "private entry") of Grand Canyon's land but did not eliminate existing claims of private ownership. (In 1920, Grand Canyon pioneer Ralph Cameron unsuccessfully tested Roosevelt's proclamation regarding land near the railroad tracks at Grand Canyon Village in *Cameron v. United States*, a case decided by the U.S. Supreme Court.) Roosevelt also declared that the mission of the monument included "recreation" and "preservation." Grand Canyon National Monument, which covered 1,279 square miles, was managed by the U.S. Forest Service until it became GCNP in 1919. (That same year, Zion and Arcadia National Parks were also created.) The Forest Service, which was the first federal agency that managed a well-known travel destination, tried to organize the area's booming tourism business at the South Rim, but the Forest Service's primary mission was to promote sustainable harvesting of timber. Congress provided little money for tourism, so the Forest Service soon formed partnerships with the Atchison, Topeka and Santa Fe Railway and the Fred Harvey Company to handle tourism at Grand Canyon.

In 1932, President Herbert Hoover used the Antiquities Act to create a new 426-square-mile Grand Canyon National Monument that included Toroweap to recognize the scientific value of western Grand Canyon. (The earlier Grand Canyon National Monument created by President Roosevelt became part of GCNP.) The legislation creating this monument included no provisions to block dams and other reclamation projects. Despite its isolation, the new Grand Canyon National Monument attracted more than 1,000 people per year by the 1960s (up from 19 in May/June of 1941). The monument remained a separate part of NPS until 1975, when it was made part of GCNP.

See also: Grand Canyon Village; Roosevelt Point

Further Reading

American Antiquities Act of 1906. 16 USC 431–433. Retrieved July 7, 2017, at https://www.nps.gov/history/local-law/anti1906.htm

B

BASS, WILLIAM AND ADA

William Wallace Bass (1849–1933), following a doctor's suggestion to move to the American West, settled at Grand Canyon's South Rim in 1884. Within a year, Bass began advertising for tourists and built tents for guests near Mystic Spring. After homesteading near Havasupai Point, Bass prospected, filed more than 20 mining claims, established a tourist facility known as Bass Camp near River Mile 108, and developed more than 50 miles of trails to support his mining and tourism operations; this was more than any other European American pioneer had done at Grand Canyon. Bass wrote many poems about Grand Canyon but never found the gold he believed had been buried in in the canyon by Lees Ferry operator John D. Lee.

Around 1891, Bass began offering the first stagecoach rides twice per week between Williams and Grand Canyon. The stagecoach brought tourists to Bass's rim-side camp, which he operated for 36 years. During this time, Bass hosted thousands of visitors, including writer Zane Grey, artist Thomas Moran, naturalist John Muir, and businessman Henry Ford. In 1914, Bass named the Dox Formation (a sandstone of the Unkar Group of the Grand Canyon Supergroup) after Virginia Dox, the first woman that Bass guided in western Grand Canyon. (Miss Dox, in turn, named a formation then known as Holy Grail Temple, Bass Tomb.) In 1892, Bass's clients also included Boston Conservatory of Music graduate and New York music teacher Ada Lenore Diefendorf (1868–1951), who accompanied Bass on a horseback trip to Havasu Canyon. (Bass played violin, so they shared a common interest in music.) The couple married two years later, raised four children (Mabel, Edith, William, and Hazel) and helped establish the first school at Grand Canyon. This made William and Ada the first non-Natives to raise a family at Grand Canyon. Ada often hiked to the river to do the family's laundry. Her life was challenging, as noted by this entry in her journal in March, 1902: "S.O.S. (Same Old Story) Cook, wash, eat, and worry." Despite these challenges, Ada later noted "I would not have turned back . . . I loved the canyon, too."

When the railroad arrived at Grand Canyon in 1901, Bill's business dropped sharply, prompting Bass to start advertising, "Don't let anyone fool you by telling you I am out of the business." Bass soon despised the railroad: "They think they can drive me out by petty tyranny, harassing treatment, perversions of the truth." Bass and Ada discontinued their stagecoach rides and moved to New York. However, they were back at Grand Canyon a few years later, after which Bill wrote a book titled *Adventures in the Canyons of the Colorado* (1920).

Bass often went to Washington, DC to lobby Congress on behalf of the Havasupai. After applying for and getting the option to open a post office at Supai in 1909

Going to School at Grand Canyon

The first school at Grand Canyon opened on July 10, 1911, in a building called the Cameron Green House at the South Rim near the Bright Angel Trailhead. The school's first teacher was Grace Miller. Thanks to funding from William and Ada Bass, the first permanent schoolhouse opened two years later; the Basses' four children comprised half of the school's students. The first student to complete all 12 grades at Grand Canyon was John Verkamp's grandson, John Verkamp III (b. 1940), who graduated in 1958. Today, Grand Canyon Unified School District serves more than 300 students and is the only K–12 school in a U.S. national park.

(at a cost of $875), Bass delivered mail to Supai for four years. In 1915, Bass had one of his best years, grossing $15,000, but he was later forced to close his camp when the South Rim concession was awarded to the Fred Harvey Company. (Bass's booklet titled *A Few Plain Truths Plainly Stated for Grand Canyon Visitors* was meant to convince tourists to avoid the Fred Harvey Company.) In 1923, Bill and Ada Bass hosted their last guest at their famous tourist camp. Soon thereafter, Bass admitted that "I have been in the struggle quite long enough so do not now retain very much desire to continue . . . So what is the use of an old man like me trying to remain in the game."

In 1926, when Bass sold his mining sites and other operations to the Santa Fe Land and Improvement Company for $20,500, Ada refused to sign the papers until she received $10,000 in her own name. Bill and Ada Bass later moved to Wickenburg, Arizona, where they managed a hotel.

Bill Bass died of a stroke at his home in 1933. Although Bass had helped end the Havasupai custom of cremating its dead, Bass was cremated and—as per his wishes—his ashes were thrown from an airplane by his and Ada's son Bill at the east end of North Bass Trail over Holy Grail Temple. This area, which is also known as Bass Tomb, includes a rock spire along Shinumo Creek where Bass had planted a garden of corn, melons, and other plants, as well as an orchard of peach and apricot trees. In 1980, Arizona governor Bruce Babbitt placed in Pioneer Cemetery a plaque honoring Bass and describing him as a "Pioneer, Guide, Miner, Poet . . . Man of Vision. Though his ashes lie on Bass Tomb, his spirit is everywhere in the Grand Canyon."

Ada Bass, who died of a stroke in Phoenix, Arizona, in 1951, is buried in Pioneer Cemetery near her oldest daughter, Edith. Ada's tombstone describes her as the "1st White Woman to Raise a Family on Rim of Grand Canyon." Bass's name adorns several sites in GCNP, including Bass Trail, Bass Canyon (which Bass called Trail Canyon), and Bass Rapid.

See also: Bass Trails and Bass Camp; Pioneer Cemetery and Other Gravesites; Tramways

Further Reading

Probst-Gipson, Debbie. 2016. *Stories the Canyon Keeps*. N.p.: Quickstep Publishing.

BASS TRAILS AND BASS CAMP

The North and South Bass Trails are named for William Wallace Bass (1849–1933), a Grand Canyon pioneer who had learned many routes into Grand Canyon from Havasupai natives. Although the North and South Bass Trails bear Bass's name, these challenging trails were later reengineered by the NPS, and parts of the original trails are often difficult to trace.

Bass built South Bass Trail in the 1880s, and in 1891 recorded this trail as the first certified Grand Canyon toll trail. This steep, rocky trail descends 1,500 feet to Mystic Spring on Darwin Plateau. Bass's original trail ended here, which is near a granary near the top of the Coconino Sandstone. Bass named the trail for a spring that he could not find; he was later shown its location by Thilwisa (Captain Burro), his Havasupai friend, in exchange for food, after which Bass named the spring Mystic Spring. When Bass extended his trail to the Colorado River, the trail became known as the South Bass Trail.

It is often difficult to get to the South Bass Trailhead (elevation 6,650 feet). The easiest access is from Rowe Well Road or Forest Service Road 328 (unpaved) from Arizona Highway 64. You have to drive 29 miles through Kaibab National Forest and the Havasupai Reservation to access the remote trailhead. (It takes most people at least 1.5 hours to get to the trailhead from Grand Canyon Village; get specific directions to the trailhead from the GCNP Backcountry Office and/or the Tusayan Ranger District of the Kaibab National Forest.) The road is rough and rocky, often not depicted accurately on maps, and impassible after heavy rain.

South Bass Trail, which is 7.8 miles long and drops 4,446 vertical feet, intersects Tonto Trail and offers views of the Grand Scenic Divide, Kaibab rim to the east, and the Supai rim to the west. It is more rugged than GCNP's Corridor Trails but easier to hike than the New Hance and Tanner Trails. South Bass Trail follows Bass Canyon and features some of the best exposures of the Unkar Group, which is absent from most of the rest of Grand Canyon because of uplift and erosion. (The Unkar Group includes Bass Limestone, which contains ancient algae that are the oldest fossils in Grand Canyon.) About a mile above the Colorado River are potholes in bedrock where Bass created Rock Camp and stored equipment. At the river, Bass's River Camp included tents and a large fire pit; the remains of that fire pit are still visible. At the river, hikers will find a historic boat (*Ross Wheeler*) and a cairn marking a spur to Ross Wheeler Beach. There are no services or drinkable water along South Bass Trail.

North Bass Trail is a challenging, steep, 13.5-mile trail from Swamp Point (elevation 7,500 feet) to the Colorado River (elevation 2,200 feet). The trail, some of which was rehabilitated in 2005, begins at the end of Swamp Ridge Road, which in and of itself can be difficult to find and navigate. (Get a map and ask for directions at the U.S. Forest Service office at Jacob Lake.) After some scenic switchbacks through the Kaibab Limestone and Toroweap Formation, hikers reach an intersection; the trail to the right goes to the two-room Muav Saddle Snowshoe Cabin, which was built by NPS in 1925 in a grove of ponderosa pine. This is the only intact building along the trail. The trail in the middle goes south to Powell Plateau, and the trail to the left continues the North Bass Trail to Muav Canyon. President Theodore Roosevelt camped near Muav Saddle Snowshoe Cabin while hunting mountain lions, thus explaining why the cabin is sometimes called Teddy's Cabin.

After about 0.3 mile, you'll reach Queen Anne Spring; hikers should purify this water before drinking it. Hikers then follow White Creek to a waterfall coming out of the Redwall Limestone, after which they continue down the drainage at the top of the Tapeats Sandstone to Shinumo Creek, which they follow for about 1.5 miles to the remains of Bass Camp. Then there is a 600-foot climb before descending to a large beach and terrace above the Colorado River.

North Bass Trail, which was originally known as the Shinumo Trail, is a continuation of the South Bass Trail. This completed the first rim-to-rim trail built by a European American in Grand Canyon. There are no services or drinkable water along North Bass Trail.

Bass linked his South and North Bass Trails with cable tramways—300 feet long and 50 feet above the river—at River Miles 107.5 and 111.4. The first of these tramways usually carried people, supplies, and animals, and the second was for Bass's mining operations. In 1902, when the USGS sent F. E. Matthes and R. T. Evans to make the first official government map of the western Grand Canyon, the surveyors used South Bass Trail to move their workers and supplies across the river. (They later used the Bright Angel Trail to access the river.) In 1906, Bass built another tram to cross the river. This tram was abandoned in the 1920s, and in 1968 the trams' cables were cut by GCNP Chief Ranger George von der Lippe because of their threat to helicopters. Today, hikers and river runners can still see the remains of a cable car and several cables at river's edge.

At the rim, the home/hotel built by Bass in 1906 became known as the White House, and another he built in 1912 closer to Grand Canyon Village became known as the Tin House. Bass's tourism operation flourished; by 1920, he had served more than 3,000 visitors, and in some years Bass grossed more than $20,000.

See also: Bass, William and Ada

Further Reading

Thybony, Scott. 2009. *Grand Canyon: South and North Bass Trail Guide*. Grand Canyon, AZ: Grand Canyon Association.

BERRY, PETER

Peter D. Berry (1858–1932) was a trail builder, guide, and prospector who came to Flagstaff in the late 1880s to avenge his brother John's murder. Although his brother's killers had already been lynched by the time Berry arrived, he stayed in the Flagstaff area and began raising sheep with Ralph and Niles Cameron. Berry married his dead brother's widow, but she left him for a traveling salesman.

Berry came to Grand Canyon in 1889, and four years later—with partners who included the Cameron brothers and Ed Gale—he founded Last Chance Mine and several other claims on Horseshoe Mesa below Grandview Point. (Visitors at Last Chance Mine can see the inscription "P. D. Berry Jan. 1st 1897" in Cave of the Domes near the mine.) Ten mules each hauled about 200 pounds of ore from the mine to the rim and then returned to the mine with food and other supplies for the crews of 4–8 miners who usually worked the site. During its busiest times, the camp could accommodate 25 workers.

Berry built a home at Grandview (he called the area "Grand View"), and in 1897 he and his second wife Martha used profits from that mine to open the two-story Grandview Hotel there. This hotel, which Berry promoted as the only first-class hotel at Grand Canyon, was the nicest hotel and the center of tourism at Grand Canyon for several years. Berry's popular hotel, which was served by a stagecoach line from Flagstaff, charged visitors $2 per night, served meals prepared by a European chef, and offered tours along local trails for $3. In 1900, the U.S. Census showed that Pete had four employees and one boarder (John Hance).

Last Chance Mine produced $75,000 worth of copper ore. In 1902, Berry and his partners sold their interests to Chicago's Henry Barbour for $1,875, who then sold it to a speculator for $40,000. However, the costs of extracting and processing the copper remained high (e.g., the closest copper smelter was in El Paso, Texas). By 1907, when the price of copper plummeted to 12.5¢ per pound, the copper mines at Horseshoe Mesa closed for good. Today, all that is left of Berry's mining operation is a roofless stone building, machinery, and nuggets of copper ore that still litter the area.

Berry filed for a homestead at Grandview Point, and his application was approved and signed by President Theodore Roosevelt on April 1, 1907. In 1913, Berry sold his homestead and hotel site to newspaper mogul William Randolph Hearst for $85,000. (In 1927, Hearst also obtained Grandview Point and Berry's mining claims at Horseshoe Mesa.) In upcoming years, Hearst often offered the site as a vacation destination for his friends. (Hearst also allegedly used the hideaway for trysts with his mistress, Marion Davies.) On most of Hearst's visits to Grand Canyon, however, he stayed at the El Tovar Hotel (Grandview was maintained by Dick Gilliland, the property's caretaker, who lived in Peter Berry's original cabin). Hearst, the inspiration for Orson Welles's *Citizen Kane* (1941), eventually owned about 200 acres at and near Grandview Point that included Horseshoe Mesa and 4.5 acres at the rim. Berry's original homestead did not reach the rim, prompting Hearst to later trade 48.9 acres and rights for a road for the 25.8 acres of land that included the rim-side strip of land that he wanted. Hearst, the most powerful media-mogul in American history, used his influence, money, and newspapers to oppose GCNP because he feared the park would take his land at Grandview. Hearst, who pioneered Sunday comic strips in newspapers, published a strip titled "Canyon Kiddies" in his *Good Housekeeping* magazine for almost two decades.

One section of land near Grandview Point included two large tanks; these so-called Hearst Tanks, which are unmarked and usually dry, include a scattered collection of building materials and other supplies from bygone days. In 1916, as tourism shifted from Grandview Point to what became Grand Canyon Village, Berry closed Grandview Hotel. Hearst told NPS director Stephen Mather that he wanted to build a public museum at Grandview for his collection of silver, but the museum was never built.

Berry died in 1932 at Grand Canyon. He is buried near Ralph Cameron in Pioneer Cemetery on the Grand Canyon's South Rim.

See also: Cameron, Ralph; Grandview Point; Grandview Trail; Pioneer Cemetery and Other Gravesites

Further Reading

Anderson, Michael. 2001. *Along the Rim: A Guide to Grand Canyon's South Rim from Hermit's Rest to Desert View.* Grand Canyon, AZ: Grand Canyon Association.

Anderson, Michael. 1998. *Living at the Edge: Explorers, Exploiters and Settlers of the Grand Canyon Region.* Grand Canyon, AZ: Grand Canyon Association.

BIRDSEYE, CLAUDE

By 1923, only 27 people (not counting James White) had boated through Grand Canyon. During the various attempts to go through the canyon, three people had drowned. There were only three books about expeditions through Grand Canyon: John Wesley Powell's account, Frederick Dellenbaugh's description of Powell's second voyage, and Ellsworth Kolb's account of his trip through Grand Canyon while making the first movie about Grand Canyon river running with his brother Emery. The first USGS expedition through Grand Canyon would follow, led by Claude Birdseye.

Claude Hale Birdseye (1878–1941) was an engineer and topographer who, after serving in World War I, led a USGS expedition in 1923 from Lees Ferry through Grand Canyon (and beyond). This expedition resulted from the growing need to control floods along the lower Colorado River and the growing number of proposals to exploit the water and power of the Colorado River (especially for agriculture in California), all of which required accurate maps of Grand Canyon. Birdseye was accompanied by boatmen, a cook, and five geologists/engineers. The 45-year-old Birdseye was the oldest member of the expedition.

Birdseye's expedition was projected to last 80 days but went on for 108. Roger Birdseye, Claude's cousin, handled the logistics of resupplying the trip. This was a formidable task; for example, for each 30 days, the expedition's ration list included 100 pounds of cornmeal, 72 cans of milk, 100 pounds of butter, 300 pounds of rice, 100 pounds of lard, 100 pounds of biscuits, 100 pounds of potatoes, 100 pounds of sugar, 240 pounds of flour, and cans of vegetables, fish, fruit, beans, coffee, tea, cocoa, spices, and dried eggs.

Birdseye, who consulted Emery Kolb before the trip began, required four boats for his trip. Three—*Glen*, *Marble*, and *Boulder*—were already at Lees Ferry, where they had been stored after a 1921 exploration of Cataract Canyon. These three boats were sturdy, Galloway-type boats that ranged from 16 to 18 feet long. Birdseye's fourth boat, *Grand*, was built for his expedition; he also took a 14-foot-long folding canvas boat for emergencies that might arise.

Birdseye was the expedition's chief topographic engineer and the first topographic engineer at USGS. Birdseye's expedition analyzed the geology, hydrology, and topography of the Colorado River through Grand Canyon to answer one question: Where could dams be built in Grand Canyon? This was the first modern survey of Grand Canyon. (At the start of the 21st century, computer-based image analysis and satellite-based platforms would replace the manual techniques pioneered by Birdseye.) Birdseye's crew, which included 37-year-old topographer Roland Whitman Burchard (who did most of the mapping), ran 84 rapids. The expedition used four 800-pound boats (stocked with two tons of equipment) that were difficult to get through the three rapids they portaged.

The 251-mile Birdseye expedition (all within today's GCNP) also carried the first radio used on a river run in Grand Canyon. This radio, specially built for the trip by the Bureau of Standards and whose long antenna also served as a clothesline, disproved the assumption that radios were useless in Grand Canyon. The radio, which was used to alert the expedition of flooding, was a public relations coup; radio station KHJ (owned and operated by the *Los Angeles Times*) in Los Angeles made special broadcasts to and about Birdseye's expedition. At Hance Rapid, Birdseye's expedition was joined by Edith Kolb (Emery's daughter), who became the first woman to run a major rapid in Grand Canyon.

One of the boatmen on the Birdseye expedition was Hawaiian native Frank Dodge. (In 1937, Dodge was also the head boatman on the Carnegie–Caltech expedition, which was the last group of river runners to run Lava Cliff Rapid before it was submerged in Lake Mead.) During the Birdseye expedition, Dodge pulled several colleagues from the river and saved them from drowning. The alcoholic Dodge described Birdseye as "the finest boss I ever had" (Boyer and Webb, 2007). In the 1930s, Dodge later worked at Lees Ferry and for hydraulic engineer Eugene Clyde La Rue on boats for the 1935–1936 Fairchild Aerial Surveys. Dodge wrote his autobiography for USGS, but it was never published. In 1965, after becoming disabled, Dodge committed suicide.

Birdseye's survey was the best equipped and best organized expedition through Grand Canyon. The expedition covered GCNP's entire 251 river miles (at the time; GCNP was later expanded to 277 miles) and identified 22 promising sites for dams between Lees Ferry and Black Canyon. Although Ellsworth Kolb had worked as a boatman on a 1921 expedition in Grand Canyon, Birdseye chose Ellsworth's brother Emery as head boatman. (Legendary Colorado River runner Bert Loper was deemed "too old.") Kolb demanded a salary of $1,000, but Birdseye—who was paid only $415—offered only $500 (this was the same salary as the director of USGS). Kolb then demanded exclusive filming rights for the trip, even claiming that "it would be self extermination for me to take part in a trip where others obtain motion pictures as it is in such pictures, more than anything else, we make our living; and I am confident that you or your Department of the Government would not knowingly want to injure us" (Boyer and Webb, 2007). Birdseye, not impressed with Kolb's self-serving claim, responded with an ultimatum: Kolb's salary would be $500, and he could take a small camera; USGS would also provide film for Kolb in exchange for copies of Kolb's film. Kolb, having no choice, agreed to the terms proposed by Birdseye.

Emery Kolb became famous as a Grand Canyon photographer, but the expedition's official photographer was hydraulic engineer Eugene Clyde La Rue. The headstrong, self-centered La Rue was one of the first scientists to use repeat photography as a scientific tool, and his film of the expedition—*The Survey of the Grand Canyon of the Colorado River in Arizona* (1923)—was one of the first black-and-white films distributed nationally. Kolb later complained he was cheated out of his self-proclaimed monopoly on film rights in Grand Canyon. (Kolb, not La Rue, became the most famous photographer at Grand Canyon; whereas La Rue's photos are stored in an obscure library operated by USGS, Kolb's photos are often republished in books about Grand Canyon.) In 1927, La Rue was a member of a filming expedition through Grand Canyon that disappeared. When Emery Kolb was asked to help with the rescue, he refused because he suspected that the disappearance was a publicity stunt. Kolb was right.

La Rue and others advocated building 13 dams in and near Grand Canyon (including one at Glen Canyon) to serve the water needs of the southwest, and many people assumed that Birdseye's famous expedition made a dam in Grand Canyon inevitable. One of the many families certain that at least one dam would be built in Grand Canyon was that of Winifred Johnson, a 21-month-old who died and was buried at Lonely Dell at Lees Ferry. Winifred's parents, after hearing about the flooding that would be produced by the dam planned for nearby Marble Canyon, reinterred Winifred at Evergreen Cemetery in Springville, Utah. Winifred's body was never returned to Lonely Dell, where her tombstone today is a cenotaph. None of the dams were ever built.

Birdseye named most of today's rapids in Grand Canyon that have mileage designation names (i.e., names referencing their location by river mile below Lees Ferry, such as 24.5 Mile Rapid). Some examples of other sites in Grand Canyon named by Birdseye are as follows:

Boulder Narrows at River Mile 18 is named for a 30-foot-wide boulder of Toroweap Formation limestone that blocks the middle of the river.

Bridge Canyon near Mile 236 is named for a small natural bridge near the river. Near here is the boundary between the Hualapai Reservation and GCNP. This was also the proposed site for a 740-foot-high dam (authorized by the Federal Power Commission) that would have flooded the Hualapai Reservation and miles of GCNP (then Grand Canyon National Monument). In 1968, Congress passed a law forbidding dams in GCNP.

House Rock Creek near River Mile 17, showcases polished, multicolored limestone. Near here in 1904, Frederick Dellenbaugh—a 27-year-old member of Powell's second expedition through Grand Canyon—found a low cliff on which "House Rock Hotel" was written in coal. Dellenbaugh later explained "from this came House Rock Valley, and the name was soon a fixture, and went on our maps. And thus easily are names established in a new country" (Brian, 2015).

President Harding Rapid at Mile 43.5 was named by Birdseye in 1923 upon hearing here by radio on August 2, 1923, of the death of President Warren

G. Harding, the 29th president of the United States. (Later in the same broadcast, the announcer acknowledged "that heroic band of engineers braving death in the rapids of the Colorado.") The relatively benign rapid had previously been referred to as Boulder Rapid because of a boulder in the center of the channel. In 1955, Grand Canyon swimmer Bill Beer named this boulder The Cheese Grater after his partner, John Daggett, was pulled under the boulder and lacerated; this was the pair's most severe accident in their entire historic swim.

Tanner Rapid at the end of Tanner Canyon at River Mile 68.5, honors Seth B. Tanner (1828–1918), a 19th-century Mormon pioneer who operated a trading post on the Hopi Reservation. Tanner claimed several mines along the Colorado River (e.g., Tanner Mine near River Mile 66) and built trails in the late 1880s to service them. Tanner Trail on the south side of the river, Nankoweap Trail on the north, and a connecting trail were often referred to collectively as Horsethief Trail and were later used to move materials and horses (and, during Prohibition, booze) from Arizona to southern Utah. Tanner Trail is one of the steepest and most challenging of the named trails into Grand Canyon from the South Rim.

Upset Rapid, which is 8 miles below Kanab Creek, is where boatman Emery Kolb capsized his boat during the Birdseye expedition. In 1967, on the day after his 44th birthday, river guide Jesse Edwin "Shorty" Burton drowned here when his boat capsized and his life jacket was caught underwater. Burton, who was famous for the pies and cobblers that he cooked for his passengers, is honored by a pie pan–shaped plaque just past this rapid at River Mile 150.5 on river right.

On August 4, on their way down the river, the Birdseye expedition etched the outline of a geologic rock pick and the letters "USGS" on river right at River Mile 16.1. River runners interested in the Colorado River's history continue to stop here and see the inscription.

Six weeks after their stay at President Harding Rapid, a flood in the canyon raised the water level of the river by more than 20 feet and increased the flow from 10,000 cubic feet per second (cfs) to 125,000 cfs in a day. Before they could continue, the expedition had to wait for three days for the river to drop enough to see their last survey point. When the group reached Diamond Creek several days later than expected, they were shown newspapers with headlines reading, "Hope Not Given Up for River Party." By this point, the expedition's radio was no longer working, so they did not get warnings about the flood.

Birdseye's famous expedition was covered by dozens of newspapers and magazines throughout North America. After running articles claiming that the canyon's flood had wiped out the expedition, newspaper headlines on November 11 then made the group heroes, reporting "Daring Scientists Conquer Grand Canyon's Wild Rapids: Hurled in Frail Boats Down 280 Miles of Savage Stream!" (*The New York Times*) and "Chief of Canyon explorers Once Reported Killed Tells of Perils: Risked Lives for 450 Miles to Make Maps" (*The Washington Post*). After the expedition, Birdseye wrote several articles about the voyage and in 1928 left the USGS to become president of an aerial photography survey company named the Aerotopograph Corporation of America; while with this company, Birdseye did a photogrammetric survey of the Boulder Dam Site. He returned to USGS in 1932, the same year that he received an honorary degree from Oberlin College, his alma mater.

Although no dams were built at Birdseye's proposed sites, Birdseye's expedition was important, for it was the last great exploration of the American West; subsequent expeditions used staged disappearances, movies, cute animals, and other gimmicks to get the public's attention. Birdseye's trip was featured in the October 1925 issue of *Scientific American*, but it was surveyor Burchard, not Birdseye, whose likeness was on the cover. The expedition produced the first topographic maps of the bottom of Grand Canyon. Thanks to these maps, subsequent expeditions into Grand Canyon were no longer considered trips into "the great unknown," for river runners knew the location and surveyed drops of every rapid. Birdseye's maps were also the basis for the first *Belknap Grand Canyon River Guide*, which became a standard resource for river runners. (This waterproof book was thought of by artist and river runner Buzz Belknap, whose maternal grandparents—Frank and Mabel Spencer—operated Hopi House for several years.) After their expedition, Birdseye's group never reconvened.

Birdseye, a distant cousin of the Birdseye family that built a frozen-foods empire, died on May 30, 1941, and was eulogized in professional journals and the popular press. He is buried at Arlington National Cemetery. In 1975, the American Society of Photogrammetry installed a memorial plaque to Birdseye and his expedition at Hopi Point, and in 1999 Birdseye was named an Engineer of the Millennium by the American Society of Civil Engineers. Birdseye is also honored by the North Rim's 6,975-foot-high Claude Birdseye Point, which is at the end of a western ridge along Shiva Temple. *Glen*, one of Birdseye's oversized Galloway-style boats, is part of GCNP's collection of historic boats; during the expedition, it was the cook's boat. In 2017, the renovated *Glen* was displayed, as part of GCNP's If Boats Could Talk conservation project, at the Grand Canyon Visitor Center on the South Rim. The *Grand* was given to the Smithsonian Institution.

See also: Thematic Essays section: Whose Water Is It?; Hopi Point; Kolb, Ellsworth and Emery; Rapids; USGS; White, James

Further Reading

Belknap, Buzz, et al. 2017. *Grand Canyon River Guide*. Evergreen, CO: Westwater Books.

Boyer, Diane E., and Webb, Robert H. 2007. *Damming Grand Canyon: The 1923 USGS Colorado River Expedition*. Logan: Utah State University Press.

Brian, Nancy. *River to Rim: A Guide to Place Names along the Colorado River in Grand Canyon from Lake Powell to Lake Mead*. Second edition. Flagstaff, AZ: Earthquest Press.

BOUCHER, LOUIS

Louis D. Boucher was a pioneering French-Canadian prospector, explorer, and guide at Grand Canyon. He came to Grand Canyon around July 1891, after which he built a tent camp and lived in a remote area at the western end of Hermit Basin at a place called Dripping Springs, where water drips from a large, overhanging cliff. Boucher worked as a guide for John Hance on Hance Trail from 1891 to 1892, after which he worked as a guide and toll collector on Bright Angel and Grandview Trails. Boucher also helped manage a camp owned by Ralph Cameron at Indian Garden. In the early 1890s, with the help of others, Boucher built what is known today as Dripping Springs Trail (then known as the Silver Bell Trail, named for Boucher's favorite mule) that went from the South Rim to Dripping Springs. At Dripping Springs, Boucher planted fruit trees and built cabins. Boucher later built Boucher Trail to the river through what is now known as Boucher Canyon. Boucher's friends called him Old Louie, and an 1892 article in *Scientific American* described him as "intelligent, obliging, and not too talkative" (Annerino, 2017). Writers and others made him famous as The Hermit.

Boucher's mining work was often funded by the Verkamp and Babbitt families. He mined copper along what is now Hermit Creek; however, unlike other prospectors, Boucher never abandoned prospecting for tourism.

In August 1909, Boucher—believing that the Atchison, Topeka and Santa Fe Railway had ruined his business—sold his trails and camps and left Grand Canyon for Mohrland, Utah, where he worked in a coal mine. After leaving Grand Canyon in 1912 following a visit to see Peter Berry at the Grandview Hotel, Boucher—the namesake of Hermits Rest—never returned. He died in obscurity. Hikers can still see the remains of Boucher's stone cabin along his trail.

See also: Hermits Rest; Hermit Trail

Further Reading

Annerino, John. 2017. *Hiking the Grand Canyon: A Detailed Guide to More Than 100 Trails.* New York: Skyhorse Publishing.

BRIDGE CANYON DAM

Bridge Canyon Dam, which was also called Hualapai Dam, was a dam proposed near River Mile 238 in Grand Canyon's Lower Granite Gorge. (Bridge Canyon is named for a small, nearby, natural bridge.) When this dam was proposed in the 1920s, the site was 80 miles downstream from GCNP's border. In 1950, an attempt to build this dam was stopped by California's delegation to the U.S. House of Representatives. Arizona wanted to use more of its allocation of water, some of which was being "loaned" to California. California opposed Arizona's claims to this water, which could have decreased the amount of water for California.

Bridge Canyon Dam, which would have been the tallest dam in the Western Hemisphere, was to be located in an extremely isolated part of Grand Canyon. The primary purpose of the 740-foot-high, 1,700-foot-long, $500 million, concrete arch-gravity dam was to produce hydroelectric power that would be sold to help pay the Central Arizona Project, an aqueduct that diverts water from the Colorado River to central and southern Arizona. The dam would have created Bridge Canyon Reservoir, which could hold 3.71 million acre-feet of water that would have flooded 93 miles of Grand Canyon (up to near the mouth Kanab Creek), including Lava Falls and Havasu Canyon. To delay the filling of the reservoir with silt, plans for building Bridge Canyon Dam required the construction of several other silt-trapping dams on tributaries such as Paria River, Little Colorado River, and Kanab Creek. The dam at Kanab Creek would have also helped fund the Central Arizona Project.

Although Bridge Canyon Dam was not as well-known as Marble Canyon Dam upstream, it would have backed up a larger reservoir. The release of water from that reservoir would have produced hydroelectric power to pump water through a canal from Lake Havasu (on the Colorado River) to Phoenix and to other areas in central Arizona. Along the way, water was to be raised more than 250 feet. The dam would have flooded lower Havasu Creek and Lava Falls, one of the iconic rapids on the Colorado River. Bridge Canyon Dam would have also flooded the Hualapai Reservation and Grand Canyon National Monument, but the Hualapai Tribe mostly supported the dam because of the money it would have received. GCNP naturalist Eddie McKee regretted that the dam would cover the area's lava flows and flood the mouth of Havasu Creek, thereby disrupting the Havasupai way of life. McKee urged building a lower dam but begrudgingly supported the plans for Bridge Canyon Dam.

When work on the dam began, workers built a small town near Bridge Canyon having "a tar-paper cook shack with tables, a walk-in refrigerator . . . sleeping accommodations for the men . . . a small generating plant which furnished lights and fans for the swamp-type air conditioners which were a necessity in this canyon" (Murdock, 1996). Workers reached this site via a 30-mile boat ride up the Colorado River from Lake Mead. After much study, the best site was determined to be the so-called Lower Gneiss site near River Mile 235 (about 19 miles northwest of Peach Springs).

In its early stages, there was relatively little resistance to Bridge Canyon Dam, largely because the site was remote and the area affected presumed unimportant. NPS objected to the dam, but their concerns were silenced by the Eisenhower administration. The Sierra Club initially supported a dam as high as 2,000 feet but later—thanks to environmental activist Martin Litton—opposed the dam. This opposition by the Sierra Club and other environmental groups grew, as did public opposition to the dam. Arizona finally gave up its plans to build dams in Grand Canyon to ensure passage of the Central Arizona Project, which promised other irrigation projects in the state.

In 1976, to replace the power that would have been generated by Bridge Canyon Dam and Marble Canyon Dam, the coal-fired Navajo Generating Station was built near Page, Arizona. In 2017, this power plant announced it would cease operations at the end of 2019.

See also: Thematic Essays section: Whose Water Is It?; Litton, Martin; Sierra Club

Further Reading

Boyer, Diane E., and Webb, Robert H. 2007. *Damming Grand Canyon: The 1923 USGS Colorado River Expedition*. Logan: Utah State University Press.

Murdock, J. Neil. 1996. Bridge Canyon Dam. In *Early History of the Colorado River Storage Project*, May 1971, U. S. Department of the Interior, Bureau of Reclamation. Grand Canyon River Guides, Winter, 1996–1997. Retrieved July 1, 2017, at http://www.gcrg.org/bqr/10-1/bcd.html

BRIGHT ANGEL LODGE

Today's Bright Angel Lodge was originally a wood-framed hotel and nearby tent camp called Bright Angel Hotel that was opened in 1896 by James Thurber. (Earlier, Thurber had operated the Tolfree Camp near Grandview Point.) In 1901, Thurber sold Bright Angel Hotel to businessman Martin Buggeln, who added several tent-cabins and raised the price of an overnight stay in a room to $3 per night. The camp was near land claimed by Ralph Cameron, who opened a two-story hotel a few dozen yards west of today's lodges.

By the 1930s, most property in this area was owned by the Atchison, Topeka and Santa Fe Railway, which hired Mary Jane Colter to design a replacement lodge. Colter's Bright Angel Lodge opened on June 22, 1935, as a cheaper alternative to the El Tovar Hotel. More than 2,000 people attended a barbeque that accompanied the opening of Bright Angel Lodge. Today, Bright Angel Lodge is a popular hotel in GCNP. The lodge's History Room, which documents the history of the Fred Harvey Company in Grand Canyon, includes a 10-foot-high fireplace made of rocks taken from the different geological layers of the canyon.

Today's Bright Angel Lodge consists of 37 rooms and 49 cabins. Suite #6101 of Bright Angel Lodge is the two-room Buckey O'Neill Cabin, which William Owen "Buckey" O'Neill (1860–1898)—a local lawyer, sheriff, miner, newspaperman, and politician—built in the 1890s and sold to Ralph Cameron. This building is the oldest continuously standing structure on the South Rim.

Bright Angel Lodge's cabin #6188 (Red Horse Station), which was once 16 miles from Grand Canyon Village, was originally the last of the three stagecoach stops between Williams and the South Rim. At the two-story Red Horse Station, stagecoaches changed horses before embarking on the final stretch to the South Rim. In 1902, Grand Canyon pioneer Ralph Cameron moved the building log by log to its present location at the South Rim and made it part of his Cameron Hotel near this site. (Cameron competed with the Santa Fe Railway's larger El Tovar Hotel by having his employees distribute flyers at the depot asking, "When you can get as good or better from a private individual, are you going to patronize a greedy, grasping corporation?") Cameron's strategy worked for a while; between 1904 and 1906, his hotel hosted nearly 2,000 guests. The Red Horse Station building also served as a post office from 1910 to 1935. Cameron's hotel (without its second story) later became the one-story, two-room cabin that was included in Bright Angel Lodge when that property was designed by Mary Colter in 1935. Whereas a room at Bright Angel Lodge cost only $19 in 1977, accommodations in 2017 started at $213 per night for a suite and $105 per night for a standard room.

See also: Grand Canyon Village

Further Reading

Anderson, Michael. 2001. *Along the Rim: A Guide to Grand Canyon's South Rim from Hermit's Rest to Desert View.* Grand Canyon, AZ: Grand Canyon Association.

BRIGHT ANGEL TRAIL

Hiking is one of the most popular ways for visitors to experience Grand Canyon. Most trails into and out of Grand Canyon were originally made by animals or Native Americans for hunting and gathering, after which they were repurposed by prospectors and early settlers for mining and tourism. Today, many of these same trails are used by hikers to explore Grand Canyon. Although there are many trails in Grand Canyon, they cover relatively little of the canyon; indeed, there are no trails along 95 percent of the North Rim nor more than 75 percent of the South Rim.

The most well-traveled trails into and out of Grand Canyon are the Bright Angel Trail (from the South Rim), South Kaibab Trail (also from the South Rim), and North Kaibab Trail (from the North Rim). These three trails, which are patrolled and maintained by GCNP year-round, are called the Corridor Trails. The origin of the modern Bright Angel Trail can be traced to 1890–1891, when Ralph and Niles Cameron and Peter Berry spent $500 to improve an old trail leading to their mines and mining claims in the canyon. These claims were extensive; for example, by about 1920, the Camerons' nearly 1,000 claims covered more than 13,000 acres of prime land at the rim and along the trail. These claims helped the Camerons and Berry control access to the trail.

In 1891, Berry registered the trail with Yavapai County as the Bright Angel Toll Road. After the Berrys transferred control of the trail to Cameron in 1901 in exchange for ownership of Grandview Toll Road (today's Grandview Trail), the trail became known as Cameron's Trail. (Interestingly, the original Bright Angel Trail at Grand Canyon started at the North Rim's Bright Angel Point.) In 1904, *Territory of Arizona v. Ralph Cameron* declared this transfer of the trail to be illegal but confirmed the right of the trail's owner to charge tolls. Although Berry remained the legal owner of the trail, Berry allowed Cameron to continue charging tolls on the trail.

Like several Grand Canyon pioneers, Cameron soon realized that he could make more money from tourism than from mining, and in 1903 he began charging tourists $1 to use the trail to go into Grand Canyon. Cameron also leased some of his land at the trailhead to Ellsworth and Emery Kolb, who opened a photography studio there. The Kolbs photographed tourists, then ran down the trail to Indian Garden, where they had built a darkroom, and used the Garden's water to develop their film and prints. (By 1932, however, a pipeline from Indian Garden brought water to the South Rim.) Today, Kolb Studio at the Bright Angel Trailhead includes a bookstore, historical exhibits, and an art gallery.

From 1905 to 1907, while chairing the Coconino County Board of Supervisors, Cameron convinced legislators to give toll operators (like himself) even more control of their trails; this enabled Cameron to more tightly control what came to be known as Bright Angel Trail. This continued until 1920, when years of litigation finally invalidated Cameron's mining claims and his ownership of Bright Angel

Indian Garden

Indian Garden is an oasis and campground 4.6 miles down Bright Angel Trail. A small creek runs through Indian Garden, which was formerly called Angel Terrace. Indian Garden is about 3,000 feet below the rim near the junction of the Bright Angel and Tonto trails. Near 1901, Ralph Cameron built a tourist camp (and sold cold beer for 5¢ per bottle) at Indian Garden that operated for nine years; the cottonwood trees at Indian Garden were planted by Cameron. Today, GCNP maintains a ranger station and a popular campground there. Indian Garden is about halfway between the South Rim and the Colorado River. Havasupai who lived at Indian Garden were forced to leave in 1919 when Grand Canyon became a national park.

Trail. The steep trail was then owned by Coconino County, during which time (in 1924) the county's residents soundly defeated a referendum to transfer the trail to NPS. Coconino County continued to charge a toll until 1928, when it gave the trail to NPS in exchange for a 54-mile road (now Route 64) from Williams, Arizona, to the South Rim. NPS then began an 11-year renovation of Bright Angel Trail that produced today's trail. Ironically, despite years of fighting to end the toll charge for use of Bright Angel Trail, NPS continued to charge a toll for use of the trail until 1933. This made Bright Angel Trail the last toll trail in NPS. Although the number of hikers did not exceed the number of mule riders on the trail until 1946, Bright Angel Trail is today the most popular trail into and out of Grand Canyon.

Bright Angel Trailhead (elevation 6,785 feet) is just west of Kolb Studio in a habitat dominated by pinyon pine (*Pinus edulis*) and Utah juniper (*Juniperus osteosperma*). For the first 4.6 miles to Indian Garden, the trail follows Bright Angel Fault. The top section of Bright Angel Trail consists of switchbacks that cut through the Kaibab Formation and then the Toroweap Formation. Hikers soon reach a short tunnel, just beyond which a panel of pictographs is etched in Kaibab limestone. (Famed architect and designer Mary Elizabeth Jane Colter reproduced some of these pictographs above the plate rail in the private dining room of the El Tovar Hotel.) By the time hikers get to the second tunnel (a good turnaround point for small children), they are standing on buff-colored Coconino Sandstone on the Bright Angel Fault. Because of this fault, rocks on the west side of the fault are 189 feet higher than those on the east side. After more steep switchbacks to the Mile-and-a-Half Resthouse, hikers are just across the Hermit Formation 1,120 feet below the rim. Here, the dominant trees are Douglas fir (*Pseudotsuga manziesii*) and Gambel oak (*Quercus gambelii*). At the rest house is an emergency phone, composting toilet, and seasonal water.

The next landmark is the Three-Mile Resthouse, another popular turnaround point. Redwall Formation below here is one of the most prominent layers in Grand Canyon; this limestone is often more than 450 feet thick. Redwall is gray but gets its reddish color from iron compounds leached from the overlying Hermit and Supai formations. A 1.5-mile trail starting behind the rest house leads to a viewpoint overlooking the Redwall Formation and out to Plateau Point, which offers excellent views of the Colorado River and Granite Gorge from river left near River Mile 89. East of this trail are the remains of an old cable car system built in 1932 during construction of the water pipeline from Indian Garden to the South Rim. The Redwall Formation beyond Three-Mile Resthouse divides the forest and desert-shrub habitats of Grand Canyon. Although the geologic formations along Bright Angel Trail are beautiful, there are few sweeping views because the trail follows Bright Angel Fault.

Past Three-Mile Resthouse, Bright Angel Trail descends through Jacobs Ladder, another series of switchbacks that follow Bright Angel Fault through the Redwall Formation to the Tonto Platform. Here, single-leaf ash (*Fraxinus anomala*), western redbud (*Cercis occidentalis*), cliffrose (*Cowania mexicana*), and mesquite (*Prosopis glandulosa*) are common. Just after intersecting the original cross-canyon pipeline, hikers reach Indian Garden on the Tonto Platform. Here, Garden Creek provides perennial water; groundwater seeping through overlying rock layers oozes out of the canyon's wall when it reaches the impervious Bright Angel Shale. Havasupai living here after 1300 CE used this water to grow crops. Water is available at Indian Garden year-round. NPS rangers regularly offer programs at Indian Garden and Phantom Ranch.

From Indian Garden, hikers can trek across the Tonto Platform along Plateau Point Trail. This three-mile (round-trip) hike takes you to Plateau Point, which overlooks the Colorado River and Inner Gorge between Pipe Creek and Monument Creek. From Plateau Point, hikers can also see the Grand Canyon Supergroup between Tapeats Sandstone and Vishnu Schist. (The red Hakatai Shale is the most prominent formation in the Supergroup.) At Plateau Point, the dominant plants are Utah agave (*Agave utahensis*) and blackbrush (*Coleogyne ramosissima*).

Trailview Overlook

West of Grand Canyon Village is Trailview Overlook (elevation 6,945 feet), which is named for its excellent views of Grand Canyon Village and Bright Angel Trail switchbacks leading from Kolb Studio to cottonwood-shaded Indian Garden. Bright Angel Trail, which follows Bright Angel Fault, was originally a path trod by bighorn sheep and other animals and later by the Havasupai; it was developed in the 1890s by Peter Berry and Ralph and Niles Cameron but was not extended to the river (to service mining claims) for another decade. Indian Garden, which is on Bright Angel Trail about halfway between the South Rim and the Colorado River, is a green oasis on the Tonto Platform (in 1900, hikers and mule riders could buy a cold beer there for 5¢). Beyond Indian Garden, Bright Angel Trail disappears into Tapeats Narrows along Garden Creek. From Trailview Overlook, visitors can also see Plateau Point (which is 3,085 feet below) and, to the south, the San Francisco Peaks, which are 55 miles away.

Bright Angel Trail then narrows as it follows Pipe Creek and Garden Creek through Tapeats Sandstone (the lowermost part of the Tonto Group that is atop Vishnu Schist). Along Garden Creek, Puebloans around 1100 CE built granaries to store their corn, yucca, beans, and squash. Bright Angel Trail then enters a series of switchbacks called Devils Corkscrew, which descend through 1,200 feet of Vishnu Schist. Here, cacti and Mormon tea (*Ephedra viridis*) become increasingly abundant.

Approximately 3.1 miles beyond Indian Garden, Bright Angel Trail intersects the Colorado River Trail. This is also where Pipe Creek joins the Colorado River. The River Rest House here has an emergency phone. Along the west side of the junction of Bright Angel Trail with Pipe Creek, hikers find water from Columbine Spring oozing from the cliff. This site, which is 7.7 miles from the trailhead, ends Bright Angel Trail.

The average grade of Bright Angel Trail is 10 percent, and at the river hikers are 4,380 feet below the South Rim. Temperatures, on average, will be 20°F warmer at the river than at the South Rim. Although the South Rim gets 16 inches of precipitation per year, the Inner Gorge gets only about half that; much of the precipitation evaporates before reaching the Colorado River. Bright Angel Trail is patrolled and maintained year-round by NPS.

See also: Cameron, Ralph; Grand Canyon Village; South Kaibab Trail

Further Reading

Thybony, Scott. 2004. *Grand Canyon: Bright Angel Trail Guide*. Grand Canyon, AZ: Grand Canyon Association.

BRIGHT ANGELS

"Bright Angel" is a common name that describes a geological fault, side canyon, trail, lodge, creek, and other places in GCNP. "Bright Angel" was first invoked by Grand Canyon explorer John Wesley Powell to honor a Methodist hymn titled "Shall We Gather at the River?" that his father (a minister) liked to sing:

> Shall we gather at the river,
> Where bright angels' feet have trod.

Powell regretted the naming of a muddy creek in Cataract Canyon (360 miles upstream)—one that expedition member Jack Sumner said was as filthy as sewage from a large, dirty city—"dirty devil," so Powell gave the clear water of Bright Angel Creek the more inspiring name: Bright Angel. "Angel Bright" is also a character in Milton's *Paradise Lost*, one of Powell's favorite books.

There are many Bright Angels in Grand Canyon. Here are some of the most famous ones:

Bright Angel Canyon is the largest tributary canyon of Grand Canyon; it runs 11 miles from the North Rim to the Colorado River. Bright Angel Canyon, with its counterpart on the south side of the Colorado River, forms a major "corridor" of GCNP that has provided access to the canyon and drinking water to Native Americans and all who have followed.

Bright Angel Creek is a perennial creek that helped erode Bright Angel Canyon. Phantom Ranch is less than a mile north of where Bright Angel Creek enters the Colorado River (Mile 88.2). Bright Angel Creek drains more than 25 square miles of land.

Bright Angel Lodge, which opened in 1935, was designed by Mary Colter and built by the Fred Harvey Company on the South Rim to replace Bright Angel Hotel. This hotel helped create Grand Canyon Village. In the lodge's History Room is a fireplace made of rocks chosen by geologist Edwin McKee to mirror a cross-section of Grand Canyon's geology. The top layer of the fireplace is made of Kaibab limestone (the youngest and uppermost layer of the Grand Canyon), with the bottom layer of the fireplace made of dark Vishnu Schist from the bottom of the Inner Gorge. In the lobby of Bright Angel Lodge is acclaimed artist William Leigh's (1866–1955) painting titled *Grand Canyon* (1908). The Atchison, Topeka and Santa Fe Railway offered Leigh free passage to the West if he would paint Grand Canyon. Leigh eventually produced six such paintings, all of which were bought by the railroad. For years, Leigh's paintings were featured on Fred Harvey Company menus at Grand Canyon. Bright Angel Lodge includes Red Horse Station and Buckey O'Neill's Cabin. In 1910, Bright Angel Lodge charged $1 for an overnight stay.

Every day, these switchbacks on Bright Angel Trail lead hundreds of hikers into and out of Grand Canyon. Bright Angel Trail, which stretches 7.7 miles from near Kolb Studio to the Colorado River, is Grand Canyon National Park's most popular Corridor Trail. This photo was taken from Trailview Overlook, which is northwest of Grand Canyon Village. (Randy Moore)

Bright Angel Point (elevation 8,148 feet, which is 5,780 feet above the Colorado River) is the most popular viewpoint on the North Rim. The famous overlook is easily accessed on the self-guided, paved, 0.25-mile Bright Angel Point Trail from Grand Canyon Lodge. The trail to Bright Angel Point begins at the corner of the back porch behind Grand Canyon Lodge. The best viewpoint on this trail is above a 10-foot-high Kaibab Limestone knoll at the end of the trail. Although the trail is smooth and paved, it is also steep in places. Bright Angel Point offers one of the most scenic views in GCNP. To the east are the 4,000-foot-deep Bright Angel Canyon, Walhalla Plateau, and Roaring Springs Canyon, and to the southeast are three peaks named for Hindu deities: Deva Temple, Brahma Temple, and Zoroaster Temple. In the west, you'll see Transept Canyon, which USGS topographer François Matthes in the early 1900s claimed was far grander than Yosemite Valley. You can also see and hear Roaring Springs, which is just above the Bright Angel Shale and more than 3,000 feet below the rim. To the south are the San Francisco Peaks (66 miles away) and the closer Red Butte volcanic cones. Most visitors' first hike at the North Rim is to Bright Angel Point, which is due north of the South Rim's Shoshone Point. The Colorado River is not visible from Bright Angel Point.

Bright Angel Trail, which descends from the South Rim into the canyon, is the most popular and heavily traveled trail into GCNP. The trail was originally a Havasupai trail that was improved by Ralph Cameron and Peter Berry. It was later named Cameron Trail for Ralph Cameron, a territorial representative who used it to reach his copper mines. (Cameron used those mines, which were not productive, to claim land.) In 1903, Cameron began charging a $1 toll to use the trail. When Cameron sold the trail to Coconino County in 1928, it was renamed Bright Angel Trail. The trail, whose trailhead was renovated in 2013, significantly influenced development of the South Rim.

See also: Cameron, Ralph

Further Reading

Aitchison, Stewart. 2008. *Grand Canyon's North Rim and Beyond: A Guide to the North Rim and the Arizona Strip*. Grand Canyon, AZ: Grand Canyon Association.

BRIGHTY, THE NORTH RIM BURRO

Marguerite Henry (1902–1997) was an American writer whose 59 children's books were loosely based on true stories of horses and other animals. One of Henry's most famous books was her now classic 222-page, Newbery Award–winning *Brighty of the Grand Canyon* (1953). Henry, the daughter of a Milwaukee book publisher, got the idea for her book after visiting a public library in Elgin, Illinois, and asking for articles about burros (small donkeys). The librarian there gave her a 31-year-old issue of *Sunset* magazine (from August 1922), in which North Rim pioneer Thomas McKee described a friendly burro named Brighty of the Grand Canyon. Brighty lived at Grand Canyon from 1890 to 1922.

Henry's book made Brighty famous. Brighty never stayed with people, but he was liked when he showed up at the Wylie Way Camp at the North Rim. The gentle and friendly Brighty spent summers carrying water from a spring 1.5 miles below the rim to the North Rim. Brighty was rewarded for his work with pancakes. Brighty was the first animal to cross Black Bridge; he also accompanied Teddy Roosevelt on one of Roosevelt's three visits to Grand Canyon. Henry's book, which did for Grand Canyon what Smokey Bear did for forest fires, ends with the following famous lines:

> [O]n moonlit nights a shaggy little form can be seen flirting along the ledges, a thin swirl of dust rising behind him. Some say it is nothing but moonbeams caught up in a cloud. But the older guides swear it is trail dust out of the past, kicked up by Brighty himself, the roving spirit of the Grand Canyon—forever wild, forever free.

Brighty was a friendly, pancake-loving burro who lived at Grand Canyon from around 1890 to 1922. This photo shows Brighty carrying water cans with Bobby McKee at the North Rim's Wylie Camp in 1918. Brighty was later immortalized in Marguerite Henry's award-winning book *Brighty of the Grand Canyon* (1953) and a subsequent movie of the same name. Today, a life-sized statue of Brighty greets visitors in the sunroom of Grand Canyon Lodge. (National Park Service)

In 1963, Betty Booth bought a copy of *Brighty of the Grand Canyon* for her three sons to read on a family vacation. Betty's husband Steve, a television producer, later read the book and made the book into an 89-minute movie titled *Brighty of the Grand Canyon* (1967). Filming of the movie took 15 weeks, of which four were at Grand Canyon (e.g., Brighty at Ribbon Falls on the North Kaibab Trail). Most of the North Rim scenes were filmed in southern Utah's Dixie National Forest. Supplies, props, and even a tamed mountain lion and people were ferried into and out of Grand Canyon via 1,200 helicopter flights and a 35-mule pack train. Brighty, who had a stunt double during filming (e.g., when he falls out of the cable car into the river), was played by Henry's own burro. The movie was produced by Stephen F. Booth Productions, and it premiered in Detroit, Booth's hometown. The movie did poorly at the box office.

Brighty was a descendant of one of the burros brought to Grand Canyon by its early miners. When these miners left Grand Canyon, they released their burros. By the 1970s, the population of feral burros exceeded 2,000, and, because burros compete for resources with bighorn sheep, the population of bighorn sheep had dwindled. These burros, which denuded parts of the Tonto Platform, had damaged Grand Canyon far more than had any prospectors and miners. In 1979, GCNP announced plans to kill some of the feral burros (as it had killed more than 3,600 burros between 1924 and 1968), noting that killing the animals would cost $30,000 but capturing and shipping them out of Grand Canyon would cost more than $350,000. Nevertheless, because *Brighty of the Grand Canyon* had romanticized the burros, there was a strong and immediate public outcry: "Don't Kill Brighty!" GCNP Superintendent Merle Stitt, an ecologist, received more than 11,000 letters protesting GCNP's plans to kill the burros.

The problem was resolved when humorist and humanitarian Cleveland Amory (1917–1998), the president of Fund for Animals, donated money that enabled GCNP to remove 577 burros from Grand Canyon on rafts and helicopters over two years (at a cost of approximately $1,000 per burro) and be given to citizens. Interestingly, NPS had stopped killing burros in 1969, and in 1972 it became a felony to kill wild burros outside NPS (but only a misdemeanor to kill bighorn sheep). After the burros were eliminated, the population of bighorn sheep began to grow. Today, GCNP is the best place in the United States to see bighorn sheep.

To help promote his movie, Booth had paid sculptor Peter Jepsen $15,000 to create a life-sized, 600-pound statue of Brighty that was originally displayed in December 1967 at the South Rim's Visitor Center. When GCNP began removing feral burros from the park, GCNP Superintendent Merle Stitt decided to place the statue in storage. In Brighty's place, GCNP installed an exhibit describing the ecological problems caused by feral burros. Tucson's Wild Burro Protection Association protested the move and threatened to take GCNP to court. Brighty was later displayed in the sun room of the North Rim's Grand Canyon Lodge, where today's visitors rub his shiny nose for good luck.

Marguerite Henry died on November 26, 1997, in Rancho Santa Fe, California. Brighty, who was Henry's most famous burro, was killed by hungry campers for food in the winter of 1922.

See also: North Rim; Roosevelt Point

Further Reading

Henry, Marguerite. 1953. *Brighty of the Grand Canyon*. New York: Rand McNally.

BROWN, FRANK

Frank Mason Brown (1845–1889) was a former politician and wealthy real estate developer/entrepreneur in Denver who, on March 25, 1889, founded the Denver, Colorado Canyon and Pacific Railroad Company. Despite his lack of experience with railroads, Brown hoped to build a water-level railroad through Grand Canyon. Brown got the idea from northern Arizona prospector Samuel S. Harper, who in 1870 spent a night on the Colorado River and believed that a railroad could be built through the canyon.

Brown's railroad, which was to stretch from Grand Junction, Colorado, to Yuma, Arizona, was meant to deliver coal from the Rocky Mountains to San Diego, after which supplies could be shipped to other areas in the West. Brown's proposed railroad was supposedly easier and cheaper to build than crossing the mountains west of the Continental Divide.

On May 25, 1889, Brown and a ragtag group of 14 men began a survey of the Colorado River from Green River, Utah, and on July 9, they were at Grand Canyon. Brown's expedition was only the third downriver expedition to enter Marble Canyon. Brown planned everything but the group's safety. Brown's chief engineer, Robert Stanton, wanted life jackets for everyone, but Brown rejected Stanton's request.

On the morning of July 10, 1889, Brown's boat—only 12 miles into Grand Canyon—flipped just past Soap Creek Rapid. The boat's two occupants, Brown and Harry McDonald, were thrown into the river. McDonald swam to safety, but Brown drowned. Brown's death resulted from his lack of a life jacket and improper clothing (e.g., heavy boots). All that was recovered was his notebook. His body was never found.

On July 10, 1889, Denver entrepreneur Frank Brown drowned when his boat capsized at River Mile 12.1 during an expedition to determine the feasibility of a railroad through Grand Canyon. Crewman Peter Hansbrough etched this inscription memorializing Brown on river left near where Brown drowned. Five days later, Hansbrough drowned a few miles downstream. (Randy Moore)

Stanton eulogized Brown with "President Brown sacrificed his life, which could so easily have been saved if he had a life preserver to keep him afloat. A noble hearted man and a true friend, he won the love of everyone associated with him" (Frank Mason Brown, 2017). Expedition member Peter Hansbrough—a long-time employee and friend of Stanton—memorialized Brown by scratching an inscription on a ledge of Coconino Sandstone near the site at River Mile 12.1 about 15 feet above the river on river left: "F. M. Brown, Pres. D, CC & PR was drowned July 10, 1889 opposite this point." River runners can see the inscription on river left. Hikers can access the site via Salt Water Wash from Highway 89A (a Navajo permit is required). The small rapid nearby is now called Browns Riffle.

The remaining members of the expedition soon pressed on; as Stanton noted, "In this world we are left but little time to mourn" (Berger, 2007). After Brown drowned, the remaining members of the expedition ran or portaged 25 more rapids. Their 25th rapid is ironically known today as 25-Mile Rapid. Over the next few days, Hansbrough spoke often of his own death and of Heaven. Five days after Brown's death, near River Mile 25, the boat carrying Hansbrough and cook's helper Henry Richards capsized, and both men drowned. Richards's body was never found, and Hansbrough's remains were not recovered until the next year almost 20 miles downstream (Hansbrough's body was identified by his boots).

Stanton noted that "the burial service [for Hansbrough] was brief and simple. We stood around the grave while one short prayer was offered, and we left him with a shaft of pure marble for his headstone, seven hundred feet high, with his name cut upon the base; and in honor of his memory we named a magnificent point opposite" (Berger, 2007), today's Point Hansbrough. Today, Hansbrough's remains rest beneath the inscription "PMH 1889" at the base of a limestone cliff on river left at River Mile 44.0 near President Harding Rapid. The crew was hesitant to cut the initials because Hansbrough had scratched Brown's inscription on the rocks at the site of the earlier drowning. As of 2017, Richards was the only African American ever to drown in Grand Canyon. Hansbrough was the first known river runner to be buried along the Colorado River.

Stanton—realizing that the expedition was undermanned and underequipped for the river's dangers—abandoned the so-called Death Canyon expedition: "Two more faithful and good men gone! Astonished and crushed by their loss, our force too small to portage our boats, and our outfit and boats unfit for such work, I decided to abandon the trip" (Berger, 2007). Stanton stored the boats and some of their food and gear in what became known as Stantons Cave (River Mile 32 on river right) and walked out of Grand Canyon just above Vaseys Paradise. Stanton returned to Grand Canyon six months later and completed the expedition. This time, he brought life jackets.

The Brown expedition set the record for boating deaths during one trip in Grand Canyon. It would be almost 40 years before anyone else would drown trying to navigate (rather than cross) the Colorado River in Grand Canyon.

Hansbrough is memorialized by Point Hansbrough, which is opposite Hansbrough's grave. However, in 1976, river runner Otis "Dock" Marston objected to naming sites for Brown and Hansbrough because those "deaths were due to stupidity . . . Continuation of this practice will soon make the course of the Colorado River a memorial to bad navigational practices" (Brian, 2015). Marston's objection was ignored.

See also: Thematic Essays section: The Colorado River's Historic Sites; Stanton, Robert

Further Reading

Berger, Todd R. 2007. *It Happened at Grand Canyon*. Guilford, CT: TwoDot Publishers.

Brian, Nancy. *River to Rim: A Guide to Place Names along the Colorado River in Grand Canyon from Lake Powell to Lake Mead*. Second edition. Flagstaff, AZ: Earthquest Press.

Brown, Frank Mason. 2017. Retrieved July 9, 2017, at https://www.findagrave.com/cgi-bin/fg.cgi?page= gr&GRid = 57608379

BUGGELN, MARTIN

Martin Buggeln Jr. (1867–1939) was a Williams, Arizona, businessman and hotelier who, after several years of operating the Grand Canyon Hotel in Williams, bought the South Rim's Bright Angel Hotel—which had opened in 1896—and a stagecoach line from James Thurber in 1901. Buggeln added several tent-cabins and changed the name of the complex to Bright Angel Camp; this was the precursor to the modern complex of cabins on the South Rim. Buggeln charged $1–$2 for a tent, $2–$3 for a room, $3 for a horse, and $5 for a guide.

Buggeln's property was in today's Grand Canyon Village, and Buggeln reasoned that the upcoming arrival of the Atchison, Topeka and Santa Fe Railway would stimulate business at the hotel. However, neither Thurber nor Buggeln ever patented the land on which the Bright Angel Hotel was located. Thus, when the Santa Fe Railway arrived, it claimed the land, thereby diminishing the hotel's value to Buggeln. Nevertheless, Buggeln made a deal with the newly arrived railroad to bring tourists to his camp. Grand Canyon's third post office, and the first post office officially listed as "Grand Canyon," opened in 1902 in Buggeln's hotel, with Buggeln serving as postmaster. (This post office closed in 1910 when it moved to the vacant Cameron Hotel nearby.) Buggeln's sister Louisa Ferrall, who also worked at the post office, helped Buggeln and Ralph Cameron fight against NPS by reading mail about GCNP that came through the post office, thereby forcing NPS officials to use code to communicate about GCNP.

After operating his hotel for five years, Buggeln—who also served as a postmaster at Grandview Point—sold the hotel to the Santa Fe Railway, which upgraded the facilities. Buggeln then bought John Hance's ranch and built a 17-room hotel by its old dining room. This property, which was one of the largest hotels on the rim, was on East Rim Drive near what is now Buggeln Hill.

Buggeln returned to Williams, after which he moved to Los Angeles, California, and opened a restaurant. Although Buggeln is buried beside his first wife Emma in Los Angeles, California, his second wife (Eva Moss Buggeln) inherited most of Buggeln's estate. Except for John Hance's asbestos mines above the northern bank of the Colorado River, Eva's sale of much of Buggeln's property to NPS eliminated the last private holding of property east of Grand Canyon Village. Buggeln's death also cancelled his permit to manage 175 head of cattle, thereby virtually ending the legal grazing of domestic animals in GCNP.

In 1957, GCNP razed Buggeln's hotel at Grandview Point. By 1960, there was little left of the properties of Hance, Thurber, Buggeln, and most other Grand Canyon pioneers except debris and trails.

See also: Grand Canyon Village; Grandview Point

Further Reading

Lago, Don. 2014. *Canyon of Dreams: Stories from Grand Canyon History.* Salt Lake City: University of Utah Press.

C

CALIFORNIA CONDORS

California condors (*Gymnogyps californianus*), with a wingspan of up to 10 feet, are the largest flying land bird in North America. They can live for more than 50 years and fly as fast as 50 miles per hour. In the wild, condors mate for life, and females lay one egg annually or every other year.

Condors were abundant for tens of millions of years across the southern United States and from Mexico to Canada. Their range and abundance shrank toward the end of the Pleistocene (about 12,000 years ago) when the scavenging of the megafauna—mastodons, saber-toothed cats, and sloths—diminished. Condors survived, however, along the coast of the Pacific Ocean from British Columbia to California.

In the late 1800s, the range and abundance of condors again shrank—this time drastically—because of habitat destruction, lead poisoning, and hunting. The last known condor in Grand Canyon was shot by a prospector in 1887, and the last sighting of a condor in Arizona occurred in 1924. Thereafter, condors lived only in California. After being placed on the Endangered Species List in 1967, the condor population continued to drop, and in 1982 only 27 condors remained on Earth. Over the next three years, condors became extinct in the wild when all of the birds were captured for captive breeding.

When captive females laid a new egg, it was immediately taken and placed in an incubator, thus prompting the female to lay another (and sometimes a third) egg. Condors were reintroduced to the wild in 1991, and they returned to the Grand Canyon area in 1996, when six birds bred in captivity were released at Vermilion Cliffs, about 30 miles north of the canyon. By 2001, condors were back at Grand Canyon. From 1996 to 2016, 189 condors were released in northern Arizona and southern Utah, and 29 chicks were hatched in the wild. By the end of 2016, however, 125 of these birds died (including 20 of the wild-hatched chicks). Chick #350, the first of these birds to hatch in the wild at Grand Canyon, did so in 2004 at The Battleship. When it launched from its cliff-side nest, the 6-month-old juvenile had a wingspan of 9 feet. In 2018, lead contamination (e.g., from ammunition used by hunters) remained the major factor hindering the recovery of condor populations.

In early 2017, 74 California condors were living in the wild in the canyons of northern Arizona and southern Utah (and more than 450 California condors were living in captivity and the wild worldwide). At Grand Canyon, the best place to see these majestic birds is at Vermilion Cliffs, the North Rim, and Navajo Bridge.

See also: Grand Canyon at a Glance

Further Reading

Nielsen, John. 2006. *Condor: To the Brink and Back—The Life and Times of One Giant Bird.* New York: Harper Perennial.

CAMERON, RALPH

Ralph Henry Cameron (1863–1953) was an American prospector, entrepreneur, and politician who pioneered early efforts to develop Grand Canyon. Cameron, who was Arizona's first Republican senator, often used his political influence as a territorial representative (1889), Coconino County sheriff (1894–1898), U.S. Congressional representative (1909–1911), and U.S. senator (1921–1927) to enhance his businesses at Grand Canyon. In 1890, Cameron, Peter Berry, and Niles Cameron opened Last Chance Mine below Grandview Point and managed a hotel there that was the terminus of the Grand Canyon Stage Line from Flagstaff.

In February 1891, Peter Berry filed the paperwork to make today's Bright Angel Trail a private toll road. There was initially relatively little traffic on the trail, and in 1901 Cameron assumed control of what was then called Bright Angel Toll Road. To control the trail, Cameron had filed 39 mining claims by 1907 along the trail and at the trailhead, including the Wizard Claim at the trail's intersection with the Colorado River, the Alder and Willow Mill Claims at Indian Garden, the Magician Claim at Devils Corkscrew, and the Gold Eagle and Cape Horn Claims atop the South Rim that included the Cameron Hotel and the trailhead. (Cameron's hotel, without its second floor, later became part of Bright Angel Lodge.) In 1919 alone, Cameron filed another 55 new claims in GCNP. Cameron claimed more than 13,000 acres at Grand Canyon, and other prospectors and entrepreneurs claimed even more, prompting a U.S. Forest Service ranger to note in 1908 that Grand Canyon "was all plastered up with fraudulent mining

Arizona politician Ralph Cameron was a Grand Canyon entrepreneur who used mining claims and political power to control Bright Angel Trail (then called "Bright Angel Toll Road"). Although Cameron is not memorialized by any structures or viewpoints in Grand Canyon National Park, his tombstone in Pioneer Cemetery tells visitors that "Arizona Can Never Forget Him." This photo of Cameron was taken around 1915. (Library of Congress)

claims so a person that wants to get anywhere or do anything cannot get any ground" (Arizona State University, 2017).

Cameron collected tolls on Bright Angel Trail for more than 20 years, during which time he often bragged that he would "make more money out of the Grand Canyon than any other man" (Lago, 2015). In 1915 alone, he collected $20,000 in tolls; people who tried to circumvent the toll were fined $50. (Cameron also sold vials of fake Grand Canyon sand.) Bright Angel Toll Road was the only toll road ever in a national park. Cameron planned a 14-mile rim-side railway that would be powered by electricity from a dam on the Colorado River, but those plans never materialized.

Pipe Creek Vista

This overlook, which is 1.3 miles from Mather Point along the Rim Trail, is a shuttle stop and popular place to avoid crowds. From Pipe Creek Vista, visitors look down onto Pipe Creek, the major canyon separating Bright Angel Trail and South Kaibab Trail. Pipe Creek Vista is also the eastern trailhead for the paved Rim Trail. Tucked just over the low wall east of the overlook you'll find several fossils of gastropods (each about an inch in diameter) embedded in the limestone slabs.

In 1894, Ralph Cameron (the namesake of Cameron, Arizona) found a meerschaum pipe along a nearby trail, carved an earlier date (as a hoax) in the pipe, and left the pipe for others to find. Others did find the pipe, for which Pipe Creek is named.

To protect his business interests, Cameron campaigned against the formation of GCNP and worked hard to have NPS director Stephen Mather fired. At one point, a lawsuit by Cameron for control of land at the South Rim (*Cameron v. United States*, 1920) reached (and was rejected by) the United States Supreme Court. Mather considered Cameron to be the worst enemy of NPS.

After he became a congressional delegate in 1909, Cameron spent relatively little time at Grand Canyon; daily operations shifted to his brother Niles and 10–12 employees. In the early 1900s, the American Mining Law required prospectors to spend at least $100 per year on assessment to maintain a valid claim. Cameron often failed to do this. In 1909, mineral examiners T. T. Swift and H. Norton Johnson of the U.S. Forest Service reported that several of Cameron's mining claims lacked minerals and tried to invalidate the claims. By 1920, Cameron's Cape Horn Claim was invalidated, and the Gold Eagle Claim was struck down the following year. Cameron, who was accustomed to using litigation, raw political power, and his well connected friends (e.g., U.S. Attorney General Harry Daughtery) to achieve personal goals, was severely criticized for using GCNP to make money. The criticism increased when newspapers reported that Cameron had filed fraudulent mining claims at the future site of Hoover Dam. Cameron lost his political power after losing elections to Congressman Carl Hayden in 1926, to Henry Ashurst in 1928, and again to Hayden in 1932.

Cameron left Arizona in 1934 but returned in the early 1950s. Cameron, who helped Arizona become a state in 1912, died at age 89 on a trip to Washington, DC, on February 7, 1953. At the time of his death, Cameron was promoting a tourist ranch at Yuma and condemning federal control of land. He was buried in Pioneer Cemetery under a plaque installed for $30 by Quincy A. Coatney, who was buried nearby 30 years later. Cameron's epitaph was chosen by John Pyle, who served as Arizona governor from 1951 to 1955: "Arizona Can Never Forget Him." Despite his influence at Grand Canyon, nothing in today's GCNP is named for Ralph Cameron.

See also: Berry, Peter; Bright Angel Trail; Grand Canyon Village; Grandview Point; Pioneer Cemetery and Other Gravesites

Further Reading

Arizona State University. 2017. Who owns the Grand Canyon? Retrieved July 9, 2017, at http://grandcanyonhistory.clas.asu.edu/history_whoownsgrandcanyon.html

Lago, Don. 2015. *Grand Canyon: A History of a Natural World and National Park*. Reno: University of Nevada Press.

CAPE FINAL AND CAPE FINAL TRAIL

This seldom-visited, easy, 4.4-mile (round-trip) rim-top trail begins at a small parking lot in a clearing of towering pines 2.5 miles north of the Cape Royal parking lot on the North Rim. The trail to Cape Final (elevation 7,916 feet), whose trailhead is 20.4 miles from Grand Canyon Lodge, follows an old fire road that climbs gently through a rim-side forest to a flat area ideal for hiking. At Cape Final, as at nearby Cape Royal, hikers are rewarded with an almost 270° view of eastern Grand Canyon from Nankoweap to the Unkar Creek Delta, including Vishnu Temple and Jupiter Temple. To the east is a stunning view of Painted Desert, which is part of the Navajo Reservation and where the elevation fluctuates less than 100 feet over 12 miles of the rim. A spectacular viewpoint suddenly appears that features views of Chuar Valley, Lava Creek, Siegfried Pyre, Gunther Castle, and Vishnu Temple. Cape Final also offers views of Grand Canyon Supergroup, which are Precambrian sediments not visible at most other viewpoints in GCNP. Cape Final is the easternmost viewpoint on the North Rim. There is no water along Cape Final Trail.

Cape Final, which is the eastern point of the Walhalla Plateau, was named in 1880 by geologist Clarence E. Dutton, when—after riding for several days—he could finally see Grand Canyon. Dutton, who accompanied the Powell Survey in 1875, described Cape Final as "doubtless the most interesting spot on the Kaibab. In pure grandeur, it is about the same as Point Sublime [but] the two differ much in the characteristics of the scenery" (Dutton, 1882).

See also: Cape Royal and Cape Royal Trail; North Rim

Further Reading

Dutton, Clarence E. 1882. *Tertiary History of the Grand Canyon*. Washington, DC: U.S. Government Printing Office. (Reprinted 2001 by the University of Arizona Press, with an introduction by Wallace Stegner and a new foreword by Stephen J. Pyne.)

CAPE ROYAL AND CAPE ROYAL TRAIL

Cape Royal (elevation 7,865 feet) is at the end of a 22.8-mile drive on a paved road from Grand Canyon Lodge. This site, which is visible from many South Rim viewpoints, is the southernmost viewpoint on the North Rim and offers one of the most beautiful views in GCNP. Clarence Dutton, who named this site in 1882, described Cape Royal as a "congregation of wonderful structures, countless and vast, profound lateral chasms" (Dutton, 1882). From here, to the south, hikers see the South Rim's Visitor Center, Sinking Ship, Horseshoe Mesa, and Grandview Point, and, in the northeast, Marble Canyon. To the west are Zoroaster Temple, Isis Temple, and Brahma Temple, which are also visible from Bright Angel Point. The flat, paved, 0.6-mile-long (round-trip) Cape Royal Trail originates at the southeast side of the Cape Royal parking lot and provides spectacular views of the canyon and Colorado River. Signs along the trail interpret the natural history of the area. A visit to Cape Royal left Teddy Roosevelt "solemn with the sense of awe" (Roosevelt, 1916).

In 1921, James T. "Uncle Jim" Owens applied to build a tourist camp at Cape Royal, but his plans were rejected by NPS director Stephen Mather. No commercial camp was ever built at Cape Royal, but NPS developed a primitive campground there in 1931. Utah Parks Company had wanted to build a lodge at Cape Royal

The North Rim's Cape Royal overlooks the spectacular Wotans Throne, a 7,733-foot-high mesa that was originally called Newberry Terrace. Wotans Throne was named by François Emile Matthes after Wotan, a deity in German mythology who is also known as Woden. (Wednesday is named after Woden.) Cape Royal, which is a 22.8-mile drive from Grand Canyon Lodge, is a popular place for watching sunsets at the North Rim. (Randy Moore)

Cliff Spring Trail

This 0.5-mile trail, which winds below the rim of the Walhalla Plateau, begins across the road from a small pullout 0.3 mile north of Cape Royal near the Angels Window Overlook. After crossing the road, hikers follow a gravel trail westward through a wooded ravine. After about 300 feet, hikers get to a chest-high boulder that hides the ruins of a Kayenta Puebloan granary that dates to 900–1150 CE. Teddy Roosevelt camped here while on a hunting expedition in the early 1900s.

From near the ruins, hikers enjoy tree-framed views of Clear Creek and Wotans Throne. Nearby is intermittent Cliff Spring beneath an overhang. Do not drink the spring's water (it may be contaminated). This is a premier place to watch forest birds as they come for a drink or to hunt food.

similar to Grand Canyon Lodge, but their plans were wiped out by the Great Depression.

Just below Cape Royal is the stunning mesa Wotans Throne (elevation 7,721 feet). In 1937, a group of explorers climbed Wotans Throne and discovered archaeological evidence that Native Americans had lived there. Anthropologists have not yet determined why these people chose such a difficult location for home.

Just east of Cape Royal is Angels Window, an often photographed natural arch in Kaibab limestone that frames a small section of the Colorado River that is 6 miles away. Angels Window is one of only two points on the North Rim with views of the river. From the trail leading over the arch, you'll see the tall, rocky Vishnu Temple to the south, as well as rapids near Unkar Delta. Far to the south, the San Francisco Peaks tower over Flagstaff. Angels Window is the only easily accessible arch in Grand Canyon.

See also: Owens, James "Jim"

Further Reading

Dutton, Clarence E. 1882. *Tertiary History of the Grand Canyon*. Washington, DC: U.S. Government Printing Office. (Reprinted 2001 by the University of Arizona Press, with an introduction by Wallace Stegner and a new foreword by Stephen J. Pyne.)

Roosevelt, Theodore. 1916. *A Book-Lover's Holidays in the Open*. New York: Charles Scribner's Sons.

CÁRDENAS, GARCÍA LÓPEZ DE

In September 1540, just 48 years after Christopher Columbus landed in the Americas, Hopi guides led 25 Spanish soldiers dispatched from 26-year-old Nueva Galicia governor Francisco Vázquez de Coronado's expedition to find the Seven Cities of Gold. (Myths about these cities arose in 1529 when Nuño Beltrán de Guzmán, a rival of Hernando Cortéz, heard from a Native American of seven rich cities in the north of New Spain.) Leading this group was Captain García López de Cárdenas. After a 20-day march, Cárdenas's group reached the canyon's South Rim at an area vaguely described as "elevated and full of low twisted pines, very cold, and lying open to the north" (Kenny, 2015), which some historians believe is between Moran Point and Desert View in the eastern part of GCNP (this area is the only part of the South Rim "open to the north"). These were the first Europeans to see Grand Canyon.

Cárdenas and his men spent three days along the rim looking for an easy way down. When they failed to find one, Cárdenas sent three men—Pablos de Melgosa, Juan Galeras, and another unnamed man—into the canyon. They got about one-third of the way to the river (to the Redwall Limestone) but returned to the rim and reported that "what seemed easy from above was not so, but instead very hard and difficult" (Kenny, 2015) and that some boulders visible from the rim (estimated to be about as tall as a man) were larger than the 342-foottall bell tower of *La Giralada*, the Great Tower of Seville. Cárdenas correctly estimated that the opposite side of the canyon was 8–10 miles away but concluded that the Colorado River was only about 6 feet wide (it's actually more than 250 feet wide where Cárdenas was). For Cárdenas and his men, Grand Canyon was an obstacle to finding the fabled Lost Cities of Gold. Although Coronado's expedition found no gold and returned to Spain in debt, it did find what is now New Mexico, Arizona, and several sections of the Rockies.

After Cárdenas's visit, the Spanish lost interest in the canyon country; no other non-Native people entered Grand Canyon for more than 200 years. Today's visitors, like those sent by Cárdenas, have similar problems comprehending Grand Canyon's size. Cárdenas is honored by Cardenas Creek, which enters from river left at River Mile 71. In 1940, the U.S. Postal Service issued a 3¢ stamp commemorating Coronado's expedition.

See also: Desert View; Moran Point

Further Reading

Kenny, Ray. 2015. A 16th-century Spanish inscription in Grand Canyon? *Park Science*, September 4, 2015. Retrieved July 1, 2017, at https://www.nature.nps.gov/ParkScience/index.cfm?ArticleID=419

CATTALO, BUFFALO, AND PEARL GRAY

The American bison (*Bison bison*), popularly known as buffalo, are common in meadows and open woodlands along the North Rim. These majestic, horned animals, which can be more than 6 feet tall and 10 feet long, usually live in herds and can run up to 35 miles per hour. Calves, who have reddish-tan fur, are always near their parents. Bison, which are the national mammal of the United States, have been on the seal of the Department of the Interior since 1912.

Buffalo once roamed the American West in vast numbers and were hunted for their prized meat and hides for decades. However, by the 1880s, overhunting had reduced the number of buffalo to only a few hundred, and many people became concerned that the species was nearing extinction. One such person was Charles

Jesse "Buffalo" Jones (1844–1919), a farmer, frontiersman, rancher, and hunter who earlier had earned a good living by hunting, killing, taming, and selling buffalo. His skills at hunting buffalo earned him the nickname "Buffalo."

Jones managed a herd of 150 buffalo at his ranch in Garden City, Kansas, a town in the southwestern part of the state that he and three others had founded on April 8, 1879. There, Jones worked with the Atchison, Topeka and Santa Fe Railway to build a depot, which brought visitors and new settlers. Jones also founded four irrigation companies that helped bring water from the Arkansas River (100 miles away) for crops. During one especially harsh winter, Jones noted that whereas many cattle froze to death, the thick-furred buffalo survived. To help conserve the buffalo (and possibly atone for his past hunting), Jones decided to begin breeding buffalo with cattle in hopes of producing "cattalo" hybrids that would have valuable fur and lean, tender meat, and which could survive harsh winters. Jones probably got this idea from Texas ranching pioneer Charles Goodnight, who was using the cross to produce what he called "beefalo." Jones also sold buffalo to parks, zoos, and other ranchers.

In 1902, Jones's friend President Theodore Roosevelt appointed Jones as the first game warden at Yellowstone National Park. Jones soon brought in a small herd of buffalo for the park. In 1906, Jones used his friendship with Roosevelt to obtain permits to fence in part of the North Rim's Kaibab Plateau and raise buffalo and other animals. (That same year, Roosevelt established Grand Canyon National Game Preserve, which included the entire Kaibab Plateau.) Jones also got permission to bring some buffalo from Yellowstone to the North Rim to hybridize with cattle, in exchange for giving the government some of his animals.

Jones began breeding buffalo bulls with Scottish Galloway cows in House Rock Valley, which is just north of Grand Canyon. (This site, which is where part of the movie *Buffalo Bill* [1944] was filmed, is where real-life Buffalo Bill [i.e., William F. Cody] wanted to raise buffalo.) Jones hoped the hybrids would have the traits of both parents, such as silky fur, tender and nutritious meat, and the ability to survive winters on meager vegetation. For the next few years, Jones's work produced mixed results; male calves from the first crosses either were sterile, were aborted, or caused the death of the cow. Other calves survived; one cattalo, with a buffalo body and spotted fur, was displayed at nearby Jacob Lake, where tourists paid 50¢ to see the strange hybrid.

Despite occasional successes, the project never became profitable, prompting Jones to abandon the project and leave the area. The herd was later managed by the U.S. Forest Service, after which the herd became the property of Arizona. The remnants of Jones's herd later escaped their pens and became a free-ranging herd, and buffalo have remained in Arizona ever since. By 1926, according to the American Bison Society, there were 80 buffalo in Arizona, and in 2018 more than 500 were living at the North Rim. These buffalo, which spend most of their time in GCNP, are descendants of Jones's experiments.

In the early 1900s, disillusioned dentist and struggling writer Pearl Gray (1872–1939) heard Jones lecture about roping mountain lions and breeding cattalo at Grand Canyon. The next year, Gray—fascinated by what he'd heard—went to see Jones in Arizona. Jones and Gray hunted and killed several lions, after which Gray went home and wrote a number of articles and a book inspired by Jones titled *Roping the Lions* (1924) and *The Last of the Plainsmen* (1908, in which the central figure is modeled after Jones). Gray became the storyteller of the American West; during his career, 56 of his 89 books—46 of which were made into movies—were set in the West. The popularity of these publications prompted Gray to change the spelling of his last name to Grey and replace his first name with his middle name. Zane Grey was inspired by what he saw at the Kaibab Plateau, and many of his stories about the American West remain popular with today's readers.

Jones died of a heart attack in 1919 in Topeka, Kansas. He is buried beside his wife and sons at Valley View Cemetery in Garden City. A statue of Jones stands at the Finney County Courthouse in Garden City, where he is also honored by Buffalo Jones Elementary School. In 1959, Jones was posthumously inducted into the Hall of Great Western Performers of the National Cowboy & Western Heritage Museum in Oklahoma City, Oklahoma.

American bison remain icons of conservation and the America West, and they are popular with visitors approaching the North Rim. However, bison are not native to Grand Canyon, and their growing population (projected to increase to more than 1,200 by 2027) has significantly damaged North Rim ecosystems by stripping areas of vegetation and polluting the area's soils, lakes, and wetlands. In 2017, GCNP announced plans to thin the North Rim's herd of bison—by roundups and killings—to no more than 200 within three to five years.

See also: Roosevelt Point

Further Reading

Easton, Robert, and Brown, Donald Mackenzie. 1961. *Lord of Beasts: The Saga of Buffalo Jones*. Tucson: University of Arizona Press.

Kersey, Ralph T. 1958. *Buffalo Jones: A True Biography*. Garden City, KS: Elliott Printers.

Grey, Zane. 1924. *Roping the Lions*. New York. Grosset & Dunlap.

Grey, Zane. 1908. *The Last of the Plainsmen*. New York: Outing Publishing.

CIVILIAN CONSERVATION CORPS

By 1933, the Great Depression had produced economic chaos and unemployment rates in the United States that exceeded 20 percent. To help remedy these problems, President Franklin D. Roosevelt—just five days after his inauguration—convened an emergency session of Congress and introduced the Emergency Conservation Work Act, which became better known as the Civilian Conservation Corps (CCC) Act. This program hired unemployed, young men for six-month assignments that included room, board, health care, and a monthly salary of $30 (of which $25 went to the worker's family and dependents). Although women were banned from the CCC, the rules were later relaxed to include older veterans of World War I and the Spanish-American War. The first inductee joined the CCC 37 days after Roosevelt's inauguration.

The first CCC workers (Company 819) came to Grand Canyon in 1933, about two months after the CCC had been created. Workers on the North Rim (Company 818), who completed the Colorado River Trail (the most difficult trail ever built within GCNP), moved to the bottom of Grand Canyon during the winter. On an average day, about 500 CCC workers were at Grand Canyon, where they worked on the North Rim, South Rim, and Inner Canyon. These workers completed several projects that today's visitors enjoy:

South Rim

Bright Angel Trailhead

Community Building, a two-story wooden building just west of the corral.

Wooden bridge, culvert, and other roadside improvements near today's Grand Canyon Association building.

GCNP boundary fence.

Rock wall, curbing, landscaping, and other improvements along scenic turnouts.

Stairs leading up from Kolb Studio.

Landscaping and paths to Kachina Lodge, El Tovar Hotel, and Yavapai Observation Station.

Rim-side wall—27 inches high and 18 inches wide—between Verkamp's Curios and Lookout Studio.

Colter Hall, a two-story women's dorm between El Tovar Hotel and Kachina Lodge.

North Rim

GCNP boundary fence.

Buildings.

Roads.

Inner Canyon

South Kaibab Trail and Bright Angel Trail (e.g., widened Bright Angel Trail to its present 4- to 6-foot width).

Ribbon Falls Trail, a 0.5-mile spur connecting the North Kaibab Trail to a waterfall. Ribbon Falls includes a sacred site of Zuni Indians.

Clear Creek Trail (9 miles).

Bright Angel Campground south of Phantom Ranch. This was the site of the camp of the CCC that worked at Phantom Ranch.

Colorado River Trail connecting Bright Angel Trail and the South Kaibab Trail on the south side of the river. Construction of this 1.7-mile trail, which was first proposed in 1929, required 20 tons of explosives and more than 10,000 jackhammer bits.

Phantom Ranch's mule coral and ranger station.

Rock and timber shelters at 1.5 miles, 3 miles, Indian Garden, and Pipe Creek along Bright Angel Trail. All of these structures have emergency telephones connected to the Trans-Canyon Telephone Line. Ironically, these telephones were only for guests; CCC workers, who were building the telephone line, were not allowed to use them (except in emergencies).

Grand Canyon was changed profoundly by the CCC's Trans-Canyon Telephone Line (maintained by Mountain Bell). The project, which was completed in only 10 months, stretched telephone lines down the North Kaibab Trail to Cottonwood Camp, Phantom Ranch, South Kaibab Trail, Bright Angel Trail, Indian Garden, and the South Rim. The project required the installation of 592 metal poles and 25 miles of cable (later reduced to 18 miles). Many of the original poles remain in place. The telephone line is commemorated by an often overlooked plaque atop the rim-side wall near the east end of the South Rim's Thunderbird Lodge, as well as a similar plaque just south of Phantom Ranch.

In 1935, at its peak, the CCC enrolled almost 500,000 men working in 2,600 camps throughout the United States that were supervised by active and reserve military officers. However, by 1941, economic conditions had improved, and there were fewer than 200,000 CCC workers in only 900 camps. The CCC, which was the most popular of Roosevelt's New Deal programs, ended in 1942—not because it was abolished but instead because Congress provided no funds for its operation. During the program's existence, more than 4,000 CCC workers (in Companies 818, 819, 847, 2543, 2833, 3318, and 4814) worked at Grand Canyon.

Nationwide, between 1933 and 1941, more than 3 million men worked in the CCC; they planted more than 2.3 billion trees (this is why the CCC was also referred to as Roosevelt's Tree Army) and built 3,470 fire towers, 97,000 miles of fire roads, 28,000 miles of new trails, more than 110,000 miles of refurbished trails, 89,000 miles of telephone lines, 800 new state parks, and more than 63,000 buildings, shelters, museums, bathrooms, and other public structures throughout the United States. GCNP's CCC workers are commemorated by a plaque at Hopi Point. Many of the buildings built by the workers are listed on the National Register of Historic Places.

See also: Grand Canyon Village; Phantom Ranch; Trans-Canyon Telephone Line

Further Reading

Audretsch, Robert W. 2011. *Shaping the Park and Saving the Boys: The Civilian Conservation Corps at Grand Canyon, 1933–1942*. Indianapolis: Dog Ear Publishing.

CLARK, GEORGIE WHITE

Georgie White Clark (1910–1992, born Bessie DeRoss) was a whitewater pioneer and, in 1952, the first woman to row the entire length of Grand Canyon. (Previous trips, often in boats with women's names, had all been guided by men.) In July 1953, while camped at Boucher Rapid (River Mile 97.1), Georgie came up with the idea of lashing three 10-person, army surplus inflatable rafts together for increased stability and to increase the number of passengers she could take through the canyon. On her trips, Georgie promised river runners that they would "get things just the way you like it."

In 1955, Clark—the first woman to lead a boating party through Grand Canyon—carried 28 passengers at once through Grand Canyon; this was the largest group ever to boat the canyon. Georgie later began her Share the Expense business model of river running, which made Grand Canyon rafting cheaper and more accessible to the public. Indeed, while other companies carried relatively few people and charged $800–$1,000 per person for their expeditions, Clark often carried more than 30 passengers per trip and charged less than $200 per person. Clark's approach opened Grand Canyon to large-scale river tourism. Georgie's configuration of boats became known as G-rigs, Triple Rigs, and Tour Buses of the River, and she proudly called her customers Georgie's Royal River Rats. However, Clark's G-rigs were often underpowered, leaving boatmen and boatwomen unable to finesse Grand Canyon's challenging rapids. Critics noted the irony of rafting the most turbulent water in America in a pontoon that neutralized the turbulence.

On a 1959 expedition led by Georgie, passenger Vernon Read injured his skull and spine and became the first injured river runner to be evacuated from the river in Grand Canyon. (Read was evacuated from near Elves Chasm.) In 1972, 64-year-old Mae Hansen was trapped beneath Georgie's flipped G-rig and drowned in House Rock Rapid. Hansen was the first commercial passenger to die on a river running trip in Grand Canyon.

Georgie, who was the first woman to own a Grand Canyon river running company, ran her rafting business for 45 years. It has been said that Georgie did more "to make this country's most dangerous rivers and canyons accessible to the average citizen than any other person living, man or woman" (Westwood, 1997). By 1961, Georgie had taken more people down the Colorado River through Grand Canyon than anyone.

Georgie's no-frills trips were known for their canned food meals, for which White offered no apologies, telling critics, "If you want to eat, go to a restaurant; if you want to see the canyon, come with me." Georgie later became a Grand Canyon celebrity who was featured in *Life*, *Sports Illustrated*, and on *The Tonight Show Starring Johnny Carson*. On the river, Georgie wore a bathing suit (often with a leopard leotard) and drank Coors beer for dinner.

In 1992, Georgie sold her company to Western River Expeditions. Later that year, only five weeks after writing her friends, "How I wish I was running that river," the 81-year-old Georgie died of cancer. When Georgie died, she had in her possession a copy of the marriage certificate of Glen and Bessie Hyde, prompting some to speculate that she *was* Bessie Hyde (a subsequent investigation showed that she was not). In 2001, the U.S. Board on Geographic Names—in a split, controversial vote—renamed 24 Mile Rapid Georgie Rapid in her honor. One of Georgie's 10-person boats is part of GCNP's collection of historic boats.

See also: Grand Canyon Museum

Further Reading

Clark, Georgie White, and Newcomb, Duane. 1977. *Georgie Clark: Thirty Years of River Running*. San Francisco: Chronicle Books.

Westwood, Richard E. 1997. *Woman of the River: Georgie White Clark, Whitewater Pioneer*. Logan: Utah State University Press.

COLTER, MARY ELIZABETH JANE

Mary Elizabeth Jane Colter (1869–1958) was born on April 4, 1869, in Pittsburgh, Pennsylvania. After short periods in Colorado and Texas, her family settled in St. Paul, Minnesota, which Colter forevermore considered home. After attending San Francisco's California School of Design in the 1890s, Colter began teaching art and drawing at Mechanic Arts High School in St. Paul.

While Colter was teaching in Minnesota, the Atchison, Topeka and Santa Fe Railway was opening routes in the American West, where the railroad partnered with the Fred Harvey Company to open hotels, restaurants, gift shops, and newsstands along its routes for its customers. Harvey (1835–1901), who began his career as a busboy and dishwasher, believed that the road to wealth in the American West was to serve good food. After opening his first two railroad restaurants in Wallace, Kansas, and Hugo, Colorado, he eventually built an empire consisting of 15 hotels, 47 restaurants, and 30 dining cars.

When Colter began her career as an architect in 1890, there were only 22 female architects in the United States. Colter's big break came in 1902, when Fred Harvey hired her to decorate the so-called Indian Building adjoining Albuquerque's new Alvarado Hotel. This job—Colter's first major job as a professional

Mary Elizabeth Jane Colter (right) was a pioneering architect who designed several iconic structures at Grand Canyon, including Desert View Watchtower, Hopi House, Hermits Rest, and Phantom Ranch. In this photo, taken ca. 1935, Colter is examining blueprints with Anna Thompson Ickes (1873–1935), the wife of Secretary of the Interior Harold LeClair Ickes. (National Park Service)

artist—began a 40-year association of Colter with the railroad and the Fred Harvey Company. Colter had no academic training as an architect, and she never got an architectural license. Nevertheless, her relationship with Fred Harvey enriched everyone involved and brought fame, respect, and admiration to Colter in the male-dominated profession of architecture and design. Despite her accomplishments, however, she was often listed as a "decorator," and her plans usually had to be approved by others before being implemented.

The South Rim's El Tovar Hotel was designed by Charles Whittlesey (1867–1941), the architect who designed the Alvarado Hotel. Whereas Whittlesey often tried to emulate European designs, Colter tried to include local culture in her decorations. When she decorated the interior of El Tovar, for example, Colter reproduced the pictographs that are just past the first tunnel of Bright Angel Trail above the plate rail of the hotel's private dining room.

Although the energetic Colter later designed and decorated buildings in Los Angeles, Chicago, Santa Fe, and elsewhere, she is best known for her masterpieces that highlight and reflect the natural beauty of Grand Canyon. The first of these iconic buildings was Hopi House, which opened on January 1, 1905, just a few days before the nearby El Tovar Hotel. Unlike architects such as Whittlesey, Colter designed her buildings at Grand Canyon to portray the cultural heritage of the area. Colter patterned Hopi House after traditional Hopi pueblos at Old Oraibi, a village (with Acoma, New Mexico) that is the oldest continuously inhabited village (since 1150) in the United States. Inside Hopi House, Colter displayed Navajo blankets that had won the Grand Prize at the St. Louis Exposition the year before. (These blankets eventually became part of the famed 4,500-piece Fred Harvey Fine Arts Collection.) Famed artists such as Navajo blanket weaver Elle of Ganado ("Red Woman," who was twice chosen to weave blankets for U.S. presidents Theodore Roosevelt and William Howard Taft), and Hopi-Tewa potter Nampeyo (famous for her distinctive, low-shoulder jars) lived in the two hogans behind Hopi House. These and other Native Americans were also included in advertisements of Grand Canyon by Fred Harvey and the Santa Fe Railway. Colter completed Hopi House as a contractor with Harvey; after finishing her work there, she returned to St. Paul to teach.

In 1910, the 41-year-old Colter became a full-time, permanent worker at the Kansas City headquarters of the Fred Harvey Company, where she was described as the foremost American architect and interior designer. Part of Colter's salary was paid by the Santa Fe Railway. Several more projects at Grand Canyon followed, including Hermits Rest (1914), The Lookout (1914, now called Lookout Studio), Phantom Ranch (1922), and Bright Angel Lodge (1935). Although these buildings are well-known, Colter's masterpiece at Grand Canyon is Desert View Watchtower, which she completed in 1932.

Before beginning work at Desert View, Colter spent months studying the towers of Ancestral Puebloans in the Four Corners area, both from the land and air. Her Watchtower includes several features of those towers that she studied, including sealed doors and a ruin beside the tower. The tower is connected to a circular kiva, and an outer wall near the staircase includes a mythical snake called *Balolookong*

("The Great Plumed Serpent"). Colter intended for the upper floor of the tower to include telescopes such as those at Lookout Studio, but they were never installed. The kiva's ceiling includes logs from the old Grandview Hotel.

Lookout Studio

The Lookout (later renamed Lookout Studio) opened on the South Rim in 1914. The building of Kaibab limestone on the rim west of El Tovar, which Mary Elizabeth Jane Colter designed and built for the Santa Fe Railway, mimics the rugged construction of 12th-century Native Americans in the Southwest. The Lookout, which has an uneven roofline that gives the appearance of a building arising from Earth, was the Santa Fe Railroad's attempt to compete with Kolb Studio; the building blocked visitors' views of Kolb Studio from the rim-side trail and was often mistaken by tourists *as* Kolb Studio. In 1954, the Santa Fe Railroad sold Lookout Studio to NPS, who used it in its ongoing (and unsuccessful) campaign to force Emery Kolb out of GCNP. (Kolb was sometimes aided by powerful friends and politicians such as Carl Hayden and Barry Goldwater.) The upper story of Lookout Studio, which features great views of Bright Angel Trail, featured a telescope for viewing Grand Canyon; this telescope is now displayed in the History Room of Bright Angel Lodge. Two small outdoor overlooks at Lookout Studio offer excellent views of Grand Canyon.

Colter wanted Desert View Watchtower to honor Native Americans, especially Hopis, who believe the canyon is a sacred site of emergence. The tower's Hopi Room includes artwork painted by then 30-year-old Hopi artist Fred Kabotie (c.1900–1986) that depicts the Hopi world, and the ceiling reveals astronomical images. (Visitors to Grand Canyon can also see Kabotie's murals in the Harvey House Café in Bright Angel Lodge and at the Painted Desert Inn in Petrified Forest National Park in eastern Arizona.) Kabotie, who illustrated the Snake Legend describing the first man (Tiyo) to navigate the Colorado River, used four colors that Hopis believe symbolize the Universe. Like The Lookout, Desert View Watchtower includes ruins that evoke images of the region's earlier inhabitants.

Colter also wanted Desert View Watchtower to honor Native Americans' connection with Grand Canyon and introduce its visitors to the depths of native culture. In addition to Kabotie's murals of Hopi life, the Desert View Watchtower hosted artists and shows featuring various tribal traditions. However, in ensuing years, this part of the tower's original purpose slowly faded, and the kiva's tribal exhibits were replaced by a giant souvenir shop managed by concessionaire Xanterra Parks & Resorts. When the Grand Canyon Inter-Tribal Advisory Council was established in 2013, one of its first suggestions was the establishment of a place in Grand Canyon that Native Americans could call home. Two years later, NPS restored the kiva's dignity by converting the souvenir shop to a cultural center that celebrates the heritage of Native American tribes traditionally associated with Grand

Canyon. The commemoration of this cultural center in 2016 was highlighted by a talk by Ed Kabotie, a Hopi musician and storyteller, and the grandson of Fred Kabotie.

Colter's choice of the name "Desert View Watchtower" likely came from her belief that some of the Native American's towers on which she modeled her work were built for astronomical and ceremonial powers. She considered her tower at Desert View to be a Native American astronomical observatory. In 1928, American astronomer George Willis Ritchey (1864–1945), who had overseen the construction of Mount Wilson Observatory (where Edwin Hubble showed that the universe is expanding), proposed to build a 250-foot-high, 10-mirror, state-of-the-art telescope and observatory on the rim at today's Desert View. *The New York Times* used its front-page headline on June 17, 1928, to announce Ritchey's plans: "Secrets of Planet Mars to Be Revealed by a Huge Telescope at Grand Canyon." A follow-up article on July 23 praised the proposed telescope planned for "somewhere in the Arizona desert on the edge of the Grand Canyon" and asked, "Will the scientists, trekking to their desert watchtower seven years hence, appreciate the significance of the anniversary?" (Lago, 2014; note the writer's use of the word "watchtower").

Ritchey's announcement in Paris was accompanied by much publicity, but when astronomer Edwin Hubble came to Grand Canyon to evaluate the worthiness of the area for such an observatory, he quickly nixed the idea; Grand Canyon's winds and updrafts made the area poor for such an observatory. Although the observatory was later built near San Diego atop Mount Palomar, Colter named her tower Desert View Watchtower when it opened in 1932. (The last of Ritchey's telescopes still in operation is at the U.S. Naval Observatory Flagstaff Station in Flagstaff, Arizona.)

Most of Colter's buildings at Grand Canyon, which look as old as the canyon itself, are tributes to Grand Canyon's tribes. However, Colter designed Hermits Rest to look as if it was designed by untrained mountain men, and its furnishings included a giant fireplace, cobwebs, and soot. Always a stickler for detail, Colter later commented, "You can't imagine what it cost to make it look this old" (Rivard, 2017). Conservationist Aldo Leopold, during a visit to Grand Canyon in 1915 to prepare his plan for the area, described Colter's buildings as "exceptionally attractive. As long as the Company's work is passed upon by Miss Colter . . . its appropriateness can be assured."

Colter, who named Hermit Trail's Santa Maria Spring to honor the Virgin Mary, retired in 1948 at the age of 79. In 1957, after watching one of her favorite buildings—the El Navajo Hotel in Gallup, New Mexico—be dismantled and sold for scrap to enable the widening of Route 66, Colter noted, "There's such a thing as living too long" (Lago, 2015). The 88-year-old Colter died less than a year later in Santa Fe, New Mexico, on January 8, 1958. Colter, who is buried in Oakland Cemetery in St. Paul, Minnesota, is remembered as one of America's greatest architects.

Colter Hall, which is north of Kachina Lodge and one of the two dorms Colter built in Grand Canyon Village in 1936–1937, is named in Colter's honor. Contrary to popular belief, Colter Butte near River Mile 56 is not named for Mary Colter but instead for James G.H. Colter (1844–1922), an early settler along the Little Colorado River in the 1870s.

See also: Desert View; Grand Canyon Village; Harvey, Fred, and the Harvey Girls; Hermits Rest; Phantom Ranch; Railroad

Further Reading

Grattan, Virginia L. 1992. *Mary Colter: Builder upon the Red Earth*. Grand Canyon, AZ: Grand Canyon Association.

Lago, Don. 2014. *Canyon of Dreams: Stories from Grand Canyon History*. Salt Lake City: University of Utah Press.

Lago, Don. 2015. *Grand Canyon: A History of a Natural World and National Park*. Reno: University of Nevada Press.

Rivard, Katherine. 2017. Aesthetic inspiration from a grand landscape. National Park Foundation, May 24, 2017. Retrieved July 9, 2017, at https://www.nationalparks.org/connect/blog/aesthetic-inspiration-grand-landscape

D

DELLENBAUGH, FREDERICK

Frederick Samuel Dellenbaugh (1853–1935) was born in McConnelsville, Ohio, educated and studied art in the United States and Europe, and the discoverer of the Escalante River, a tributary of the Colorado River that was the last unknown river in the United States. When he was only 17 years old, Dellenbaugh was an artist and the assistant topographer with the second expedition through Grand Canyon by John Wesley Powell.

In 1884, Dellenbaugh lived with the Hopi Indians near the Four Corners area. Five years later, Dellenbaugh accompanied the Harriman expedition that charted the coasts of Siberia and Alaska. During the summer of 1903, Dellenbaugh painted *Zion Canyon*, which was displayed at Zion National Park and the World's Fair in St. Louis the following year. (Today it hangs in the Zion Human History Museum, and other paintings by Dellenbaugh are displayed at Yale University.) In 1904, Dellenbaugh helped found the Explorers Club in New York City, his base of operations.

Dellenbaugh wrote several books about the American West, including *The Romance of the Colorado River: The Story of Its Discovery in 1540, with an Account of the Later Explorations* (1902). This book became an influential history of Grand Canyon that described Powell's first trek through Grand Canyon as "one of the distinguished feats of history" (Dellenbaugh, 1902). Dellenbaugh overlooked Powell's mistakes, exaggerations, and self-dramatizations and rejected James White's story of traversing Grand Canyon as "flimsy." The book included photographs by James Fennemore and J. K. Hillers, and its frontispiece was a watercolor sketch by Thomas Moran titled "Looking Up the Bright Angel Trail."

Six years later, Dellenbaugh's *A Canyon Voyage: The Narrative of the Second Powell Expedition* (1908) was another uncritical homage to Powell. In that book, Dellenbaugh described the Colorado River's water as sometimes being "well-nigh solid, not water at all" and noted that "the power of the river can hardly be described. It is fierce, unrelenting, demonical" (Dellenbaugh 1908). Dellenbaugh strongly supported later efforts to dam the Colorado River, but those dams were never built in Grand Canyon.

Dellenbaugh, who considered the Colorado River to be "the enemy" that his expedition "was vanquishing" in "another day's battle" (Lago, 2015), later became a self-appointed consultant to a committee charged with designing a memorial to John Wesley Powell's two expeditions through Grand Canyon. Dellenbaugh argued that Powell would not have approved of including the names of William Dunn, Seneca Howland, and Oramel Howland on the memorial because they abandoned the 1869 expedition at Separation Canyon. Although Powell never disparaged Dunn and the Howlands, Dellenbaugh branded them "deserters." Thanks to Dellenbaugh, their names do not appear on the memorial at Powell Point. This omission was criticized harshly by Robert Stanton, who claimed that it was unfair and attached an "undeserved stigma of cowardice" to the three explorers.

After a walk through snow to an Explorers Club meeting, the 81-year-old Dellenbaugh died of pneumonia on January 29, 1935, in his New York City apartment. He is buried in the Fantinekill Cemetery in Ellenville, New York, and is commemorated by Mount Dellenbaugh, a remote peak (elevation 7,072 feet) on the Shivwits Plateau of the Arizona Strip.

See also: Thematic Essays section: John Wesley Powell

Further Reading

Dellenbaugh, Frederick, S. 1908. *A Canyon Voyage: The Narrative of the Second Powell Expedition*. New York: G. P. Putnam's Sons.

Dellenbaugh, Frederick S. 1902. *The Romance of the Colorado River: The Story of Its Discovery in 1540, with an Account of the Later Explorations*. New York: G. P. Putnam's Sons.

Lago, Don. 2015. *Grand Canyon: A History of a Natural World and National Park*. Reno: University of Nevada Press.

DESERT VIEW

Desert View and Desert View Watchtower (elevation 7,438 feet), which overlook the eastern end of Grand Canyon and the adjacent Painted Desert, offer the first view of Grand Canyon for visitors arriving on Arizona State Route 64 from the east (i.e., from Cameron, Arizona, which is 39 miles away). In 1919, El Tovar Hotel offered three-day trips to Desert View along the El Tovar-Desert View Road (later called East Rim Drive) for $5 per day. For $10, visitors could buy a ticket for a one-day motorized trip to Desert View; drivers of these vehicles often took passenger pigeons to summon help in case they had trouble along the way. The original speed limit was 8–12 miles per hour, but it was increased to 45 miles per hour in 1955 when the road was widened and realigned.

Below Desert View, the Colorado River turns west and continues into the deepest parts of Grand Canyon. You can hike but not drive east along the canyon from Desert View because the land drops sharply toward Cedar Canyon, a tributary of the Little Colorado River. Campsites near Desert View are available on a first-come/first-served, self-registration basis. Desert View is 23 miles by road and 18.5 miles (by air) from Grand Canyon Village. Directly across the canyon from Desert View and 8 miles away (as the condor flies) is the North Rim's Cape Final.

Desert View is dominated by the rustic, 70-foot-tall Desert View Watchtower, which was designed in the early 1930s by Mary Jane Colter and built by the Bridge Department of the Atchison, Topeka and Santa Fe Railway. Desert View Watchtower is visible from the Colorado River at River Mile 65 and for several miles thereafter, as well as from several other viewpoints along the South Rim. This tower is not a copy of a tower elsewhere but instead is what Colter described as a "re-creation" of an Indian watchtower (at 70 feet, it is taller than any known Native American

The eastern entrance to Grand Canyon National Park's South Rim is dominated by Mary Elizabeth Jane Colter's masterpiece, Desert View Watchtower. Colter wanted the tower to honor Native Americans, for whom Grand Canyon is sacred. (Randy Moore)

tower). The ground floor of the steel-reinforced tower is a large, round room modeled after a Native American kiva (with additions such as large windows for observing the canyon). Kivas were used for religious ceremonies. People entered the kiva through the ceiling by using a ladder (the opening also was an escape hole for smoke from the kiva's fireplace). Some of the logs for the kiva's ceiling came from the Grandview Hotel, the first hotel built on the South Rim. The tower, whose concrete/steel superstructure is covered by a veneer of native Kaibab Limestone, opened in 1932 as a shrine for Native Americans, especially the Hopis, whose ancestors lived in Grand Canyon and for whom the canyon is a sacred place of emergence. When building the tower, Colter meticulously strove to give the illusion of age; as she told workers, "any tool mark [would be] a conspicuous scar on the face of our walls" (National Park Service, 2017). Desert View was formally dedicated on May 13, 1933, at a ceremony that included dignitaries, artists, and ceremonial dancers. Navajo refer to the watchtower as *Yahehahi* ("standing tower").

From the kiva, visitors ascend stairs to the tower's Hopi Room, where they find a large, circular painting depicting the Hopi Snake Legend painted by Hopi artist and educator Fred Kabotie (c.1900–1986). Also shown are images of Muyingwa (the god of germination), Baloongahoya (the god of echo), and a Hopi wedding scene. Kabotie, who managed the gift shop at Hopi House in 1935–1937, was assisted by rock art enthusiast Fred Geary, who rendered drawings found at ancient Hopi sites, and by Hopi artist Chester Dennis, who produced petroglyphs along the wall around the balcony. Kabotie's art was featured on playing cards sold by the Fred Harvey Company, and one of Kabotie's murals, which features a Hopi man named Tiyo floating down the Colorado River in a hollowed cottonwood log, describes the cultural connection between Grand Canyon and the Hopis. (More of Kabotie's art, which was commissioned in 1958 by Fred Harvey's son Byron Harvey Jr. to depict the many tourists whom Harvey met who were indifferent to Native American culture, is displayed in the lounge of Bright Angel Lodge.) Tiyo is honored by Tiyo Point near River Mile 91.5.

From Desert View, visitors have a great view of the 150-mile-long East Kaibab Monocline, which accounts for the 3,060-foot-drop in elevation between Desert View and Cameron. Desert View also offers dramatic views of Painted Desert, which stretches toward the Hopi and Navajo Reservations along the Colorado River. Desert View offers an excellent view of the Grand Canyon Supergroup, which is visible as the low, tilted layers of sediment near the river. When Desert View Watchtower opened, tourists were charged 25¢ to ascend the tower's 85 steps with a Hopi guide.

Desert View includes a memorial to the midair crash in 1956 of two commercial planes high above nearby Chuar Butte, which killed all 128 people aboard both planes. This was the worst peacetime civilian airline disaster in commercial aviation history and remains the worst of several disasters at Grand Canyon. This crash (one of 65 midair collisions between 1950 and 1955) led to the development of anticollision radar and the creation of the Federal Aviation Administration in 1958. A plaque commemorating the crash was installed at Desert View in 2014.

For many years, the kiva was a giant gift shop operated by a GCNP concessionaire. However, GCNP restored the dignity of the kiva by shrinking the gift shop so that the room could be used to display Native American exhibits. Also at Desert View are a gift shop, gas station (established in 1939), and restaurant. Although the 39-mile road from Cameron to Desert View was completed in 1935, most visitors to the South Rim continue to enter GCNP from Tusayan.

See also: Colter, Mary Elizabeth Jane; Tiyo

Further Reading

Colter, Mary Elizabeth Jane. 2015 (1933). *Manual for Drivers and Guides Descriptive of the Indian Watchtower at Desert View and Its Relation, Architecturally, to the Prehistoric Ruins of the Southwest*. Flagstaff, AZ: Grand Canyon Association (Fred Harvey Company).

Kabotie, Fred, and Bill Belknap. 1977. *Fred Kabotie: Hopi Indian Artist*. Flagstaff: Museum of Northern Arizona Press.

National Park Service. 2017. Desert View. Retrieved July 2, 2017, at https://www.nps.gov/grca/planyourvisit/upload/Desert_View-b.pdf

DINOSAURS

In late 1924, Samuel Hubbard—the curator of archaeology at Oakland Museum of California—led an expedition into Havasu Canyon to "bring before the scientific world, certain discoveries relating to prehistoric man" (Hubbard, 1925) that was made by museum patron Edward L. Doheny in three previous trips to the area. On one of those trips, Doheny—an oil tycoon who sponsored the trip—had accompanied prospector James Mooney, who fell to his death at the beautiful Havasu Creek waterfall that was later named in his honor (Mooney Falls).

In 1925, Oakland Museum of California published the expedition's findings in *The Doheny Scientific Expedition to the Hava Supai Canyon*. That report included reproductions and interpretations of "wall pictures" from the canyon, as well as drawings depicting elephants, ibexes, a petrified giant human, and a bipedal, tail-dragging, then conventional "dinosaur" resembling *Diplodocus* rearing up on its hind legs. The booklet's second printing, which was designed to discredit the teaching of evolution, included a "Supplement" that reported a human tooth from the Eocene and the heel of a shoe from Triassic limestone and claimed that the art "proved" that dinosaurs had lived with humans at Grand Canyon. The authors suggested that "if the observations are true, then the theory of Evolution . . . is not true," that "humans appeared on Earth before apes," and that "Man was the first mammal to appear and not the last one" (Hubbard, 1925). In a supplement published later, the authors claimed that life's history would not be understood until people took into consideration the evolution of man's soul. This report prompted *The New York Times* to claim in a headline on April 20, 1925, that the expedition had "Put Man's Age Back Ten Million Years." However, expedition member Charles Gilmore, a famed paleontologist and dinosaur expert of the U.S. National Museum, refused to endorse the expedition's conclusions. No mainstream scientist has endorsed the expedition's report since then. Nevertheless, claims that humans lived with dinosaurs a few thousand years ago remain popular and are a foundation of Young-Earth creationism.

The fossils of dinosaurs nearest Grand Canyon are in Mesozoic rocks of Vermilion and Echo Cliffs. The Mesozoic sediments at Grand Canyon that contained dinosaur fossils have eroded away to reveal Kaibab limestone, much like mud washed off pavement.

See also: Thematic Essays section: Grand Canyon and Modern Christian Fundamentalism; "Lost Worlds"; Religious Books and Scripture Plaques

Further Reading

Doheny, E. L., et al. 1927. *Discoveries Relating to Prehistoric Man by the Doheny Scientific Expedition in the Hava Supai Canyon, Northern Arizona, with Supplement*. Oakland, CA: Oakland Museum.

Hubbard, Samuel. 1925. *The Doheny Scientific Expedition to the Hava Supai Canyon, Northern Arizona, October and November, 1924, with Supplement*. Oakland, CA: Oakland Museum.

DOMÍNGUEZ AND ESCALANTE

In 1776, only a month after Francisco Tomás Garcés visited Havasu Canyon, Spanish Franciscan priests Atanasio Domínguez and Silvestre Vélez de Escalante, accompanied by cartographer Bernardo Miera y Pacheco and eight men from Santa Fe, were exploring the Colorado Basin for a route from Santa Fe, New Mexico, to the Catholic mission in Monterey, California. These were the first Europeans to reach Lees Ferry and to see and cross Glen Canyon. Vélez described "lush, mountainous land filled with game and timber, strange ruins of stone cities and villages, and rivers showing signs of precious metals" and "a great canyon . . . [having] on both sides extremely high cliffs and rocks." On its way back to Santa Fe, the expedition camped from October 26 to November 7, 1776, at the site of their crossing in Glen Canyon, which became known as Crossing of the Fathers. This site was visited almost 100 years later by John Wesley Powell. The site and its plaque are now under Lake Powell's Padre Bay. The expedition was one of the last major explorations by the Spanish in the modern-day American West.

The Domínguez–Escalante expedition did not reach the mission, but its maps and notes were important for later explorers. The route used by the Domínguez–Escalante expedition became part of a trade route named the Old Spanish Trail.

See also: Lake Powell and Glen Canyon Dam

Further Reading

Warner, Ted. 1976. *The Dominguez–Escalante Journal: Their Expedition through Colorado, Utah, Arizona, and New Mexico in 1776.* Provo, UT: Brigham Young University Press.

DUTTON, CLARENCE

Clarence Edward Dutton (1841–1912) was a Union soldier in the 21st Connecticut Volunteers, a Yale College graduate (including a stint in Yale Divinity School), and a geologist who worked for USGS from 1875 to 1891. Dutton had trained for the ministry but was diverted to the army. While in the army and thanks to an intervention by Ulysses S. Grant, Dutton was "loaned" to agencies such as the Public Lands Commission and USGS, during which time he never relinquished his commission. Dutton wrote several influential books about the Colorado Plateau region and studied seismology and volcanism in Hawaii, California, and Oregon (e.g., he introduced the terms *'a'a* and *pahoehoe* for clinky and ropey lava flows, respectively). Unlike other prominent geologists such as Charles Lyell, Dutton's work was often speculative.

Dutton was a protégé and close friend of John Wesley Powell, who persuaded Dutton to become a geologist and lead a geological expedition out West. Powell helped secure money and approval (by Congress and the War Department) for a "Geological and Geographical Survey of the Rocky Mountains, J. W. Powell, commanding," an important survey also known simply as the Powell Survey. In the 1870s, this survey targeted the Colorado Plateau, whose vast, rugged geography was the last part of the continental United States not yet systematically explored and surveyed. Powell, Dutton, mapmaker Almon Thompson, and scientist Grove Karl Gilbert also gathered large amounts of ethnological material about the area's native inhabitants. Dutton helped document that Grand Canyon was formed by the uplift of the Colorado Plateau and down-cut of the Colorado

River. Theories about continental drift were based partly on Dutton's work at Grand Canyon.

Grand Canyon's significance to geology was obvious to Dutton, who noted:

> It would be difficult to find anywhere else in the world a spot yielding so much subject matter for the contemplation of the geologist, certainly there is none situated in the midst of such dramatic and inspiring surroundings. Dimension means nothing to the senses [at Grand Canyon], and all we are left with is a troubled sense of immensity. (Dutton, 1882)

However, whereas Gilbert measured Grand Canyon with numbers, Dutton described it with words. Dutton's writing helped readers appreciate the aesthetic significance of Grand Canyon, which he labeled "a great innovation in modern ideas of scenery and in our conceptions of the grandeur, beauty, and power of nature. It must be dwelt upon and studied, and the study must comprise the slow acquisition of the meaning and spirit of that marvelous scenery" (Dutton, 1882).

Dutton's *The Tertiary History of the Grand Canyon District* is primarily a scientific treatise about geology and landscapes. However, Dutton knew that Grand Canyon was more than *just* science, geology, and landscapes. To help readers appreciate this, Dutton knew that he needed far more than science's "severe ascetic style" to "exalt the mind sufficiently to comprehend the sublimity of the subjects" (Dutton, 1882). So instead of invoking the jargon of science, Dutton wrote more like Thoreau and Muir as he described Grand Canyon's "superlative manifestation" of marvelous scenery as he told readers what they were looking at and why it was important. For example, in Chapter 7 of *The Tertiary History of the Grand Canyon District*, Dutton leads readers on a suspenseful "imaginary journey . . . to the brink of the chasm" at Point Sublime. The chapter describes "very sweet and touching" ravines and playful "woodland nymphs . . . peeping at us from the flowery banks . . ." As Dutton approaches the rim, it "seems to glow with an unwonted wealth of floral beauty. It is scarcely a hundred yards distant; let us pluck a bouquet from it. We ride up the slope." The chapter's final sentence then stuns readers: "The earth suddenly sinks at our feet to illimitable depths. In an instant, in the twinkling of an eye, the awful scene is before us." Whereas John Wesley Powell's scientific writing was dry, Dutton's distinctive, evocative prose describes a feeling that has been encountered by countless people when they've walked up to the rim of Grand Canyon. Dutton's creative, poetic language taught geologists and others how to see Grand Canyon and helped transform Grand Canyon from an American place to an American symbol. Dutton was Grand Canyon's most eloquent writer.

Dutton's style of writing, which would not be published today, was new and effective. According to American writer and conservationist Wallace Stegner (1909–1993), the author of *Beyond the Hundredth Meridian* (1954) whose doctoral thesis was about Dutton, "Dutton first taught the world to look at [Grand Canyon] and see it as it was . . . Dutton is almost as much the *genius loci* of the Grand Canyon as Muir is of Yosemite" (Stegner, cited in Dutton, 1882). For Dutton, Grand Canyon "first bewilders, and at length overpowers" and should only "be described in blank verse and painted on canvas." To again quote Stegner, "Although it is Powell's monument to which the tourists walk after dinner to watch the sunset from the South Rim, it is with Dutton's eyes, as often as not, that they see" (Stegner, cited in Dutton, 1882).

Dutton began *The Tertiary History of the Grand Canyon District* by claiming that "this work is chiefly devoted to a description of the methods and results of EROSION upon a grand scale" (Dutton, 1882). However, much of his work involved naming sites trying to convey their meaning, for Dutton knew that "it is not enough to know the facts; their meaning is of much greater significance" (Dutton, 1882). Dutton disliked many of the original, utilitarian names of sites at Grand Canyon (e.g., Dripping Springs). Grand Canyon, Dutton claimed, provides "the visible grandeur a deep significance" where "the geologist finds himself a poet." This inspiration is obvious in many of the sites at Grand Canyon that Dutton named for Asian religions or ancient Roman and Greek mythology, including:

Brahma Temple is a pyramidal butte north of Zoroaster Temple named for Brahma, the Hindu creator of the universe. Brahma Temple overlooks Bright Angel Canyon.

Cape Final is a 7,919-foot peak on Cape Royal Road named by Dutton in 1880 when, after riding for several miles, he could finally could see Grand Canyon. Dutton called Cape Final, the eastern point of the Walhalla Plateau, "the most interesting spot on the Kaibab." Many people consider Cape Final, which overlooks several inactive volcanoes, to be the best of the North Rim's overlooks.

Cape Solitude on the South Rim near River Mile 62 towers 3,400 feet above the confluence of the Colorado and Little Colorado Rivers. The point was so named by Dutton because "it stands solitary and alone." From Cape Solitude, visitors can look across the river to see the Temple and Chuar Buttes.

Powell Plateau is a seldom visited North Rim plateau at the eastern end of North Bass Trail that Dutton named in 1882. The plateau, which is visible in the west from Grand Canyon Village, is surrounded by formations named for Grand Canyon surveyors (e.g., Dutton Point, Ives Point, Walthenberg Canyon, Newberry Point). Powell Plateau honors John Wesley Powell, who named tens of places in Grand Canyon (and hundreds more in the American West) but named none for himself. Powell is also honored by Lake Powell, which formed behind Glen Canyon Dam after it was completed in 1963.

Shiva Temple is a wooded butte named for Shiva, the Hindu destroyer of the ego and universe. Shiva Temple, the largest of the North Rim buttes, rises 4,000 feet from its base; it was named because it represented "the grandest of all the buttes, and the most majestic in aspect, though not the most ornate." In 1937, scientists from American Museum of Natural History examined the "lost worlds" atop Shiva Temple and Wotans Throne to see if plants and animals growing atop these buttes—which had been isolated from the North Rim for 100,000 years—had evolved new species. They had not; the species there were like those of the surrounding land. However, the scientists found Anasazi ruins, deer antlers, empty boxes of film, and tissues left during an earlier visit by South Rim pioneer Emery Kolb and his friends. Today, Shiva Temple—whose top spans six square miles—is a popular place for climbers.

Vishnu Temple, which is south of Angels Window, was named by Dutton in 1880 after Vishnu, one of the supreme trinity (with Brahma and Shiva) of the Hindu deity. Vishnu the Redeemer helped men against evil and misfortune. According to Grand Canyon explorer Frederick Dellenbaugh, Vishnu Temple is the most impressive geologic formation in the world.

Vulcans Throne, which overlooks Lava Falls, is a cinder cone at River Mile 179 (about 1 mile west of the North Rim's Toroweap Overlook) named for Vulcan, the Roman god of fire. (This site on river right is different from Vulcans Anvil, which is a 50-foot-tall cinder cone in the middle of the river at River Mile 177.8, at the approach to Lava Falls.) Dutton was inspired by the geological diversity at Vulcans Throne, noting that "apart from the merely scenic effects it would be hard to find anywhere in the world a spot presenting so much material for the contemplation of the geologist" (Dutton, 1882).

Dutton, who said that being at Grand Canyon made "the heart ache and the throat tighten," began the tradition of naming sites in Grand Canyon for mythological deities (e.g., Vishnu Temple, Isis Temple, Horus Temple). This was a new and inspiring way to view geology that differed from the prospector mentality that had dominated Grand Canyon and other parks until that time (and which produced utilitarian names such as Breezy Point and Crystal Creek). Some people objected to Dutton's names; for example, in 1933, Frederick Dellenbaugh noted, "I had several arguments with [Dutton] on the subject as I objected violently to Oriental and Egyptian nomenclature," later adding that it was:

> a pedantic and ridiculous thing to saddle onto that country a lot of names from European legends like the Arthurian, or the Greek, or Egyptian mythology. Nothing could be more inappropriate to my mind than "Holy Grail" Temple in Arizona . . . This matter of names worries me greatly, not that I want any romantic or high-sounding names, that I want to avoid the commonplaceness of pedantry. (Brian, 2015)

Similarly, nature writer Mary Austin warned Grand Canyon's visitors to:

> be careful whom you ask to point the place out [for] you, lest you be answered by one of the silly names cut out of a mythological dictionary and shaken in a hat before they were applied to the Grand Cañon for the amazing number of Americans who can never see anything unless it is supposed to look like something else. (Lago, 2015)

In 1916, Aldo Leopold proposed replacing Dutton's names with those from Native Americans, but Dutton's colorful names remain in place.

In 1880, Dutton led the first detailed geological survey of Grand Canyon for USGS. Dutton's report was first published as part of the USGS *Annual Report of 1881* and was republished the following year as an independent monograph. Dutton published three monographs about his work at Grand Canyon, and the first and only edition of his influential 264-page *The Tertiary History of the Grand Canyon District, with Atlas* (1882) was the second (and first published) book of the trilogy. USGS monographs were meant for use by USGS alone, so the Survey printed only 3,000 copies; this is why Dutton's monograph has always been a rare book.

When Dutton's monograph was published in 1882, the director of USGS was Grand Canyon explorer John Wesley Powell. Unlike Powell's earlier book, which had described Grand Canyon from the river, Dutton's scientific, comprehensive, and romanticized book described Grand Canyon from the rim and linked the formation of Grand Canyon with erosion.

The original artist with Powell's expedition in 1869 was E. O. Beaman, but he left the group at Lees Ferry to photograph Hopi Indians. He was replaced by young James Fennemore, who had lived with John Lee at Lees Ferry and had photographed Lee at his execution. When the frail Fennemore got sick and left for Salt Lake City, Jack Hillers became the expedition's photographer. Hillers, who had been with John Wesley Powell's second expedition through Grand Canyon, became a great photographer of the American West. For his book, Dutton—who later was Powell's chief assistant in Powell's irrigation surveys—also used the work of artists Thomas Moran and William Henry Holmes; these artists relied heavily on Hillers's photographs and their direct observations when preparing illustrations for Dutton's book. Moran, who cared little for literal transcriptions of nature, painted landscapes that captured emotions, while Holmes drew the canyon with precision so that other geologists could recognize the rock layers.

Dutton's book was the first USGS book about Grand Canyon, and it remains a definitive and classic book about Grand Canyon. The 3,000 copies that were published sold well in the United States and abroad, and Dutton's book quickly established itself as a model of modern natural history. Dutton, with geologists G. K. Gilbert and John Strong Newberry, made Grand Canyon a centerpiece of American geology. Dutton appreciated Grand Canyon's ever changing beauty, noting:

> It is never the same, even from day to day, or even from hour to hour. In the early morning its mood and subjective influences are usually calmer and more full of repose than at other times, but as the sun rises higher the whole scene is so changed that we cannot recall our first impressions . . . the sensitiveness of the picture to the slightest variations in these conditions is very wonderful. (Dutton, 1882)

With stunning art and remarkable prose, Dutton bonded Grand Canyon to art and linked the area to bigger themes. More than any other scientist or explorer, Dutton placed Grand Canyon outside of typical conventions of perception. Powell made Grand Canyon famous, and Gilbert described it with math and reason, but Dutton made Grand Canyon grand.

Dutton, who in 1888 was one of the 33 founders of the National Geographic Society, retired in 1901. He died on January 4, 1912, in Englewood, New Jersey, and is buried in the Memoriam Cemetery in Wallingford, Connecticut. After his death, Dutton's massive collection of personal papers was lost. In Grand Canyon, Dutton is memorialized by Dutton Point on the east side of Powell Plateau.

See also: Thematic Essays section: John Wesley Powell; United States Geological Survey

Further Reading

Brian, Nancy. 2015. *River to Rim: A Guide to Place Names along the Colorado River in Grand Canyon from Lake Powell to Lake Mead*. Second edition. Flagstaff, AZ: Earthquest Press.

Dutton, Clarence E. 1882. *Tertiary History of the Grand Canyon District*. Washington, DC: U.S. Government Printing Office. (Reprinted 2001 by the University of Arizona Press, with an introduction by Wallace Stegner and a new foreword by Stephen J. Pyne.)

Lago, Don. 2015. *Grand Canyon: A History of a Natural World and National Park*. Reno: University of Nevada Press.

Pyne, Stephen J. 1983. *Dutton's Point: A Natural History of the Grand Canyon*. Grand Canyon, AZ: Grand Canyon Natural History Association.

Stegner, Wallace Earle. 1982. *Beyond the Hundredth Meridian: John Wesley Powell and the Second Opening of the West*. Lincoln: University of Nebraska Press.

E

EDDY, CLYDE

Clyde L. Eddy (1889–1954) first saw Grand Canyon in 1919 while on his honeymoon. Eight years later, during the summer of 1927, Eddy led his own expedition through Grand Canyon, most of whose members were recruited with "Explorers Wanted" advertisements posted in college newspapers and on bulletin boards in college fraternity houses. (Eddy chose three men from Harvard, two from Coe College, two from Notre Dame, and one from Northwestern University.) As a result of his experiences in the military, Eddy believed that his crew "need not be 'tough whiskered' to be brave, and that a 'pink wristed' college boy will stand up as well in the face of long continued danger as, for instance, the average 'hard boiled' army sergeant" (Eddy, 1929). Eddy, who gave preference "to men who have had outdoor experience," rejected people "who cannot swim, handle a boat, do his share of camp 'chores,' and handle himself in the woods."

Eddy's expedition, which was guided by Parley Galloway (Nathan Galloway's son), relied on Frederick Dellenbaugh's books and notes to get through Grand Canyon. The expedition consisted of 13 men, a bear cub named Cataract that Eddy bought at a New York zoo, and a rescued Airedale dog named Rags, which Eddy believed was a "picturesque addition" to the group. The group was carried through Grand Canyon on three oversized Powell-type Whitehall boats named *Coronado*, *Powell*, and *Dellenbaugh*, each made of Mexican mahogany and otherwise constructed like those used by John Wesley Powell in 1869. During the expedition, which was replenished with food and supplies at Lees Ferry and the foot of Hermit Trail, Eddy noted water-piled driftwood 100 feet above the river from the flood in 1884, in which Eddy speculated that the Colorado River flowed at 250,000 cubic feet per second (cfs).

The purpose of Eddy's expedition was adventure, and Eddy found it. At Hance Rapid, Eddy "looked over the side of the boat and into the whirling, churning water and knew that I had looked into the face of . . . death," but added that the constant danger along the river rewarded him with a "richness of life—where the daily threat of death makes clear how good it is to live" (Eddy, 1929). Some of Eddy's crew left the expedition during the voyage. Nine of the 13 crew who started the trip finished it. Rags, was taken by crewmember Robert Bartl to Wisconsin, and Cataract was sold to an Atchison, Topeka and Santa Fe Railway engineer and taken to San Francisco. Later in 1927, Eddy returned to Grand Canyon and rafted from Lees Ferry to Bright Angel Creek. This expedition of 13 men in six boats was meant to make a movie of Grand Canyon. Eddy described his first expedition in his book titled *Down the World's Most Dangerous River* (1929, featuring an introduction by Frederick S. Dellenbaugh) and with an 8-minute-long film (with no sound) titled *Down the Colorado River in a Rowboat* (1927). Eddy, who described the Colorado as "a mad, crazy stream" (Eddy, 1929), later said the trip through Grand Canyon was the bravest thing he ever did.

In 1934, Eddy returned to Grand Canyon with seven others: Russell "Doc" Frazier, Fred Jaine, Bus Hatch, Alton Hatch, Bill Fahrni, Frank Swain, and Royce "Cap" Mowrey. This trip was run at nearly record low water (1,600 cfs), and the area was dusty, causing the expedition to become known as the Dusty Dozen. Of this crew, only Eddy had been through Grand Canyon before. The group started their expedition with 16 oars and found five more along the way (e.g., that had been left during the failed rescue of Glen and Bessie Hyde in 1928) but ended the trip with only six oars. Eddy learned that running the river at low flow rates reduced the risk but increased the work because the group had to portage more rapids. During the trip, the Dusty Dozen abandoned a boat at Lava Falls. On August 3, 1934, at the downstream end of Separation Canyon, Frazier wired a bronze plaque (that he brought from Salt Lake City) to the canyon wall honoring the three men who left Powell's 1869 expedition through Grand Canyon. Nearby the men also hung a copper plaque (made by Hatch) listing the date and their names. (This plaque fell off between 1937–1939, was found in the mid-1960s when Lake Mead was at low levels, and now is in the offices of Hatch River Expeditions in Vernal, Utah.) When the Dusty Dozen reached Boulder Dam, they traded one of their boats for a case of beer.

Eddy, who considered Separation Rapid the most dangerous rapid in Grand Canyon, is honored in Grand Canyon by Clyde Eddy Cave on river right at River Mile 31.5.

See also: Dellenbaugh, Frederick; Rapids

Further Reading

Eddy, Clyde L. 1929. *Down the World's Most Dangerous River.* New York: Frederick A. Stokes Company. (Reprinted, with a foreword by Peter Miller, by University of New Mexico Press and Avanyu Press in 2012.)

EL TOVAR HOTEL

The luxurious, four-story, 100-room El Tovar Hotel—a "Hotel in the Wilderness" that architect Charles Frederick Whittlesey (1867–1941) promised would be "the finest hotel in Arizona"—opened on the South Rim in 1905. This hotel, which combined "Swiss chateaux" with "castles of the Rhine" (or, as some claimed, "a European hunting lodge with Western chic"), doubled the number of hotel rooms at the village area of the South Rim; it also put several nearby hotels out of business when it opened as "the most expensively constructed and appointed log house in America" and "the most unique, the most comfortable, and one of the costliest resorts in the Southwest." Each room at El Tovar was different; for example, the Charles Whittlesey Suite was decorated with the hotel's original blueprints, and the Zane Grey Suite included art related to Grey and his books. The only hotel at

In 1905, the Santa Fe Railway opened El Tovar Hotel on the South Rim. This was one of the most luxurious and expensive hotels in America; indeed, each room was unique, and guests enjoyed amenities such as electric lights, hot and cold water, and fresh milk provided by a nearby herd of cows. In this photo taken in 1906, the carriage is facing the canyon. (Grand Canyon National Park Museum Collection)

the South Rim not run out of business by El Tovar was the Bright Angel Hotel, which was replaced by the Mary Colter–designed Bright Angel Lodge in 1935.

Fred Harvey often named his hotels after Spaniards. "Cárdenas" was the first choice for the name for El Tovar; this would have honored Don García López de Cárdenas, the first non–Native American to see Grand Canyon. However, that name was already used by another Fred Harvey hotel, as were Alvarado (one of Coronado's artillery officers) and Castañeda (the historian of Coronado's expedition). So the new hotel was named for López's boss, Don Pedro de Tovar, who had visited Hopi villages in 1540 but had never seen Grand Canyon. Harvey described El Tovar this way: "Not a Waldorf-Astoria—admirable as that type is for the city—but a big country clubhouse, where the traveler seeking high-class accommodation also finds freedom from ultrafashionable restrictions" (Harvey, 1905). Sadly, the Spaniards had committed atrocities against several groups of Native Americans; Harvey paid tribute to the conquistadors, but architect Mary Colter paid homage to the Hopis and other Native Americans.

The Fred Harvey Company operated El Tovar Hotel, but it was owned by Santa Fe Railway. The 327-by-218-foot El Tovar, which was "built of native boulders and pine logs from far-off Oregon," cost $308,000 ($8.1 million in 2017) and could house more than 200 guests. (Despite, its high cost, El Tovar was not the most expensive trackside hotel built by the Santa Fe Railway; that designation belonged to the La Posada Hotel in Winslow, Arizona.) The hotel, which promised high-class accommodations "free from the ultrafashionable restrictions" of city hotels, had electric lights, a chicken house for eggs, hot and cold water, a barber shop, a solarium, laundry room, a herd of cows that provided fresh milk, a greenhouse that provided fresh fruits and vegetables, and rooms dedicated to music, wine, and billiards. The hotel's art room displayed "some of Thomas Moran's masterpieces" (Harvey, 1905), as El Tovar does today. Water was brought from Del Rio, Arizona, more than 110 miles away, and stored at El Tovar in a 162,000-gallon tank. There was one bathroom on each floor, for which guests could make reservations. Whereas some guests at Bright Angel Hotel paid as little as 75¢ per night, guests at El Tovar paid $3.50 to $6 per night. The nearby El Tovar stables housed horses and mules for transportation in the park before the arrival of automobiles. Today, these stables house mules used to take visitors and supplies to Phantom Ranch. Despite its cost, El Tovar was popular; by 1945, there was a seven-day limit on stays at the hotel, and the following year the limit was reduced to three days.

Since opening more than a century ago, El Tovar—which is easily identified by the wooden turret atop the building—has hosted countless celebrities, including Albert Einstein, Zane Grey, George Bernard Shaw, and Presidents William Howard Taft and Teddy Roosevelt. (Roosevelt, who in 1912 ran for a third term as president as the candidate of the Bull Moose Party, donated the smaller of the two moose heads in El Tovar's lobby.) In El Tovar's early days, boats that successfully ran through Grand Canyon (e.g., *Utah*, from the Charles Russell–Edwin Monett expedition in 1907–1908) were often displayed on El Tovar's porch, usually accompanied by signs noting, "This Boat Came Through the Grand Canyon." A sign near the entry of the hotel displays the Tovar coat of arms.

Pirl Ward

Visitors at Grand Canyon Village often notice a flat, tombstone-like marker in a grove of trees behind Hopi House and 180 feet in front of El Tovar Hotel. The marker is inscribed "Pirl A. Ward 1879–1934." Most visitors assume the marker denotes a gravesite, and lore has long claimed that the site is the grave of a Harvey Girl whose life ended poorly at Grand Canyon. This speculation makes for an intriguing story, but it's not true.

Pirl Adelbert Ward (1879–1934), who was born in Peabody, Kansas, in 1879, was a carpenter and contractor who helped build El Tovar Hotel. Ward, who married Lillian Williams in Williams, Arizona, in 1905, also worked on the Coronado Hotel in San Diego, California, and on the high school in Winslow, Arizona. He died of cirrhosis in 1934 and is buried in Artesia, California. The monument near El Tovar, placed there without permission from NPS, honors his memory and contributions to the hotel.

El Tovar Hotel, which now has 78 rooms, remains one of the grandest lodges in NPS. It is open to the public and includes a gift shop, dining room, and lounge. The Fred Harvey Company, which is now Xanterra Parks & Resorts, has managed El Tovar (and many other South Rim facilities) since it opened. El Tovar Hotel remains the most expensive hotel at the South Rim. Whereas in 1940 a room cost $7 and $65 in 1977, prices for a suite at El Tovar in 2017 started at $401 per night and at $217 per night for a standard room.

See also: Grand Canyon Village

Further Reading

Harvey, Fred. 1905. *El Tovar: A New Hotel at Grand Canyon*. Chicago: Santa Fe and W. H. Simpson.

EVOLUTION

Although Grand Canyon has become an icon of religious fundamentalism in the past few decades, several of its sites honor pioneers of evolution. Many of these relatively obscure sites (all near Bass Canyon) were named by British journalist, preacher, and publisher George Wharton James (1858–1923) in the late 1800s while James worked as a guide for William Wallace Bass. In 1902, USGS topographer François Emile Matthes (1874–1948) completed the evolution-related set of landmarks when he named Evolution Amphitheatre:

Evolution Amphitheater near River Mile 110.5 is at the western base of Bass Canyon between Huxley Terrace and Spencer Terrace.

Darwin Plateau, near River Mile 107.5 and west of Ruby Rapids, is named for Charles Robert Darwin (1809–1882), the codiscoverer of evolution by natural selection. Darwin Plateau, which can be accessed along the Bass Trail, has three "fingers" that jut north: Spencer Terrace to the east, Huxley Terrace in the middle, and Grand Scenic Divide to the west. Darwin, like most of his British supporters (e.g., Wallace, Huxley), never visited Grand Canyon.

Huxley Terrace, near River Mile 107.5, is named for Thomas Henry Huxley (1825–1895), a prominent defender of Charles Darwin's theory of evolution by natural selection. Huxley Terrace, which was originally called Observation Plateau, is between Darwin Plateau and Wallace Butte.

Wallace Butte is a 5,204-foot-high formation near River Mile 107.5 that honors British naturalist and explorer Alfred Russel Wallace (1823–1913), the codiscoverer (with Charles Darwin) of evolution by natural selection. Across the Colorado River from this area is Holy Grail Temple, which is named for the Last Supper chalice.

Lyell Butte is a 5,368-foot-high formation near River Mile 81.5 that honors Charles Lyell (1797–1875), a British lawyer and the foremost geologist of his era. Lyell, who hoped to "free the science from Moses" because he believed biblical literalists were "wholly destitute of geological knowledge," documented Earth's vast age. Lyell's *Principles of Geology* (1830) profoundly influenced Darwin and helped establish uniformitarianism, which is summarized by Lyell's phrase "the present is the key to the past"—that is, that Earth's geological features have developed over long periods through a variety of slow geologic processes "now in operation," including common events such as rain, volcanic activity, wind, and earthquakes. Lyell initially resisted Charles Darwin's conclusions about evolution but later accepted them when he realized how it answered questions such as how invasive species could often outcompete indigenous species (this was a problem if species had arisen perfectly adapted to local conditions).

Cope Butte, which is near River Mile 93.5 below Pima Point and just north of Cathedral Stairs along Hermit Trail, is named for American paleontologist Edward Drinker Cope (1840–1897). The deeply religious Cope, an expert on animals of the Cretaceous and Tertiary strata, studied the paleontology of Grand Canyon in the 1870s. During Cope's bitter 20-year feud with Othniel Marsh (The Bone War), Cope discovered, described, and named more than 1,000 species of vertebrates that he believed would vindicate his belief in evolutionary "progress" and perfectability. During their "war," Grand Canyon pioneer and USGS director John Wesley Powell sided with Marsh and often blocked Cope's publications.

Dana Butte, which is a 4,964-foot-high formation near River Mile 90.5 directly north of Powell Memorial, honors prominent Yale University geologist James Wright Dana (1813–1895). Dana reconciled his religious beliefs with science by championing evolution and an old Earth. Dana Butte, which is below Hopi Point near Indian Garden, Battleship, and Cheops Pyramid, can be seen from Maricopa Point and is upstream from Granite Rapids. Dana Butte is made mostly of Redwall Limestone.

Spencer Terrace (originally called Mystic Springs Plateau), near River Mile 107.5 honors Herbert Spencer (1820–1903), a British philosopher who studied evolution. Spencer, a Social Darwinist, was the first to use the phrase "survival of the fittest."

Marsh Butte, which is west of Hermits Rest and near River Mile 96.5, honors American paleontologist Othniel Charles Marsh (1831–1899). During John Wesley Powell's tenure as director of USGS, Marsh guided vertebrate paleontology for the USGS from 1882 to 1892. Marsh, who described and named more than 500 species of vertebrates, convinced many people of the validity of Charles Darwin's ideas about evolution.

Tyndall Dome near River Mile 107.5 honors John Tyndall (1820–1893), a British physicist who was an outspoken advocate of evolution and opponent of religious orthodoxy. At the annual meeting of the British Association for the Advancement of Science in Belfast, Ireland, in 1874, Tyndall created controversy by claiming that the church's long monopoly on issues related to human origins would soon be replaced by those based on evolution.

Shaler Plateau, near River Mile 102, honors American geologist Nathaniel Southgate Shaler (1841–1906), who was among the first scientists in the United

States to promote Charles Darwin's ideas about evolution. Shaler Plateau is below Hualapai Point and just downstream from Turquoise Canyon.

Fiske Butte is a 4,692-foot-high formation near River Mile 107.5 that honors John Fiske (1842–1901), an American philosopher who promoted the ideas of Charles Darwin and Thomas Henry Huxley while claiming "there has never been any conflict between religion and science."

There are also more subtle recognitions of evolutionary biologists in Grand Canyon. For example, Charles Doolittle Walcott (1850–1927), who succeeded John Wesley Powell as director of USGS from 1894 to 1907, discovered Canada's fossil-laden Burgess Shale. (Before he found the Burgess Shale in Canada, Walcott saw it in Grand Canyon. Walcott's discovery of stromatolites was the first discovery of preserved Precambrian cells.) At Grand Canyon, Walcott helped supervise construction of Nankoweap Trail at the eastern edge of Grand Canyon; he also studied Grand Canyon rocks in 1882–1883 and named the Chuar Group (1883) and Temple Butte Formation (1889). At Grand Canyon, Walcott is honored by the Walcott Member Formation, which is one of three parts of the Kwagunt Formation.

See also: James, George Wharton

Further Reading

Larson, Edward L. 2006. *Evolution: The Remarkable History of a Scientific Theory.* New York: Modern Library.

F

FIRE TOWERS

Near the beginning of the 20th century, it was common to install fire towers in large forests to detect and stop forest fires. The first fire tower in Arizona was installed in 1909 at Grand Canyon near Hopi Point. Towers were built later at Bright Angel Point (1919), Grandview Point (1919), and Signal Point (1929). In 1927, the wooden tower at Hopi Point was replaced by a 40-foot-tall steel tower, and in 1953 the top of that tower was removed to lower the tower's height to 24 feet. The towers were effective. Although several fires are started each year in GCNP by the more than 25,000 lightning strikes, few of these fires burn more than 100 acres.

The 75-foot-high tower originally at Bright Angel Point was moved in 1933 by the CCC to the entrance of the North Rim. That tower is where environmentalist and writer Edward Abbey worked for four seasons as a lookout; the tedious work there inspired him to write his novel titled *Black Sun* (1971).

By the 1950s, some of the fire towers in GCNP were used to monitor air quality, as part of the Public Health Service National Air Sampling Network. The first tower to do this was at Hopi Point, which in 1957 was equipped with a device to measure benzol-soluble and total particulate matter.

NPS stopped staffing the fire towers in GCNP in the late 1970s, when it began monitoring fires aerially. In 1978, while Smokey Bear implored visitors to Kaibab National Forest to prevent forest fires, GCNP began prescribed burns, which recognized fires as a natural part of the ecological health of a forest. In 2009, the four historic fire towers in GCNP were listed on the National Historic Lookout Register.

By 1936, GCNP and Kaibab National Forest were using radios to fight fires. Today, most fires in GCNP are discovered and monitored by satellite imagery, airplanes, helicopters, and drones.

See also: Hopi Point

Further Reading

Abbey, Edward. 1971. *Black Sun*. New York: Simon & Schuster.

Pyne, Stephen. 1995. *Fire on the Rim: A Firefighter's Season at the Grand Canyon*. Seattle: University of Washington Press.

FISH

The Colorado River in Grand Canyon was once teeming with fish. Indeed, in 1892, Grand Canyon prospector Ben Beamer described 20-inch-long humpback chub and how he could "lean over water's edge and pull them out [of the Colorado River] by the tail two at a time." (National Park Service, 2017b). These fish were important sources of food for Native Americans and early explorers who lived in and ventured through Grand Canyon. Today, few species of native fish are left in Grand Canyon, and some are gone.

Eight species of fish were native to Grand Canyon, the status of which now ranges from abundant to extirpated:

Humpback chub (*Gila cypha*)—This almost scaleless fish, whose striking appearance results from a large hump behind its head, is the only chub still living in GCNP. The largest population of these fish resides near the confluence of the Little Colorado and Colorado Rivers. Humpback chub, which was the last species of native fish in the Colorado River to be described by scientists (in 1946), was on the first federal list of endangered species issued in 1967.

Razorback sucker (*Xyrauchen texanus*, also known as the humpback sucker)—These fish, which can grow to more than 3 feet long, can live more than 35 years and are now rare in GCNP. The fish were listed as critically endangered in 1991. In 2014, a population of razorback suckers was found upriver from Lake Mead in GCNP.

Bluehead sucker (*Catostomus discobolus*)—This species, which is usually less than 20 inches long, is now restricted to populations in the Little Colorado River, Havasu Creek, Shinumo Creek, and Bright Angel Creek.

Flannelmouth sucker (*Catostomus latipinnis*)—This species, which is characterized by large lower lips, is a native fish that grows up to 25 inches long and weighs more than 5 pounds. Flannelmouth sucker, which is the largest species of native fish still relatively abundant in Grand Canyon, is the only endemic species of fish native to Grand Canyon that is not endangered.

Speckled dace (*Rhinichthys osculus*, also known as spotted dace)—This species is common in Grand Canyon; it is the most abundant native fish remaining in GCNP.

Colorado pikeminnow (*Ptychocheilus lucius*)—This fish, which was once the top predator of the Colorado River, grew up to 6 feet long and weighed more than 100 pounds. Once called the white salmon of the Colorado River, Colorado pikeminnow were listed as endangered in 1967. Today, natural populations of this fish are gone from Grand Canyon.

Roundtail chub (*Gila robusta*)—This sport fish is a candidate for the federal Endangered Species List. It is no longer found in GCNP.

Bonytail chub (*Gila elegans*, also known as bonytail)—These fish were once abundant but are now endangered and are among the rarest of endemic big-river fishes of the Colorado River. Virtually no juveniles have been found in recent years, leading many to assume that the species is functionally extinct.

Of these species, five remain in the park (humpback chub, razorback sucker, bluehead sucker, flannelmouth sucker, speckled dace) and four are endangered (humpback chub, razorback sucker, bonytail chub, and Colorado pikeminnow). Three species—that is, Colorado pikeminnow, roundtail chub, and bonytail—have been extirpated. In 2014, the Zuni bluehead sucker (*Catostomus discobolus yarrow*)—a fish that lives only in the Little Colorado River watershed—was declared endangered by the U.S. Fish and Wildlife Service. The fish was first proposed as a candidate for the endangered species designation in 1985.

A variety of factors have contributed to the demise of most of Grand Canyon's native fish:

Habitat alteration and destruction by dams. Hoover Dam, which flooded the lower Grand Canyon, blocked the migration of the Colorado pikeminnow. The closing of Glen Canyon Dam in 1963 had an even greater impact by drastically clearing and lowering the temperature of the Colorado River's water in Grand Canyon, thereby producing conditions in which many native fish could not spawn. Water released from Lake Powell is near 49°F, and the water warms only a few degrees by the time it reaches Lake Mead. These cold temperatures are detrimental to several native fish; for example, water must be at least 61°F for humpback chub to spawn.

Competition and predation. There are now at least 13 species of exotic (i.e., non-native) fish in GCNP, the first of which was channel catfish (*Ictalurus punctatus*), which was introduced in the Colorado River in 1919. By 1923, GCNP officials had put trout in tributaries of GCNP and in 1931–1932 stocked several tributaries (e.g., Clear Creek) with tens of thousands of trout eggs, along with shrimplike amphipods (e.g., *Gammarus lacustris*) that the trout could eat. Here are some of the most prominent invasive species of fish now living in Grand Canyon:

Rainbow trout (*Oncorhynchus mykiss*) are often released from a fishery between Glen Canyon Dam and Lees Ferry. Rainbow trout are found throughout Grand Canyon and are especially abundant in Upper Grand Canyon.

Common carp (*Cyprinus carpio*), which were introduced in rivers of the United States by the U.S. Fish Commission in 1872, are now the most abundant and widespread non-native fish in North America. Carp are common throughout Grand Canyon.

Mosquitofish (*Gambusia affinis*), which were introduced in the 1930s, are the only fish in Grand Canyon that bear live young. All other species of fish in Grand Canyon are oviparous, meaning that their eggs are fertilized after leaving the females' bodies.

Striped bass (*Morone saxatilis*) were introduced in the 1970s and are now a prized trophy fish. However, striped bass often eat juvenile bonytails.

Red shiners (*Cyprinella lutrensis*) are rare in Grand Canyon but common beyond Separation Canyon.

Fathead minnows (*Pimephales promelas*) seldom grow beyond 4 inches long but eat juveniles of some native fish.

Brown trout (*Salmo trutta*) are voracious predators common in parts of Grand Canyon.

Attempts to remove exotic species in Grand Canyon have had mixed results. For example, efforts to reintroduce boneytails into GCNP have been slowed by concerns about spreading quagga mussel, an invasive species.

Fishing is allowed in GCNP, but you must have a license if you are 14 years old or older. The best fishing is usually (1) in the dry season, when the water is clear, and (2) between Lees Ferry and the confluence of the Little Colorado and Colorado Rivers.

See also: Thematic Essays section: The Future of Grand Canyon; Lake Powell and Glen Canyon Dam

Further Reading

National Park Service. 2017a. Fish. Retrieved July 7, 2017, at https://www.nps.gov/grca/learn/nature/fish.htm

National Park Service. 2017b. Humpback chub. Retrieved July 7, 2017, at https://www.nps.gov/grca/learn/nature/fish-humpback-chub.htm

FLAVELL, GEORGE

On August 27, 1896, trapper, hunter, and tattoo artist George F. Flavell (who also called himself George Clark) and his partner/passenger Ramón Montéz (some sources give the name Montos) set out from Green River, Wyoming, in a homemade, 15.5-foot-long, flat-bottomed boat named *Panthon*. Unlike earlier explorers, who were surveying the river or doing science-related work, Flavell and Montéz were trappers intent on gathering pelts to sell. Flavell rowed bow-first, facing downstream. Their little publicized expedition was done, "First, for the adventure; second, to see what so few people have seen; third, to hunt and trap; fourth, to examine the perpendicular walls of rocks for gold" (Historical Marker Database, 2017).

Flavell and Montéz eventually went through Grand Canyon, running all of its rapids except Soap Creek. As Flavell noted:

> I feel confident we will get through. We must expect some accidents and expect to hit some rocks. There is only one stone we must not hit. That we must miss at all hazard: our Tomb Stone!" (Arizona State University, 2017).

Montéz was the first Mexican and the 15th person through Grand Canyon.

Although it received little publicity, the Flavell–Montéz trip was only the third successful downriver run through Grand Canyon (and the first in a dory). The expedition included the first successful runs of Lava Falls, Hance, and several other rapids, which had been portaged by earlier river runners. Flavell and Montéz reached Needles, California, in 65 days. After the trip, Flavell's vivid account of the trip was published as *Log of the Panthon* (1987), which dispelled the exaggerated claims by John Wesley Powell that Grand Canyon and the Colorado River were extremely dangerous. North Rim entrepreneur David Rust was one of the few witnesses of Flavell and Montéz's run through Grand Canyon. During the trip, Flavell faced the bow and looked downstream; this popularized the rowing style that is used in Grand Canyon today. Neither Flavell nor Montéz wore a life jacket.

See also: Thematic Essays section: John Wesley Powell; Rapids

Further Reading

Arizona State University. 2017. Lava Falls. Retrieved July 1, 2017, at http://grandcanyonhistory.clas.asu.edu/sites_coloradorivercorridor_lavafalls.html

Flavell, George F., Carmony, Neil B., and Brown, David E. 1987. *The Log of the Panthon: An Account of an 1896 River Voyage from Green River, Wyoming to Yuma, Arizona through the Grand Canyon*. Boulder, CO: Pruett Publishers.

Historical Marker Database. 2017. Green River, Wyoming Historical Markers. Retrieved July 1, 2017, at https://www.hmdb.org/results.asp?Town=Green%20River&State=Wyoming

FRAZIER, RUSSELL

Russell G. "Doc" Frazier (1893–1968) was a physician and northern Utah river runner who, in 1934, financed a trip to place a plaque at Separation Canyon (River Mile 239.8) commemorating the three men who left John Wesley Powell's 1869 expedition there through Grand Canyon. Frazier, who was a member (with seven others) of the 1934 Dusty Dozen expedition (completed by Frazier and only six others), was accompanied by Willis Johnson, Julius F. Stone, and Stone's son. The plaque read as follows: "In Memory of Seneca Howland, O. G. Howland, Wm. H. Dunn, who left the first Powell party at this point Aug. 28, 1869 & were killed by Indians on the Shivwits Plateau. Erected by Colorado River Club 1934." After that plaque was submerged in Lake Mead in 1935, Frazier helped finance another trip—this time in 1939 across Lake Mead with 80-year-old Stone and others—to mount a new plaque at Separation Canyon. In 1943, that plaque was moved to its current location higher on the cliff on river right. This plaque, one of the most famous in GCNP, reads as follows:

> Here on August 28, 1869, Seneca Howland, O.G. Howland and William H. Dunn separated from the original Powell party, climbed to the North Rim and were killed by the Indians. For further authentic information, see 'Colorado River Controversies' Obtainable from University Libraries. This Cenotaph was placed and dedicated in 1939 by later Colorado River Voyagers.

Frazier became famous after being selected from 1,100 applicants to be the chief physician and surgeon on Admiral Richard E. Byrd's third expedition to Antarctica in 1939–1941. Frazier won a Congressional Medal for his service in 1947. He died of cancer at the age of 74 on January 14, 1968, in Salt Lake City, Utah.

See also: Thematic Essays section: John Wesley Powell

Further Reading

Utah History. 2017. Russell G. Frazier. Retrieved March 2, 2017, at http://historytogo.utah.gov/people/utahns_of_achievement/russellgfrazier.html

G

GALLOWAY, NATHAN

Nathan "Than" Galloway (1853–1913) was born in Lehi, Utah, and was raised on a farm. He later became a trapper and today is considered the father of modern whitewater rowing techniques. He was among the first river runners to popularize the stern-first method of running rapids; instead of pulling downstream with his back to the rapids (as oarsmen of the Powell and Stanton expeditions had done), Galloway faced downstream and pulled upstream to slow his momentum, thus enabling him to dodge upcoming obstacles. In 1898, Galloway passed along his new way of boating to his friend David Rust, who in 1907 passed it to Ellsworth Kolb. The system became known as drifting; Norm Nevills called it "facing your danger." Galloway, who is often considered to be Grand Canyon's first great boatman, was also the first person to traverse the entire Grand Canyon twice.

Realizing that no boat could be stronger than a rock, Galloway designed lighter boats that he could maneuver to avoid crashes with rocks. Galloway modeled his boats after the shapes of ducks and geese (which face downstream to ride riffles), for he wanted his boats to function like a duck floating on the water. Instead of using the heavy, deep-keeled boats of John Wesley Powell, Galloway built upturned, narrow, lightweight, flat-bottomed boats that he could pivot easily for greater safety in Grand Canyon's rapids. Galloway boats dominated river running in Grand Canyon for almost four decades before new materials (e.g., neoprene) produced new boats that were cheaper, safer, easier to store, easier to transport, and could carry more people than wooden boats.

Galloway spent a few months dredging for gold with Robert Stanton of the Hoskinnini Company in Glen Canyon. There, he met the company's president, mining financier Julius Stone. Galloway left Hoskinnini by 1900 and in 1909 was the boatman for Stone's 10-week trip from Green River, Wyoming, through Grand Canyon, to Needles, California. This trip, in which Stone and Galloway were joined by 17-year-old boatman Seymour Dubendorff, photographer Raymond Cogswell, and Charles Sharp, used four Galloway-style boats whose construction (of white pine and elm) in Detroit was supervised by Galloway. This was the first pleasure trip through Grand Canyon. The boat rowed by Stone, which was unnamed and referred to as "the Stone boat," is the oldest extant boat to have run the Colorado River through Grand Canyon and is the centerpiece of GCNP's collection of historic boats. Galloway believed that many people would attempt to navigate the Grand Canyon, but few would survive the whole trip. Stone described his expedition trip in his *Canyon Country: The Romance of a Drop of Water and a Grain of Sand* (1932). Galloway was a widely admired and sought-after expert on running Grand Canyon. For example, when the Kolb brothers were planning their 1911–1912 expedition through Grand Canyon, they sought boat-related advice from Galloway.

On December 22, 1913, Galloway fell across a bar on his wagon, crushed his throat, and suffocated. He is buried in Vernal Memorial Park in Vernal, Utah. Galloway is commemorated by Galloway Canyon near River Mile 134. At Inscription Cave near Elves Chasm (River Mile 117.2), Galloway etched his name and date in an overhang; this faint inscription is one of the oldest known river inscriptions in Grand Canyon.

See also: Kolb, Ellsworth and Emery; Stanton, Robert; Stone, Julius

Further Reading

Arizona State University. 2017. Running the river. Retrieved June 2, 2017, at http://grandcanyonhistory.clas.asu.edu/history_coloradoriver_runningtheriver.html

Stone, Julius F. 1932. *Canyon Country: The Romance of a Drop of Water and a Grain of Sand.* New York: G. P. Putnam's Sons.

Sutton, Brook. 2015. Whitewater pioneer Nathaniel Galloway. *Adventure Journal*, June 18, 2015. Retrieved July 8, 2017, at https://www.adventure-journal.com/2015/06/historical-badass-whitewater-pioneer-nathaniel-galloway/

GANNETT, HENRY

Henry Gannett (1846–1914) was an American explorer and geographer who was a pioneer at making topographic maps. By 1879, Gannett was lobbying for centralizing the government's mapmaking; the new organization for which he lobbied became the USGS. Gannett's work with USGS is considered the beginning of true topographic work in the United States. At USGS in 1880, Gannett's friend John Wesley Powell promoted Gannett to the position of chief geographer, where Gannett served until his death. During this time, Gannett's dedication to resolve problems caused by duplicate, confusing names led to the formation of the U.S. Board on Geographic Names in 1890, which he chaired until 1910. In 1888, Gannett was a founding member of the National Geographic Society, for which he served as president from 1910 to 1914. Examples of sites in Grand Canyon named by Gannet include the following:

Buddha Temple is a 7,218-feet peak named for the Buddha (563–483 BCE), who renounced material things and attained enlightenment by meditating under a tree near a desert river. The formation resembles a Buddhist temple.

Deva Temple is a 7,339-foot peak southeast of Cottonwood Camp that Gannet named after Durga, the wife of Shiva in Hindu mythology.

Gannett was a founding member of Washington, DC's Cosmos Club, which was founded in 1878. He is honored by Wyoming's 13,804-foot-high Gannett Peak and by the Henry Gannett Award, which is sponsored by USGS.

See also: Powell, John Wesley

Further Reading

Jenkins, Mark Collins. 2012. Henry Gannett: National Geographic founder and "The Man Whose Brain Conceived It." *Explorers Journal*, May 22, 2012. Retrieved July 13, 2017, at http://voices.nationalgeographic.com/2012/05/22/henry-gannett-the-man-whose-brain-conceived-it/

GARCÉS, FRANCISCO TOMÁS

In 1776, Franciscan friar Francisco Tomás Garcés's journal describing his solo expedition from the lower Colorado River to the Hopi pueblos became the earliest written record of travel immediately south of Grand Canyon. Garcés, who was exploring the Colorado River on his fifth missionary trip with Indian guides, was

Naming the Colorado River

The Colorado River has been christened with many names, most of which have referred to the river's red, sediment-laden water. In 1540, Hernando de Alarcón named the river *El Rio de Buena Guía* ("River of Good Guidance" or "Good Guide River"). (*Buena Guia* was the motto in the coat of arms of Viceroy Mendoza, who sent Alarcón with supplies to Francisco Vásquez de Coronado.) In 1605, Spanish conquistador and New Mexico governor Don Juan de Oñate—in his last major expedition in America—named the river *Rio Grande de Buena Esperanza* ("Grand River of Good Hope.") (Oñate was also the first to use the name *Rio Colorado*, but he was referring to the Little Colorado River.) Near the same time, Spanish explorer Melchior Diaz—after seeing Yuma Indians with torches near the mouth of the Colorado River—named the river *Rio Tizon* ("Firebrand River"). In 1774, Spanish explorer and missionary priest Francisco Tomás Garcés became the first writer to consistently refer to the river as *Rio Colorado*. Francisco Silvestre Vélez called the river *Rio Mysterioso* ("River of Mystery"), and in the 1830s, fur trapper James Ohio Pattie called the river Red River. Native Americans also had their own names for the Colorado River; for example,

Navajo: *Rio Poketto* ("Red River") and *Bits'iis Ninéez* ("River of Never Ending Life")
Paiute: *Pahaweap* ("Water Deep in the Earth," "A Long Way to the Water") or *Pahgaiv* ("Mighty River")
Pima: *Buqui Aquimuti* ("Red River")
Hopi: *Pisisvaiyu* or *Pisishbaiyu* ("Where Water Flows")
Coco-Maricopa: *Gritetho* ("Grand or Great River")
Yuma: *Javill* or *Hahweal* ("Red River")
Havasupai: *Ha Tay G'am I*, *Sil Gsvgov*, and *Chicamimi Hakatai* ("Large Canyon with Loud Sound")

Despite these many earlier names, Grand Canyon explorer John Wesley Powell usually gets credit for naming the river the Colorado River, primarily because of his stature and consistent use of the name.

searching for new patrons for his church and his king. Like Cárdenas before him, Garcés saw nothing special about Grand Canyon except its ruggedness, noting that he was "astonished at the roughness of this country, and at the barrier which nature had fixed therein" (Pyne, 1999). The intrepid Garcés, the first European since 1540 to see Grand Canyon, and his Hopi guides spent a night just east of today's Grand Canyon Village.

Garcés was the first person to record seeing the Colorado River in the canyon area and to consistently refer to Grand Canyon's river as the Rio Colorado ("colored river," a reference to the red sediment that was abundant in the water before construction of Glen Canyon Dam). Garcés was the first non-Native person to visit the Havasupai, whom he failed to convert to Christianity. On July 4, 1776, Garcés was expelled by the Hopi Indians, and in 1781 he was killed by Yuma Indians along the lower Colorado River.

Garcés was the first known non-Native to name Grand Canyon; he named it *Puerto de Bucareli*, but this name did not become common. Garcés's group, who were declared martyrs by the Catholic Church, were the last Spanish explorers to visit Grand Canyon; American explorers would not see the canyon for another 80 years.

See also: Supai and the Waterfalls of the Havasupai

Further Reading

Pyne, Stephen J. 1999. *How the Grand Canyon Became Grand: A Short History*. New York: Penguin.

GATEWAYS TO GRAND CANYON

Most entrances to Grand Canyon and GCNP are near towns whose economies are based primarily on tourism generated by the Canyon. Here are the most familiar of these "gateways" to Grand Canyon:

Cameron, Arizona (population 885), which was originally called Tanner's Crossing, is at the intersection of U.S. Highway 89 and Arizona State Route 64, just south of the Little Colorado River on the Navajo Reservation. Cameron, which offers cabins, gas stations, an RV park, an art gallery, gift shops, and the famous Cameron Trading Post and accompanying lodge (in business for more than 100 years), is 32 miles from GCNP's South Rim East Entrance. Prices for rooms at the lodge start at $79/$119, depending on the season. Ancient Hopi mesas and Navajo lands (including Painted Desert, Canyon de Chelly, and Monument Valley) are easily accessed from Cameron. From the lodge, visitors can watch sunset over the Little Colorado Gorge.

Cameron was founded in 1911 after Ralph Cameron obtained federal money to build a bridge there across the Little Colorado River. The Cameron Suspension Bridge near Cameron Trading Post and lodge carries U.S. Highway 89 over the Little Colorado River. The road from Cameron (elevation 4,370 feet) to Desert View (7,434 feet) was completed in 1935. Cameron is the gateway for people visiting the South Rim from the east.

Fredonia, Arizona (population 1,314), the northernmost town in Arizona, is the town closest to Toroweap. Fredonia, which bills itself as "The Gateway to the Canyons," covers 7.4 square miles and is at an elevation of 4,671 feet. The town, which is on the Arizona Strip, was founded in 1885 by five men hoping to build a dam to enable irrigation of farmland.

Green River, Wyoming (population 12,515) was the gateway to Grand Canyon for many of the early river runners, including John Wesley Powell (in Powell's days, the town was called Green River Station). Because Green River was a stop on the Transcontinental Railroad, Powell and others could ship their boats and supplies there and travel on the Green River (a chief tributary of the Colorado River) to Grand Canyon. Powell had arrived in Green River on May 11, 1869, to prepare for his expedition, one day after the Golden Spike had united the Transcontinental Railroad at Promontory Summit in Utah Territory. Green River's Island Park (the northwest end of Expedition Island), which is a National Historic Landmark, displays historical markers honoring a variety of Grand Canyon icons, including Powell, Buzz Holmstrom, Norman Nevills, George Flavell, Julius Stone, the Kolb brothers, and "The Great Unknown."

Jacob Lake, Arizona (unincorporated), which is 44 miles from the North Rim, is at the intersection of U.S. Highway 89A and Arizona State Route 67. Arizona State Route 67 (Kaibab Plateau–North Rim Parkway, a National Scenic Byway), which began as a two-lane road cut through the Kaibab National Forest in the 1920s, is the only paved road to the North Rim. Jacob Lake includes a lodge, restaurant, campground, gas station, and a U.S. Forest Service visitor center. Services at Jacob Lake are open year-round, but Arizona State Route 67 closes soon after the first large snowfall (usually November). Between Jacob Lake and the North Rim entrance station is beautiful De Motte Park, an open meadow named by John Wesley Powell for Harvey De Motte, a mathematics professor at Wesleyan University who accompanied Powell when he surveyed the North Rim. Some local residents called the area V. T. Park and V. T. Tourist Park for Mormon ranchers Van Slack and Thompson.

Jacob Lake (elevation 7,925 feet) is named for Jacob Hamblin ("The Buckskin Apostle"), who explored the area from the 1850s to the 1870s. In 1862, Hamblin made the first recorded trip around Grand Canyon. The U.S. Forestry Service station at Jacob Lake features a plaque honoring Theodore Roosevelt.

Kingman, Arizona (population 28,068), which advertises itself as "The Heart of Route 66," is the town closest to the Skywalk at Grand Canyon West. Kingman is 70 miles from the Skywalk.

Marble Canyon, Arizona (unincorporated), which is 12 miles southwest of Page, Arizona, is a 10-minute drive from Lees Ferry, which is designated as River Mile 0 in GCNP. Marble Canyon is on U.S. Highway 89A and is a 3-minute walk from the two Navajo Bridges that cross the Colorado River near River Mile 4. At Marble Canyon you'll find Marble Canyon Lodge (built by David "Buck" Lowrey in 1927 and rebuilt following a fire in 2013) and an airport, restaurant, post office, gas station, laundromat, general store, and gift shop. Other lodges just down the road include Lees Ferry Lodge at Vermilion Cliffs (opened by bootlegger Maggie

Johnson) and Cliff Dwellers Lodge, built in 1949 by Art Greene. A sign at the Marble Canyon airport notifies pilots that the landing fee is $5.

Peach Springs, Arizona (population 1,090), is the headquarters of the Hualapai Nation and the town nearest Hualapai Hilltop, the trailhead to Supai. At Peach Springs, which is 50 miles east of Kingman, you can sign up for Grand Canyon's only one-day rafting trips.

Tusayan, Arizona (population 558)—the busiest and most developed gateway to Grand Canyon—is on Arizona State Route 64 7 miles south of the South Rim and 2 miles south the South Rim's entrance station. Here visitors find hotels, stores, campgrounds, RV parks, restaurants, a gas station, hiking trails, and the National Geographic Visitor Center. Grand Canyon Airport at the southern edge of Tusayan was dedicated in 1967 and today is the fourth-busiest airport in Arizona. Just north of Tusayan at the boundary of Kaibab National Forest and GCNP is an often photographed statue of Smokey Bear, a popular advertising mascot who has been imploring Americans to prevent forest fires since 1944.

The oldest hotel in Tusayan is the Red Feather Lodge, which was opened in 1964 by Tusayan pioneer Robert P. "R.P." Thurston (1893–1979). Just south of Red Feather Lodge is the 318-room Best Western Premier Grand Canyon Squire Inn, which in 2017 *TripAdvisor* named the most booked hotel in the world.

Tusayan (elevation 6,540 feet) is a phonetic version of the Hopi word meaning "country of isolated buttes." You can hike from Tusayan to the South Rim via the Arizona Trail, which then traverses Grand Canyon via the South Kaibab Trail to the North Rim. Tusayan, which covers only 144 acres (0.225 square mile), is the smallest town (by area) in Arizona. Nevertheless, more than 2 million cars carrying more than 4 million tourists pass through Tusayan every year. Services in Tusayan are open year-round, but snow and ice sometimes close roads.

Williams, Arizona (population 3,023), which advertises itself as "The Gateway to the Grand Canyon," is the closest town to Grand Canyon on an interstate highway (in this case, Interstate 40). On October 13, 1984, Williams—which is named for Old Bill Williams, whose statue stands at the west end of town—became the last town along Historic Route 66 to be bypassed by an interstate highway. In Williams, visitors can board the Grand Canyon Railway for the 64-mile trip to the South Rim. Prices of a round-trip ticket start at $65.

See also: Cameron, Ralph

Further Reading

Historical Marker Database. 2017. Green River, Wyoming Historical Markers. Retrieved July 1, 2017, at https://www.hmdb.org/results.asp?Town=Green%20River&State=Wyoming

GEOLOGY ALONG THE RIVER

River runners in Grand Canyon "go back in time" as the Colorado River gradually cuts through progressively older geologic formations and reveals almost 2 billion years of Earth's geologic history. Early prospectors had a variety of names for these geological features. For example, the top layers of each rim (i.e., today's Kaibab Limestone, Coconino Sandstone, and Hermit and Supai formations) were called Limestone, Gray Sandstone, and Red Sandstone, respectively. Since those days, geologists have studied and standardized the names of Grand Canyon's geological formations.

Listed here, in their order of appearance on a trip downstream, are those rock formations and the River Miles (i.e., the distance, in miles, from Lees Ferry) at which they first appear along the river. Geologic descriptions of these formations are provided in the Thematic Essay, The "Postcard Rocks" of Grand Canyon: The Paleozoic Era.

River Mile	Rock Formation
0.0	Soft Shinarump Conglomerate and Moenkopi Formation (Downstream you'll see the Vermilion Cliffs.)
1.0	Kaibab Limestone
1.9	Toroweap Formation
4.3	Coconino Sandstone (just upstream from Navajo Bridges)
5.1	Hermit Formation
11.4	Esplanade Sandstone (This begins the Supai Group.)
15.2	Wescogame Formation
18.2	Manakacha Formation
20.4	Watahomigi Formation
23	Redwall Limestone (Early prospectors called this layer Blue Lime.)
24.2	Surprise Canyon Formation
36.8	Muav Limestone
38	Temple Butte Formation
49.4	Bright Angel Shale (Early prospectors called this layer Serpentine Shale.)
58.8	Tapeats Sandstone (Early prospectors called this layer Tonto Sandstone.)
61.5	The view downstream from here toward Chuar Butte features the full Paleozoic sequence of rocks.
63.4	Dox Sandstone (This begins the Unkar Group.)
63.4	Great Unconformity
63.9	Here is a great view of the Grand Canyon Supergroup. This formation extends to River Mile 78 and then again from River Miles 131–139.
64.1	Here, 3,580 feet below the rim, is another excellent view of the entire Paleozoic sequence of rocks.

65.7	The Butte Fault crosses the Colorado River here. This fault raised land to the west 1,400 feet. Tapeats Sandstone was last seen at river level 2.5 miles upstream; because of the uplift here, Tapeats Sandstone will again appear at the river 50 river-miles downstream.
66.1	Cardenas Basalt
68.6	Nankoweap Formation above black lava
75.5	Shinumo Quartzite
76.4	The view downstream here (i.e., just before Hance Rapid) features another excellent view of the Grand Canyon Supergroup.
76.5	Hakatai Shale
77.7	Bass Formation
81	All formations of Grand Canyon rock have appeared, except for lava flows (which are most prominent near Miles 179, 181, and 184).
86.5	Zoroaster Temple, which towers 4,700 feet above the river, is prominent in the northeast.
90	A view upriver shows faulting and the granite dikes of Grand Canyon's Precambrian rocks, as well as small blocks of the younger Grand Canyon Supergroup.
97	Point Sublime dominates the skyline downstream here.
107.7	The 4,000-foot-high stack of Grand Canyon's Paleozoic rocks is prominent downstream here. These rocks have been sequentially exposed along the river down the canyon and are capped by Wheeler Point on the Powell Plateau.
120.6	Conquistador Aisle, which begins here, is the longest, straightest part of the Colorado River in GCNP.
127.0	Randy's Rock is a boulder of Tapeats Sandstone in the left part of the river.
135.5	Granite Narrows is the narrowest and deepest point of the Colorado River in Grand Canyon. Here, the river is 76 feet wide.
177.5	Toroweap Overlook is high above on river right.
185	Mount Trumbull is visible downstream.
189.9	Hurricane Fault crosses the river between here and River Mile 191.5.
266.8	The full set of Grand Canyon rocks is visible by looking at the North Rim.

See also: Rapids

Further Reading

Belknap, Buzz, et al. 2017. *Grand Canyon River Guide*. Evergreen, CO: Westwater Books.

GOLDWATER, BARRY

Barry Morris Goldwater (1909–1998) was an American businessman, a five-term U.S. Senator from Arizona, and the Republican Party nominee for president of the United States in 1964. (Goldwater lost that election in a landslide to incumbent Lyndon Baines Johnson). Goldwater, an avid river runner and amateur radio operator, announced the first radio program from the bottom of Grand Canyon in 1940 on a trip in which he was accompanied by Ellsworth Kolb. Robert L. Ripley of *Believe It or Not*, who pioneered "on location" broadcasts from strange places (including the first ship-to-shore broadcast and the first broadcast from Australia to America), was on the South Rim. During Goldwater's broadcast, Arizona governor Robert Taylor Jones described Grand Canyon as the greatest believe-it-or-not wonder of all time.

In 1940, Goldwater ran the Colorado River through Grand Canyon with Norman Nevills's second commercial river trip. Goldwater began giving talks about Grand Canyon and, after realizing he had a talent for public speaking, became a successful politician. Because of his position as U.S. senator and his interest in Grand Canyon, Goldwater was an influential voice in matters regarding GCNP. Goldwater was among the first 80 people to run the Colorado River through Grand Canyon, and he claimed he knew more about the Colorado River in Grand Canyon and the bottom of Grand Canyon than any living person. He fought hard for funds for Glen Canyon Dam but later admitted the dam was a mistake.

Barry Goldwater was a river runner in Grand Canyon National Park who became a successful politician. Goldwater, who was the Republican nominee for president in 1964, was among the first 80 people to run the Colorado River through Grand Canyon. Goldwater's popular talks about Grand Canyon revealed his talents as a public speaker, which later helped fuel his political career. (Library of Congress)

In 1970, the Arizona Historical Foundation republished Goldwater's journal (originally published in 1940 under a different title) from a Grand Canyon trip as the 209-page *Delightful Journey*. Goldwater, who had visited Hopi reservations as a young man, eventually amassed a world-class collection of 437 kachina dolls. Goldwater, who served 30 years in the U.S. Senate, died on May 29, 1998, at his home in Paradise Valley, Arizona. He is buried in Christ Church of the Ascension in Paradise Valley, Arizona.

See also: Kolb, Ellsworth and Emery

Further Reading

Goldwater, Barry Morris. 1970. *Delightful Journey: Down the Green and Colorado Rivers.* Tempe: Arizona Historical Foundation.

GRAND CANYON DEPOT

Most of the early visitors to Grand Canyon came to the South Rim by taking a train to Williams, Arizona, and then taking a stagecoach for the remaining 64 miles to the canyon. (A model of these stagecoaches is above the fireplace in El Tovar Hotel.) In the late 1800s, however, the Atchison, Topeka and Santa Fe Railway began laying track to the South Rim to serve the area's copper mines. This project failed, leaving the tracks 8 miles short of the rim. In 1901, the Grand Canyon Railway (a subsidiary of the Santa Fe) bought the project and finished building the rails to the South Rim. The final construction included a small, shack-like terminal near the rim.

The first steam-powered train (on a spur line of the Santa Fe Railway) arrived at the South Rim on September 17, 1901. This ended the pioneer era at Grand Canyon and dramatically increased tourism at the South Rim. This, in turn, spurred development of Grand Canyon Village. (The Navajo word for Grand Canyon Village is *Bidáá' Ha'azt'i'*, which means "railroad coming to the edge.") Indeed, for the next 50 years, the Santa Fe Railway built nearly all of Grand Canyon Village's hotels, restaurants, lookouts, curio shops, and public utilities. Whereas a ticket for the stagecoach ride from Williams to the South Rim had cost $20 and involved a jarring 10-hour journey, the relatively short, smooth ride aboard the Santa Fe Railway cost only $3.95. More and more people began coming to Grand Canyon. The original terminal was near the Bright Angel Hotel, which paid the railroad half of its profits in exchange for the railroad delivering people, water, and supplies to its door.

After the opening of El Tovar Hotel in 1905, the terminal was moved east, away from the Bright Angel Hotel and directly adjacent to the El Tovar Hotel. In 1909–1910, the railroad built a new depot. That depot, which was designed by architect Francis W. Wilson to complement El Tovar Hotel, was built of locally harvested ponderosa pine. When it opened, the copper letters on the building spelled "Grand Cañon," but six months later the Santa Fe Railway Company ordered all vestiges of Spanish language to be removed from company signage. In the 1960s, when crews tired of polishing the depot's copper letters, the letters were painted white, but this coating of white paint was removed in 1989 when the depot was restored. Grand Canyon Depot, the only log structure ever designed by Wilson, is one of only three remaining railroad depots built of logs serving active railroads in the United States. The Santa Fe Railway was also a patron of the arts, often hosting artists such as Thomas Moran, William Leigh, and Louis Akin at Grand Canyon

The arrival of the Santa Fe Railway at Grand Canyon in 1901 stimulated tourism and, in the process, transformed the economy and landscape of the South Rim. Grand Canyon Depot, shown here, was built in 1909. Today, the railroad carries more than 200,000 visitors per year to and from Grand Canyon from Williams, Arizona. In the background on the left is the El Tovar Hotel, which was owned by the railroad. (Randy Moore)

in exchange for paintings that were featured in advertisements, brochures, and menus across the country.

In 1954, Santa Fe Railway donated its buildings to NPS and sold its tourism-related property to the Fred Harvey Company. The railroad continued to bring passengers to the South Rim until 1968, when it could no longer compete with automobile traffic (the last trip by the Santa Fe Railway carried only three paying passengers). The railroad bowed out entirely from carrying passengers to Grand Canyon in 1972 but continued to carry freight until 1974. The depot was obtained by the NPS in 1982, and—following failed attempts in 1977, 1980, and 1984 to revive service—passengers again began arriving at the South Rim aboard Grand Canyon Railway in 1989. This occurred a year after the line was bought by Max and Thelma Biegert of Phoenix. The first run of the restored railroad was on September 17, 1989, exactly 88 years after the arrival of the original train at the South Rim.

Grand Canyon Railway Station, which looks much as it did when it was built in 1909, is the only rail station in NPS built primarily of logs. The 2.5-story wood structure (about 3,000 square feet) faces south (i.e., away from the canyon). Inside the first floor are four main rooms: a ticket office and waiting room, an agent's office, a storage room, and a baggage room. Upstairs is a two-bedroom apartment used by the original railroad agent. Much of the depot's original hardware remains, all of it stamped "GC." The depot also includes historic photos and information about the railway. The depot, which is a National Historic Landmark, is within 300 feet of El Tovar Hotel, which was also built by the railroad.

Today, the train to Grand Canyon is a nostalgic but convenient way to get to Grand Canyon from Williams, Arizona. The train carries more than 200,000 people per year to the South Rim, thereby eliminating approximately 70,000 vehicles from the park. The one daily trip to Grand Canyon leaves Williams at 9:30 a.m. and arrives at Grand Canyon near 11:45 a.m.; it departs at 3:30 and arrives in Williams at 5:45 p.m. Grand Canyon Railway, which in 2017 commissioned two luxury, dome-topped cars named in honor of Fred Harvey and Mary Colter, remains the only railroad line to a destination within a national park.

See also: Grand Canyon Village; Harvey, Fred and the Harvey Girls

Further Reading

Anderson, Michael. 2001. *Along the Rim: A Guide to Grand Canyon's South Rim from Hermit's Rest to Desert View.* Grand Canyon, AZ: Grand Canyon Association.

GRAND CANYON MUSEUM

GCNP includes Grand Canyon Museum Collection, a 6,000-foot research facility on the South Rim near the Albright Training Center. The facility, which opened in 1999, houses almost a million objects related to Grand Canyon, including archaeological specimens (e.g., pottery, baskets, rugs, split-twig figurines from Stantons Cave), mineral specimens, fossils, news articles, brochures, herbarium specimens, boats, and historical objects such as signs from the Kolb brothers' business and movies, Walter Clement Powell's diaries (from John Wesley Powell's second expedition through Grand Canyon), and John Wesley Powell's pocket watch. The museum also includes the pen used by President Woodrow Wilson to sign the legislation to create GCNP, but this pen is often displayed while on loan at Verkamps Visitor Center. Other information about Grand Canyon and John Wesley Powell is available in the John Wesley Powell Memorial Museum in Page, Arizona, and in the John Wesley Powell River History Museum in Green River, Utah.

Grand Canyon Museum also includes a collection of boats used in famous events in Grand Canyon's history. In some instances, the boats became as famous as the accomplishments that they made possible.

Since the early days of river running, famous boats have been displayed at Grand Canyon's South Rim. For example, in 1911, the South Rim's El Tovar Hotel displayed the boat *Utah* used during the 1908 Russell–Monett expedition through Grand Canyon; this was the first metal boat to go through Grand Canyon. During that same year, Nathan Galloway tried to convince North Rim pioneer David Rust to display the boat used in 1909 by Galloway and Julius Stone to go through Grand Canyon, but El Tovar—claiming that their porch could not reasonably accommodate another boat—refused the offer.

Some of Grand Canyon's most famous boats remain along the river; for example, the abandoned *Ross Wheeler* at River Mile 108.5, from the Russell expedition of 1915, is a popular stop for river runners, as are the remains at River Mile 41.6 of *Grand Canyon*, in which Colorado River legend Bert Loper died in 1949. The John Wesley Powell River History Museum in Green River, Utah, houses Ellsworth Kolb's *Defiance* (from the Kolb brothers' moviemaking expedition in 1911), an unnamed canoe used by North Rim pioneer David Rust in the 1920s, and *Mexican Hat* used by Norm Nevills and Don Harris in 1938.

In 1949, GCNP began collecting historic boats and storing them near the Visitor Center Courtyard. Although such boats had been lost or abandoned, GCNP eventually gathered an impressive collection that includes these historic boats:

Boat	User	Year
"The Stone boat"	Julius Stone	1909
Edith	Emery Kolb	1911
Glen	USGS Survey	1921–1923
WEN	Norman Nevills	1938
Esmeralda II	Ed Hudson, Dock Marston	1949
Woman of the River	Georgie White Clark	1954
Music Temple	P. T. Reilly, Martin Litton	1962
Sport Yak II	Dock Marston	1963

The GCNP collection also includes a piece of the hull of *Nellie Powell* from John Wesley Powell's 1871–1872 expedition through Grand Canyon.

GCNP's work on its collection of historic boats escalated in 1996, when a consortium of river running companies created the Colorado River Fund that provided funds to restore and preserve the historic boats. Seven years later, the GCNP Foundation raised an additional $150,000 to preserve the boats and begin its Grand Canyon Historic Boat Project. That same year, the foundation began developing its *If Boats Could Talk* pamphlet that was distributed at the South Rim and Marble Canyon. In 2009, the nonprofit Grand Canyon River Heritage Coalition was created to help NPS and other interested groups raise funds and plan facilities for historic boats and related artifacts associated with Grand Canyon.

See also: Birdseye, Claude; Clark, Georgie White; Kolb, Ellsworth and Emery; Loper, Bert; Marston, Otis "Dock"; Nevills, Norman

Further Reading

National Park Service. 2017. Grand Canyon Museum Collection. Retrieved May 3, 2017, at https://www.nps.gov/grca/learn/historyculture/muscol.htm

GRAND CANYON NATURAL HISTORY ASSOCIATION

In February 1932, led by GCNP naturalist Edwin D. "Eddie" McKee, 132 residents of GCNP and the surrounding area formed Grand Canyon Natural History Association (GCNHA, known today as Grand Canyon Association) "to stimulate and encourage scientific research and investigation in the fields of geology, botany, zoology, ethnology, archaeology, and related subjects in the Grand Canyon region." This was the fourth such association in the United States (others were at Yosemite, Zion, and Rocky Mountain National Parks).

The original board of directors of GCNHA included Grand Canyon pioneer Emery Kolb and GCNP Superintendent Miner Tillotson. McKee was the association's first executive secretary, and membership was $1 per year. The association also raised money by selling books and holding dances at the Grand Canyon Community Hall (today's Community Building). Membership reached 650 in 1995, 2,500 in 2000, and 8,500 in 2007.

Earlier, in 1925, GCNP park naturalist Glen Sturdevant had started publishing *Grand Canyon Nature Notes* to disperse natural history research done at Grand Canyon. With help from GCNHA, Sturdevant continued to publish the newsletter, as did his successor, Edwin McKee. In 1935, the newsletter was replaced by bulletins addressing specific subjects. The first of these bulletins was Vernon Bailey's *Mammals of the Grand Canyon Region*, which was published in 1935. In 1942, GCNHA hired the park's first librarian, Barbara Bryant, and in 1989 the association moved into the old hospital on the South Rim. The following year, GCNHA renovated and reopened Kolb Studio, which had been vacant for more than a decade.

During the 1930s, the association's interpretive programs were supplemented with CCC workers. These projects included exhibits at several sites in GCNP, including on the rims and at places such as Cedar Ridge, where visitors could see *in situ* exhibits of fern fossils. CCC workers also worked in the library and as guides in GCNP.

In October 1994, Grand Canyon Natural History Association became the Grand Canyon Association (GCA), whose mission is "to cultivate knowledge, discovery, and stewardship for the benefit of GCNP and its visitors." The Association, which has 17,000 members and annual revenues of $14-million, remains active at the North and South Rims of GCNP. Grand Canyon Field Institute, a nonprofit educational branch of GCA, offers guided tours in GCNP.

See also: McKee, Eddie; Sturdevant, Glen

Further Reading

Berger, Todd R., ed. 2008. *Reflections of Grand Canyon Historians: Ideas, Arguments, and First-Person Accounts*. Grand Canyon, AZ: Grand Canyon Association.

GRAND CANYON VILLAGE

At the South Rim, the most concentrated area of development is the Historic District (also called Grand Canyon Village), the largest example of town planning in NPS. Grand Canyon Village began to take form on the South Rim with the arrival of tourists in stagecoaches from Williams, Arizona, in the late 1800s. When Grand Canyon Railway arrived at Grand Canyon Village in 1901, tourism increased dramatically, and with the tourists came concessionaires hoping to serve and exploit those tourists with hotels, tours, shops, and toll trails. Nearly all of Grand Canyon Village was built, owned, and operated by concessionaires such as the Fred Harvey

Company and the Atchison, Topeka and Santa Fe Railway. A master plan by NPS landscape architect Daniel Ray Hull envisioned a "village square" where the railroad intersected the road just below El Tovar Hotel. Grand Canyon Village, which is the primary administrative and tourist visitation area at the South Rim, is also the largest and most developed area in GCNP.

Mule Barn, Blacksmith Shop, and Stables

These facilities, which were collectively referred to as the Transportation Department, were built across the railroad tracks from Bright Angel Lodge in 1907–1908 to house and care for the animals used to carry supplies and people around the park. These buildings were active social centers in the early days of Grand Canyon Village, for far more people then than now rode mules to Indian Garden and beyond. The blacksmith shop, whose history is described on a historical plaque at the east end of the barn, is the oldest such continually operating facility in the United States.

Grand Canyon Village, in which many of the buildings and businesses predate GCNP, was added to the National Register of Historic Places in 1975. It is a highlight of a visit to Grand Canyon and can be reached by car, shuttle bus, or foot from Grand Canyon Visitors Center along the Rim Trail.

See also: Bright Angel Lodge; El Tovar Hotel; Hopi House; Kolb Studio

Further Reading

Anderson, Michael. 2001. *Along the Rim: A Guide to Grand Canyon's South Rim from Hermit's Rest to Desert View.* Grand Canyon, AZ: Grand Canyon Association.

GRAND CANYON WEST

The Hualapai ("People of the Tall Pines") are Native people who, until the late 1800s, inhabited more than 5 million acres of the American Southwest. In 1883, however, the Hualapai were gathered onto the Hualapai Indian Reservation, which today covers almost 1 million acres that border more than 100 miles of the Grand Canyon and Colorado River. About half of the more than 2,200 enrolled members of the Hualapai Tribe live in Peach Springs, Arizona, a town on Historic Route 66 that includes the tribal headquarters of the Hualapai Nation. In 1998, the Hualapai decided to open their land (River Miles 165–273) to visitors, the first of which were 30 visitors from the United Kingdom. Today, their most successful tourist attraction is Grand Canyon West.

Grand Canyon West, also known as The West Rim, is located just outside of GCNP in a desolate area on the Hualapai Reservation in northwestern Arizona. This reservation stretches from Grand Canyon to Bill Williams River and from the Black Mountains to the San Francisco Peaks. The Hualapai have lived in and south

Bat Cave

In the early 1950s, George Steinke, Beal Masterson, and Murl Emery began vacuum-mining 12,000-year-old bat guano from a cave in western Grand Canyon 800 feet above the river on river right near River Mile 267. These entrepreneurs later sold the rights to the guano to a mining company that began planning how to increase production at the site. The cave was estimated to contain 100,000 tons of guano worth about $15 million.

After spending $3 million to build a 9,400-foot-long tram linking Bat Cave with Guano Point on the opposite side of the Colorado River, the investors discovered that the cave contained only about 1,000 tons of guano. Moreover, by the late 1950s, the commercial production of nitrate fertilizers made the mining of guano uneconomical, and in 1960 the mine at Bat Cave closed. Bat Cave, which is closed to visitors, is across the river and near the Haulapai Skywalk.

of Grand Canyon since at least 1200 CE, and today their economy is based on ranching, timber sales, hunting, arts and crafts, and tourism.

Grand Canyon West is the brainchild of Las Vegas developer David Jin (1962–2013), who invested $30 million to develop the site. Jin, who was originally from Shanghai, also owned several tour companies that brought tourists from Asia to the United States. Because most tourists at Grand Canyon West are Asian, signs and menus at Grand Canyon West include Chinese translations. Legal disputes between Jin and the Hualapai were not resolved until 2014. Most visitors to Grand Canyon West come from Las Vegas (120 miles away) on tour buses and are deposited at a welcoming area that features shops, food, and an airport from which they can arrange helicopter tours and rafting trips. A two-day trip camps overnight at Spencer Canyon, a site the Hualapai believe is their source of origin. (The Hualapai River Runners are the only Native American–owned river rafting company in Grand Canyon.) Grand Canyon West, which is north of Kingman, Arizona, is a 230-mile drive from Grand Canyon Village on the South Rim, a 264-mile drive from Phoenix, a 386-mile drive from the North Rim, and a 130-mile drive from Flagstaff.

Grand Canyon West consists of three major sites, all of which are linked by shuttle buses:

Hualapai Ranch is an "Old West" area featuring food, entertainment, wagon rides, exhibits, horseback rides, and overnight ranch-style cabins. This site does not offer a view of the canyon's rim, which is a 30-minute horseback ride away.

Guano Point, which is northwest of Eagle Point, features the best views of the canyon at Grand Canyon West. From the peninsula there, visitors have nearly a 360-degree view of the canyon. The short Highpoint Hike leads to the remnants of a cable tramway that, in the 1950s, took workers and materials

8,800 feet to and from Bat Cave, a guano mine 800 feet above the opposite side of the Colorado River. In the 1940s and 1950s, the U.S. Guano Corporation excavated bat guano from Bat Cave and lifted it along a 9,400-foot cable to Quartermaster Point (and up more than 2,550 feet) to the opposite side of the river. By the 1960s, however, the mine was depleted of guano and abandoned. (Today's fertilizers are made of nitrates rather than guano.) Soon after the mine closed, a plane cut the transriver cable but landed safely. Bat Cave continues to house a large colony of Mexican free-tailed bats. Visitors to Guano Point can buy food and Native American crafts and jewelry.

Eagle Point, which is named for a rock formation across the canyon that looks like a bird with outstretched wings, is north of Hualapai Ranch and is the most popular stop at Grand Canyon West because it features the Grand Canyon Skywalk. This horseshoe-shaped, transparent walkway is the most recognized cantilever structure in the world. It extends 70 feet over the canyon, and its floor of reinforced glass enables visitors to look directly down at the canyon floor, which is more than 3,600 feet below. It is this experience of "floating on air," rather than its general view of the canyon, that attracts people to the Skywalk. The walkway is 10 feet wide and has railings 5 feet 2 inches high.

Grand Canyon Skywalk is a popular tourist attraction along a side canyon on the Hualapai Reservation in western Grand Canyon. The Skywalk, which is more than a mile from the Colorado River, can be seen on river left from around River Mile 265. The attraction towers 3,600 feet above the Colorado River, and 500–800 feet above the canyon floor directly below. (Veselina Dzhingarova)

> Visitors must buy separate tickets ($35 per person in 2017) to access the Skyway, and they are not allowed to take cameras, cell phones, or other photographic devices onto the Skywalk. (However, you can pay employees of the Skywalk to take your picture there; packages range from $25 to $90.) At Eagle Point, you'll also find food, souvenirs, Native American dwellings, an amphitheater that features performances, and jewelry and crafts.
>
> Construction of Grand Canyon Skywalk took four years (only about a year less time than it took to build Hoover Dam). The Skywalk, which opened in 2007, is made of 83,000 pounds of glass and nearly 1 million pounds of steel. It can support 71 million pounds (the equivalent of 71 loaded 747 airplanes), but no more than 120 people are allowed on the Skywalk at one time. Opponents of the Skywalk complained that the attraction would deface sacred land, while advocates of the Skywalk argued that the site would boost the tribe's finances. When the Skywalk was built, engineers were allowed to drill only into the top (and not into the face) of the canyon wall; this is why the Skywalk is cantilevered. The first official visitor to walk on the Skywalk was astronaut Buzz Aldrin.
>
> Contrary to popular belief, Grand Canyon Skywalk does not overlook Grand Canyon; instead, it overlooks a side canyon. River runners see Grand Canyon Skywalk in the distance on river left from near River Mile 265.5 of the Colorado River.

The opening of Skywalk increased attendance at Grand Canyon West from about 300 people per day to several thousand people per day. More than 5 million people have visited Grand Canyon West since it opened in 2007, and attendance rates are increasing rapidly (e.g., more than a million people have visited Grand Canyon West each year since 2015). Future plans for the area include the construction of a museum, zipline, theater, golf course, restaurants, and a casino. Grand Canyon West, which is one of the region's biggest economic engines, is also the only place in Grand Canyon at which visitors can ride a helicopter to 4,000 feet below the rim for a boat ride on the Colorado River.

See also: Airplanes and Helicopters

Further Reading

National Park Service. 2017. The Hualapai Tribe and Skywalk. Retrieved July 3, 2017, at https://www.nps.gov/grca/planyourvisit/skywalk.htm

GRANDVIEW POINT

Grandview Point (also called Grand View; elevation 7,400 feet) was settled by Ralph Cameron and Peter Berry in the 1890s because it offered the best access to their Last Chance Mine (and seven adjacent mining claims) at the U-shaped Horseshoe Mesa 2,600 feet below the rim. That mine, where five to seven miners each earned $4 per day for their work, contained exceedingly pure copper; a 600-pound nugget, which was 70 percent copper carbonate, won awards at Chicago's World Fair in

1893. (Hopi Indians used the ore to color paint.) Most ore from Last Chance Mine was only about 35 percent pure.

When innovations in processing copper ore dropped the price of copper to 12.5¢ per pound, it was no longer profitable to haul and process ore from Berry's mine. Berry and his wife Martha then used profits from the mine to build (from ponderosa pine) the two-story Grand View Hotel, which for several years was the only first-class hotel at the Grand Canyon. When this hotel opened near Grandview Point in 1897, it became the new terminus for Grand Canyon Stage Line from Flagstaff, Arizona. During the next few years, tourism at Grand Canyon was focused at Grandview; this is why there was a post office here from 1902 to 1913. (Cameron used his profits from Last Chance Mine to develop an area west of Grandview Point at what became Grand Canyon Village.) Stalactites and stalagmites from Cave of the Domes near Last Chance Mine supported the front porch of Grand View Hotel.

In the 1890s, Grandview was the center of visitation and the most popular destination at Grand Canyon. Tourism here thrived until the arrival of the Santa Fe Railway in 1901 and the accompanying growth of Grand Canyon Village. To save his business, Berry offered free stagecoach rides to his hotel. However, he sold the hotel and his mining interests in 1902 to Henry P. Barber, who quickly sold the mine to the Canyon Copper Company. By 1908, however, both the hotel and mine had closed.

In 1913, Berry sold his homestead to newspaper magnate William Randolph Hearst. (Berry stayed as the property's caretaker until 1919.) To protect his property and interests in Grand Canyon, Hearst used his influence and friends (including Arizona Senator Henry Ashurst) to insert a provision into the U.S. Code that "where privately owned lands within [GCNP] lie within three hundred feet of the rim of the Grand Canyon, no building, tent, fence, or other structure shall be erected on the park lands lying between said privately owned lands and the rim." (Hearst's cabin was 300 feet from the rim.) Despite threats by Hearst to develop the land, the NPS—in the only condemnation lawsuit ever filed by GCNP—bought Hearst's land at Grandview in 1939 for $85,000, plus 6 percent interest for the initial underpayment for the land.

Today, there are few visual signs of Grandview Point's colorful past; all of the original buildings at Grandview Point are gone. However, at Horseshoe Mesa, you'll find the mine's openings, ruins of a stone building, and some campsites. Several caves penetrate the Redwall Limestone on the west and northwest sides of the mesa. One of these caves is Cave of the Domes, a multichambered cave that is the only cave in GCNP open to the public. (This cave, however, is closed periodically due to threats posed by disease-carrying rodents.) Inside Cave of the Domes are several historic inscriptions, including "P. D. Berry Jan. 1st 1897" and "E. L. Kolb E. C. Kolb Pittsburg, Pa." The trail from Grandview Point to Horseshoe Mesa doesn't reach the Colorado River; beyond the mesa, the cliffs of Grand Canyon's Inner Gorge are too steep for hiking.

From Grandview Point, views include (west to east) Grapevine Creek, Horseshoe Mesa, Vishnu Temple, Palisades of the Desert, and Desert View Watchtower

(9.5 miles away). West of Horseshoe Mesa is Cottonwood Creek, and to the east is Hance Creek. In the far northeast, the Colorado River (near the mouth of Tanner Canyon) is visible. A look north from Grandview Point features (left to right) Zoroaster Temple, Brahma Temple, Angels Gate, Wotans Throne, Cape Royal, and Jupiter Temple.

Grandview Point is one of the highest overlooks on the South Rim, is the southernmost point of the South Rim, and is approximately halfway between Desert View and Grand Canyon Village. Views from Grandview Point are sometimes restricted by trees; to enjoy better views, hike a short way down Grandview Trail.

See also: Berry, Peter; Cameron, Ralph; Grandview Trail

Further Reading

Anderson, Michael. 2001. *Along the Rim: A Guide to Grand Canyon's South Rim from Hermit's Rest to Desert View.* Grand Canyon, AZ: Grand Canyon Association.

GRANDVIEW TRAIL

Unlike most trails in GCNP, which were developed from paths originally forged by animals or Native Americans, Grandview Trail was built in 1892–1893 to transfer people and materials to and from Last Chance Mine. This 19th-century mine, which was owned by Peter Berry and Ralph and Niles Cameron, is in Redwall Limestone about halfway between the South Rim and Colorado River. Construction of the $12,000 trail to the mine took eight months. The original trail, which was known as the Berry Trail, began about 1 mile east of the modern trailhead; hikers can still see holes drilled into the Coconino Sandstone for explosives used to blast the trail.

Today's 3.1-mile (one-way) Grandview Trail begins on the canyon side of Grandview Point, which is about 11 miles west of Grand Canyon Village. The trail ends 2,500 feet below the rim atop Redwall Limestone at Last Chance Mine on the Horseshoe Mesa. At Horseshoe Mesa, the remnants of Last Chance Mine include the 6-foot-high walls, chimney, and pipes of a cookhouse that had adjoined a bunkhouse, corral, and blacksmith shop. Visitors also find greenish-blue fragments of copper ore extracted from the mine. Workers at the mine were paid $4 per day.

A spur trail to the north leads hikers to a camping area and switchbacks down 1,000 feet to the Tonto Trail; it also crosses the Surprise Canyon Formation, an exposure found only in a few places in GCNP. The sheer drop along a spur trail to the east leading to Miners Spring (a basin in Bright Angel Shale) is where notorious murderer Robert Spangler killed his third wife in 1993.

In 1897, Berry used some of the profits from his Last Chance Mine to open the two-story Grand View Hotel at Grandview Point. This hotel, combined with stagecoach service from Flagstaff, Arizona, that delivered tourists to Grandview Point, made this area the primary destination for Grand Canyon's tourists. Berry's hotel provided the nicest accommodations at the South Rim until 1905, when the El Tovar Hotel opened 11 miles to the west at what became Grand Canyon Village. The mine was sold to investors in 1902, who closed the mine in 1908 after copper prices plummeted. The arrival of the Atchison, Topeka and Santa Fe Railway at the South Rim permanently shifted tourism away from Grandview Point to Grand Canyon Village.

Berry's hotel was empty by 1912. William Randolph Hearst bought Last Chance Mine in 1913, and GCNP took control of the property in 1940. The Grandview Trailhead, which is at an elevation of 7,400 feet, is the highest trailhead on the South Rim.

More than any other trail in GCNP, Grandview Trail recalls the legacy of mining in Grand Canyon. Grandview Trail, which was restored by NPS in the 1970s, is unique among GCNP inner-canyon trails in being partially made of cobblestone and log cribbing. These features were included when the trail was restored in 2005 after a rockfall destroyed some sections of the trail's switchbacks. The trail also includes an 8-foot-long wooden beam bridge. On the west side of Horseshoe Mesa is Cave of the Domes, a series of domed chambers (some connected with crawlways) discovered by the cook at Last Chance Mine and extending about 0.5 mile under the mesa. Stalactites and stalagmites from this cave were used to support steps at Berry's Grand View Hotel. Cave of the Domes is one of the few accessible caves in GCNP open to visitors. There is a guest register on the left of the first chamber that hikers enter.

Grandview Trail includes 31 switchbacks and is often steep and rocky. About 2 miles down the trail is the unmarked Coconino Saddle, which is a convenient turnaround point (1,200 feet below the rim) for many hikers. Grandview Trail does not reach the Colorado River but does link to Tonto Trail, which links with other trails taking hikers to the river and elsewhere. Grandview Trail, which is not regularly maintained, features great views of Vishnu Temple and Zoroaster Temple. Grandview Trail has no water.

See also: Berry, Peter; Cameron, Ralph; Grandview Point

Further Reading

Thybony, Scott. 2009. *Grand Canyon: Grandview Trail Guide*. Grand Canyon, AZ: Grand Canyon Association.

GREAT DEER RUN

In 1906, President Theodore Roosevelt created the Grand Canyon Game Preserve, thereby protecting mule deer and other game animals at Grand Canyon. This protection also began an eradication of the area's deer predators; by 1931, more than 800 mountain lions, 500 bobcats, 30 wolves, and 4,000 coyotes had been killed in the area. The preserve's warden and the area's most famous hunter was James T. "Uncle Jim" Owens, who is memorialized by the 5-mile (round-trip) Uncle Jim Trail on the North Rim. Over the next 13 years, Owens killed hundreds of mountain lions, bobcats, wolves, bears, and other predators. This reduction in the populations of predators increased the number of deer from about 4,000 in 1906 to more than 25,000 by 1924. By this time, Grand Canyon was overrun by deer. This exploding population of deer—especially along the North Rim—produced Grand Canyon's first (and only) Great Deer Run of 1924.

To address the problem, Secretary of Agriculture Henry Wallace appointed a Kaibab Deer Investigating Committee, which recommended that NPS reduce the size of the deer population by half. After a short hunting season in the fall of 1924 killed only about 650 deer, Flagstaff-based cattle rancher George McCormick—supported by Governor George W. P. Hunt—offered to herd as many as 10,000 deer along the Nankoweap and Tanner Trails from the North Rim to the South Rim. (McCormick apparently believed that deer could be herded like his cattle.) On December 16, McCormick—aided by 25 local residents, 72 Navajos, and 24 Paiutes, who would share $2.50 (paid by the state of Arizona) for each animal that reached the South Rim—began their herding. Famed movie director D. W. Griffith arrived to film the roundup, knowing that footage of deer swimming across the Colorado River would be "a million-dollar film."

The so-called Great Deer Run didn't turn out as planned. A local newspaper announced the fiasco with the headline "Kaibab Deer Refuse to Herd," and Mark Musgrave, a representative of the U.S. Biological Survey, reported that "there was no organization whatsoever" (Berger, 2007). Not a single deer made it to the South Rim, and the population of deer remained far too large to be supported by the resources available (thanks to the near lack of predators). In subsequent years, thousands of deer died of starvation, and today's population of deer at the North Rim is only slightly larger than it was in the early 1900s.

The futile attempt to herd the deer was witnessed by writer Zane Grey, who had first visited Grand Canyon 19 years earlier on his honeymoon. Grey fictionalized the event in his *The Deer Stalker* (1925). Grey's criticisms of antipredator policies and his advocacy of natural ecosystems helped promote modern ideas about wildlife management. Although GCNP continued to offer bounties for predators until 1927, GCNP superintendent Miner Tillotson ended GCNP's bounty system in 1931. Three years later, GCNP stopped feeding its deer.

See also: Owens, James "Jim"

Further Reading

Berger, Todd R. 2007. *It Happened at Grand Canyon*. Guilford, CT: TwoDot Publishers.

H

HAMBLIN, JACOB

Jacob Hamblin (1819–1886) was a Mormon missionary who was sent by Brigham Young to work with Native Americans and explore the area around Grand Canyon. While on his way to Hopi villages, Hamblin—who was known as "The Buckskin Apostle"—was one of the few witnesses of Joseph Christmas Ives's 1858 exploration of Grand Canyon. Hamblin tried for years to establish missions at the Hopi villages but failed. Hamblin's testimony against Lees Ferry ferryman and namesake John D. Lee helped produce the conviction that led to Lee's execution in 1877. (Lee referred to Hamblin as "Dirty Fingered Jake" and "the fiend of Hell.")

Hamblin's travels enabled him to discover a variety of important places around Grand Canyon, including Pearce Ferry, Lees Ferry, and Crossing of the Fathers (which is now under Lake Powell). In 1862, Hamblin completed the first recorded trip around Grand Canyon, completing the 500-mile trip in 60 days. The following year, Hamblin repeated the trip. In 1864, Hamblin built a log raft and, for the first time, crossed the Colorado River at the mouth of Paria Creek in Grand Canyon. In 1870, Hamblin advised John Wesley Powell about interactions with Native Americans.

When polygamy was outlawed in 1882 with passage of the Edmunds Act, Hamblin became a fugitive who moved to several different locations to avoid capture. Hamblin died a fugitive in Pleasanton, New Mexico, on August 31, 1886. He is buried in Alpine, Arizona. Hamblin, who helped found Kanab, Utah, in 1870, is honored by Jacob Lake and Hamblin Wash along U.S. Highway 89 in northern Arizona. Hamblin's home in Salt Lake City, Utah, is a museum operated by the Mormon Church.

See also: Lees Ferry

Further Reading

Brooks, Juanita. 1980. *Jacob Hamblin: Mormon Apostle to the Indians*. Salt Lake City, UT: Westwater.

Lago, Don. 2015. *Grand Canyon: A History of a Natural World and National Park*. Reno: University of Nevada Press.

HANCE, JOHN

John Hance (c.1840–1919) was the first non–Native American resident at Grand Canyon. Hance, who in 1859 had been part of the Pikes Peak gold rush, first saw Grand Canyon in 1883 and the following year claimed a 160-acre homestead just east of Grandview Point. Hance, with William Hull, built a small lodge that was one of Grand Canyon's first tourist accommodations.

By 1886, Hance was advertising his services as a tour guide with ads such as this one that appeared in a newspaper in Flagstaff, Arizona:

> Being thoroughly conversant with all the trails leading to the Grand Canyon of the Colorado, I am prepared to conduct parties thereto at any time. I have a fine spring of water near my house on the rim of the Canyon, and can furnish accommodations for tourists and their animals. (Brown, 2009)

By 1894, the Atchison, Topeka and Santa Fe Railway was taking tourists from Flagstaff to Hance's Cañon Hotel via stagecoach for less than $20. Hance also converted an ancient trail into the first tourist trail into Grand Canyon in the late 19th century. Soon thereafter, with John Marshall, C. H. McClure, and William Ashurst, Hance began mining asbestos from what was called Asbestos Canyon near River Mile 78.5. (This site is near the mouth of Red Canyon near Sockdolager Rapid.) Asbestos was important as insulation and fireproofing for cities, whose indoor lighting was based on gas flames and whose buildings were made primarily of wood. Some of the asbestos from Grand Canyon was shipped to France, where it was made into fireproof fabrics. Hance's mine, which produced asbestos fibers almost 4 feet long, is an Archived Site in the Environmental Protection Agency's Superfund program, which means that it has been evaluated and requires no further remedial action. Hance's mine is closed to visitors.

When expenses for removing ore from Grand Canyon became prohibitive, Hance shifted his work to tourism. He first advertised for tourists in the September 18, 1886, issue of Flagstaff's *Arizona Champion*. By the next year, Hance was also advertising accommodations and mule rides into Grand Canyon (from Grandview Point).

By the early 1900s, Hance was a raconteur living in Grand Canyon Village, where he was given free room and board to entertain tourists. The hospitable "Captain" Hance, who claimed to have served in the Union and Confederate armies, often regaled tourists from the porch of El Tovar Hotel. He was known for his tall tales (e.g., that he dug Grand Canyon while looking for precious metals, and piled the dirt to form the San Francisco Peaks). As one visitor wrote in Hance's guestbook, "To see the Canyon only, and not to see Captain John Hance, is to miss half the show" (Campbell, 2015). On January 25, 1893, Grand Canyon pioneer William "Buckey" O'Neill wrote similar sentiments in Hance's book: "God made the cañon, John Hance the trails. Without the other, neither would be complete" (Campbell, 2015). Hance became an important part of experiencing Grand Canyon; as American writer Hamlin Garland (1860–1940) noted, "The man who ought to be remembered with the Grand Canyon of the Colorado is not a scientist, nor a painter, nor a poet . . . His name is John Hance."

In 1895, Hance sold his property at Grand Canyon for $1,500 to Lyman Tolfree and James W. Thurber, who took over the Atlantic and Pacific Railroad's stagecoach line from Flagstaff. When a post office was opened on what had been his land in

1897, Hance named the area "Tourist, Arizona" and served as the first postmaster. In 1906, Thurber sold the Hance homestead to Martin Buggeln, who built a two-story, 17-room hotel there the next year. Hance's mine eventually was sold to the Hance Asbestos Mining Company of Kittery, Maine. Although that company's property at Grand Canyon eventually included 16 claims covering 325 acres, Hance never bought any stock in the company. Hance Asbestos Mining Company was out of business by February 1915.

Hance worked as a guide until 1915. In 1918, he suffered a stroke and moved to Flagstaff, where he lived as an indigent on North Leroux Street in the (now iconic) Weatherford Hotel (est. 1897). He died in the Coconino County Indigent Hospital (today's Arizona Historical Society's Pioneer Museum) on the afternoon of January 26, 1919 (the same day as Teddy Roosevelt's death), just four weeks before Grand Canyon became a national park (*Arizona Republic* listed his age as 84). Hance's death certificate, which lists his cause of death as "Senility [and] Endocarditis," notes he was to be buried in Flagstaff, but—as he wished—his body was returned to the South Rim. At the time of his death, Hance's only possessions were three mining claims and some household furniture.

After a funeral service at the railroad platform, Hance was one of the first people to be interred at the South Rim site that would later become Pioneer Cemetery. Hance's tombstone describes Hance as the "First Locator on Grand Canyon, Arizona Trail Blazer, And Guide." Hance is memorialized by Hance Rapid at River Mile 77.1 and by Hance Creek, which enters the Colorado River from river left east of Horseshoe Mesa at River Mile 78.5.

See also: Grandview Point; Hance Trails; Pioneer Cemetery and Other Gravesites

Further Reading

Brown, Ron. 2009. John Hance tells how he arrived at Grand Canyon. Insider's look at Grand Canyon #14, October 30, 2009. Retrieved March 15, 2017, at https://www.nps.gov/grca/learn/photosmultimedia/inside-look-014.htm

Campbell, Mike. 2015. The Hance Guestbook: In praise of John Hance and his tall tales. *Canyonology Treks*, April 15 2015. Retrieved April 21, 2017, at https://www.canyonology.com/hance-guestbook-praise-john-hance-tall-tales/

Murphy, Shane. 2017. The death of John Hance. *The Ol' Pioneer*, 28 (4), 3–7.

HANCE TRAILS

In 1883, Grand Canyon pioneer John Hance built a home between Grandview and Moran Points. Hance's cabin was near a tank and spring, and nearby Hance developed an old Havasupai trail into what became known as the Old Hance Trail. This was the first developed trail into Grand Canyon. The following year, Hance began leading tourists down his trail, making Hance the canyon's first tour-guide. The next year, Hance erected tents by his cabin for tourists, thereby becoming the first "hotel" operator at Grandview Point.

When a flashflood and rockslides damaged his trail in the 1880s, Hance abandoned the trail and in 1894 opened the Red Canyon Trail, which soon became known as the New Hance Trail (parts of Old Hance Trail remain visible, but they are not maintained). Hance's mines have been off-limits to tourists since 1988. This trail enabled Hance to reach the Tonto Trail and his mines in Red Canyon. Hance charged tourists $1 to walk on the trail, $2 to ride a mule on the trail, and $10 to use Hance's pack animals and guide services. By 1895, Hance sold his property and the rights to his trail to William Thurber and J. H. Tolfree and promised he'd not open another trail or work as a guide within 30 miles of their site. By 1900, Thurber and Tolfree ended their tourism ventures, and the trail reverted to Hance. Hance patented his homestead in 1907 and sold his property to Martin Buggeln, after which Hance moved to Grand Canyon Village, where he worked as a greeter for the Fred Harvey Company.

New Hance Trail, which is 6.5 miles long and descends 4,400 feet, was condemned by geologist Eddie McKee in 1933 as one of the two worst trails in GCNP. The trail begins about a mile west of Moran Point. In 1988, GCNP began minimal maintenance of the trail, but it remains a steep, challenging hike—some places are only 2 feet wide. There is no water along New Hance Trail.

See also: Hance, John

Further Reading

Abbott, Lon, and Cook, Teri. 2004. *Hiking the Grand Canyon's Geology*. Seattle, WA: Mountaineers Books.

HARVEY, FRED, AND THE HARVEY GIRLS

Frederick Henry Harvey (1835–1901) began his career in the restaurant business as a busboy and dishwasher and later headed a food service empire that changed how people eat when they travel. His association with railroads and the American West began in 1876, when he agreed to a handshake deal with the Atchison, Topeka and Santa Fe Railway Company to provide food and other services in a second-floor lunchroom at the railroad depot in Topeka, Kansas. Two years later, Harvey opened his restaurant-hotel in Florence, Kansas. By the late 1880s, Harvey's Harvey House restaurants—the nation's first restaurant chain—were about every 100 miles along the railroads. By 1893, Harvey owned the exclusive dining rights along the railroad west of Kansas City. At each of his restaurants, the railroad paid for the buildings, utilities (including lights, heating, and ice), and transportation for Harvey's employees. Harvey, in turn, provided all the furniture, linens, and tableware for the restaurants, and got all of the money generated by food sales.

Harvey House restaurants, hotels, and other hospitality-based businesses set new standards for quality and civility in the West. Delicious meals—for example, featuring oysters, sea turtle, garden-fresh salads, and other such delicacies for 75¢—were served on fine china and linen tablecloths. Harvey's partnership with the railroad was one of the great success stories of the American West.

Harvey's restaurants were famous for so-called Harvey Girl waitresses, who were young, white women "18 to 30 years of age, of good character, attractive, and intelligent" (Howard and Pardue, 2016). These women, who worked 10-hour shifts six or seven days per week (with a 10:00 p.m. curfew), earned up to $50 per month (in addition to free room/board, laundry, and gratuities) and enjoyed rail passes. Harvey hired and chaperoned more than 100,000 Harvey Girls, each of whom wore starched uniforms and agreed not to get married during their first six months on the job.

Fred Harvey's first concessions arrived at Grand Canyon in the early 1900s and had been active for 15 years when Grand Canyon was made a national park in 1919. Given this history, it was not surprising that Harvey became GCNP's first NPS-approved concessionaire at the South Rim in 1920. There, Harvey—with the Santa Fe Railway—enjoyed a virtual monopoly (the U.S. Forest Service referred to this arrangement as "regulated competition"). The first Harvey Girls at Grand Canyon worked at El Tovar Hotel, and they later staffed Bright Angel Lodge. From El Tovar and elsewhere on the South Rim, Harvey took tourists to various sites via touring cars and buses (one of Harvey's 1929 Cadillac 341-B touring cars is displayed in Williams, Arizona). The cars and buses, which broke down often, included cages for six homing pigeons that were used to summon help.

Today, Bright Angel Lodge's History Room features several informative exhibits about Harvey, Harvey Girls, and Harvey's food empire, which "civilized the American West" (Howard and Pardue, 2016). The Fred Harvey Company remained in the Harvey family until the death of Fred's grandson in 1965. In 1968, the Fred Harvey Company merged with American Factors Company (AmFac), a Hawaii-based company that was later renamed Xanterra Parks & Resorts. In 2006, Xanterra bought the Grand Canyon Railroad, and in 2015 was awarded a 15-year contract for concessions at the South Rim that could be worth more than $1 billion. This contract, one of the most lucrative in NPS, will bring mobile food trucks and a valet parking service to the South Rim, and a $12 million renovation of 30 cabins at Maswik Lodge.

Harvey Girls were made famous in 1946 by *The Harvey Girls*, an MGM musical starring Judy Garland based on the 1942 novel by Samuel Adams of the same name. This movie, which was among the top 15 grossing movies of 1946, won an Academy Award for Best Song for "On the Atchison, Topeka and the Santa Fe." Many of Grand Canyon's Harvey Girls stayed at or returned to Grand Canyon after retiring. Several are buried in Pioneer Cemetery, including Bette Rich Swanson, whose epitaph tells visitors that she was "Just an Old Harvey Girl at Rest in the Place She Loved."

When Harvey died on February 9, 1901, in Leavenworth, Kansas, his company was operated by his two sons, Byron S. and Ford F. Harvey. In its heyday, the Fred Harvey Company owned 12 large hotels, 65 restaurants and lunch counters, and 60 dining cars. A large portrait honoring Harvey decorates the stairwell at El Tovar Hotel.

See also: Colter, Mary Elizabeth Jane; Grand Canyon Village; Railroad

Further Reading

Howard, Kathleen L., and Pardue, Diana F. 2016. *Over the Edge: Fred Harvey at the Grand Canyon and in the Great Southwest*. Tucson, AZ: Rio Nuevo Publishers.

Poling-Kempes, Lesley. 1989. *The Harvey Girls: Women Who Opened the West*. New York: Paragon House.

HATCH, ROBERT "BUS"

Robert Raphael "Bus" Hatch (1902–1967) was a carpenter and contractor in Vernal, Utah, who began running rivers in northeastern Utah in 1929. He later operated the first commercial rafting company in the United States and made his first trip through Grand Canyon in 1934. That same year, Hatch was a member of the so-called Dusty Dozen, which also included Royce "Cap" Mowrey, Russell G. Frazier, W.J. "Bill" Fahrni, Clyde Eddy, Alton Hatch (Bus's brother), and head boatman Franklin Enos "Frank" Swain. The expedition occurred at low-flow rates (less than 1,800 cubic feet per second) during the Dust Bowl period, and the river was "dusty."

The Dusty Dozen expedition was still relatively novel, for fewer than 50 people had gone through Grand Canyon. During the expedition, Bus Hatch found 5-inch split-twig figurines at Stantons Cave (River Mile 32.0), some of which were later determined to be 4,000 years old. During the trip, the group reported a skeleton at South Canyon, the site where Robert Stanton and his remaining crew hiked out of Grand Canyon following the drowning deaths of three of their colleagues. Bus Hatch was later part of the first commercial expedition that rowed a pontoon through Grand Canyon.

Hatch founded Hatch Expeditions and, in the process, became the first person to go through Grand Canyon who owned an outfitting business on the Colorado River. Hatch's sons Don and Ted made Hatch River Expeditions a famous commercial river-running company that continues to operate at Grand Canyon. Hatch ran the Colorado River through Grand Canyon until he died in 1967, after which his family and descendants took over the business. In 2018, the fourth generation of Hatch's descendants was operating Hatch Expeditions, one of the only family-owned and operated river running companies in the world. Bus is buried in Vernal Memorial Park in Vernal, Utah.

See also: Eddy, Clyde

Further Reading

Hatch River Expeditions. 2017. Grand Canyon whitewater rafting. Retrieved June 1, 2017, at http://hatchriverexpeditions.com

Webb, Roy. 1989. *Riverman: The Story of Bus Hatch*. Third edition. flagstaff, AZ: Fretwater Press.

HERMIT TRAIL

In the late 1890s, settlers at Grand Canyon's South Rim improved an Indian trail into Hermit Basin, which is 8 miles west of what became Grand Canyon Village. This trail reached water but not arable land and therefore was used for hunting rather than agriculture. In 1902, just a few months after the Santa Fe Railway arrived at the South Rim, prospector Louis D. Boucher recorded the deed for his Silver Bell Trail nearby. Boucher, a French-Canadian prospector who lived alone and operated several mines in the area for 20 years, became famous as Grand Canyon's "hermit," despite the fact that he was not a loner and often socialized with neighbors and visitors. At the time, the most popular trail into the canyon from the South Rim was Ralph Cameron's Bright Angel Toll Trail, which was the only trail into the canyon from what is today's Grand Canyon Village. In contrast to Cameron, Boucher did not charge people to use his trail. Interestingly, more features of Grand Canyon are named for Boucher than any other pioneer. For example, three are named Boucher (Rapids, Trail, Creek), and nine are named Hermit (Rest, Road, Rapids, Camp, Shale, Trail, Fault, Creek, Basin). Eremita (Spanish for "hermit") Mesa and Hermit Tank also bear his name.

In 1909, the Atchison, Topeka and Santa Fe Railway—needing the unrestricted use of a trail into Grand Canyon for its customers—enhanced its tours at Grand Canyon by buying rights to Boucher's trail. The railroad built a new Hermit Trail in 1911–1912 that NPS described as "wide, comfortable, and constructed on the most approved engineering lines" that would make inexperienced hikers "conclude that they are engaged in a truly hazardous enterprise." Hermit Trail to the (relatively) luxurious Hermit Camp—which GCNP described as "[t]he last word in gracious tourism below the rim" (Annerino, 2017)—was the only such trail designated exclusively for tourists. Much of the new 4-foot-wide trail was created with bordering stonewalls and hand-fitted slabs of sandstone laid end to end for miles, making it the closest thing to a paved trail in Grand Canyon. In the 1920s and 1930s, Hermit Trail was the model trail in NPS.

The railroad complemented the trail by building Hermit Camp near Hermit Creek on the Tonto Platform, 1,000 feet above the Colorado River and 3,600 feet below Pima Point. From El Tovar Hotel, tourists could pay $18.25 for an overnight trip to the camp. The camp consisted of 11 canvas-topped tents, showers, restrooms, dining hall, and other buildings, which were in one of the few areas with water. The $100,000 Hermit Camp competed with other camps at Grand Canyon owned by William Bass, Ralph Cameron, and David Rust. Hermit Camp was operated by the Fred Harvey Company and included an on-site chef, rest areas, horse stables, and cottages for visitors.

Hermit Trail, which descends 4,240 feet along its 8.1-mile length, begins 500 feet west of Hermits Rest. The trail is moderately well maintained and begins with switchbacks through 600 feet of Kaibab Limestone, Toroweap Formation, and Coconino Sandstone into the Hermit Creek drainage. The first part of the trail is steep, dropping 2,000 feet over the first 2.5 miles. Hermit Trail joins the 2-mile-long, seldom used Waldron Trail atop the Hermit Formation, where you'll see the remains of a camp. (Waldron Trail is named for John L. Waldron, who was hired by Bass to build trails and buildings at Grand Canyon in the 1890s.) After intersecting the now deteriorated Dripping Springs Trail, Hermit Trail then drops into Hermit Gorge in the Supai Formation. The Tonto and Hermit Trails overlap for approximately a mile. Approximately 2.3 miles down Hermit Trail is its junction with Boucher Trail; from here, it's 9.3 miles along this trail to the river. Hermit Trail is also the most exposed rim-to-river trail in GCNP; some places are barely 3 feet wide. Experienced hikers who brave this trail get to enjoy a spectacular view from Yuma Point. Several of the trailside slabs along Hermit Trail in the Coconino Formation display fossilized tracks of reptiles.

Four miles down Hermit Trail from the rim is Santa Maria Spring, which was named by architect Mary Elizabeth Colter to honor the Virgin Mary. Nearby is a stone shelter (built in 1913) and a metal water trough and hitching rail; these are remnants of the days when mules took tourists to Hermit Camp. Hermit Trail also displays beautiful fossils of marine invertebrates from 260–350 million years ago. Beyond Santa Maria Spring, the trail levels off atop Redwall Limestone and then reaches Lookout Point, which is directly above Hermit Camp. The trail then drops to the Tonto Platform through the Cathedral Stairs, which is a break in Redwall Limestone that includes a series of switchbacks from which hikers can see the remnants of the camp and nearby mining operations. From here, hikers see a beautiful downriver panorama. To the east is Cope Butte, which is named for the famous paleontologist, Edward Drinker Cope, who never visited Grand Canyon. Hikers who continue to the river can watch river runners navigate Hermit Rapid; most boatmen and boatwomen run the middle of this rapid.

Hermit Trail ends at Hermit Camp, which is below Pima Point at Hermit Creek (about 1 mile past the junction with Tonto Trail). This camp began as a group of tents, each with its own stove, beds, and wooden floor, and later grew to include more than 10 buildings. The remains of the camp include a cold cellar built into a hill that was later used by NPS as a storage room. Today there are campsites at the remains of Hermit Camp. Hermit Trail, with its rest stops, wide trail, gentle grades, and retaining walls made of the major rock formations that the trail passes through, became a model for trails in NPS.

In 1914, Mary Colter designed Hermits Rest at the head of Hermit Trail, and in 1925, the railroad began building a tramway from Pima Point to Hermit Camp. This tramway was the longest single-span tram in the world. The 6,000-foot ride along the 7/8-inch cable took 18 minutes for adventurous tourists to get from the rim to Hermit Camp. The tramway was later torn down, and on August 15, 1930, NPS decided to abandon the camp. Valuable items were salvaged and removed from the camp during the fall of 1936, and on November 10, 1936, residents and visitors watched from Pima Point as the NPS burned the remains of Hermit Camp. Today, all that's left of the camp are buildings' foundations, the tramway's concrete foundation, a corral, and the tramway's wheels and cables. (From the west side of Pima Point, visitors can see the foundations of buildings that were part of Hermit Camp.) In the 1940s, GCNP rangers installed a barrier along the neglected trail about 1.5 miles down the trail (at the Supai Formation), but this barrier was later removed when more hikers returned to the trail.

Hermit Trail is strenuous; it is neither maintained nor patrolled as much as Corridor Trails. Hermit Trail is also steeper and rockier than these trails, and hikers must often find the route in washed-out areas.

See also: Boucher, Louis; Colter, Mary Elizabeth Jane; Hermits Rest; Railroad

Further Reading

Annerino, John. 2017. *Hiking the Grand Canyon: A Detailed Guide to More Than 100 Trails*. New York: Skyhorse Publishing.

Historic Structure Report. 2007. Hermits Rest. Retrieved January 19, 2018, at http://capla.arizona.edu/sites/default/files/projects/Hermit%27s%20Rest%20Historic%20Structure%20Report.pdf

Thybony, Scott. 2005. *Grand Canyon: Hermit Trail Guide*. Grand Canyon, AZ: Grand Canyon Association.

HERMITS REST

Approximately 1.5 miles west of Pima Point and 8.9 miles west of Grand Canyon Village is Hermits Rest (elevation 6,650 feet), a rustic, rocky building designed by Mary Jane Colter in 1914 for Santa Fe Railway. Admission to Hermits Rest originally required a ticket sold at El Tovar Hotel or Bright Angel Camp for 50¢, but today admission is free. Visitors are welcomed to Hermits Rest by a stone arch and a broken mission bell; in earlier days, a lantern here helped guide travelers to the site. Colter added soot, smoke residue, and cobwebs in Hermits Rest above the 20-foot-high vaulted fireplace before the site opened in 1914 to provide a rustic, lived-in appearance. Colter wanted the building to look like the rocky den of a hermit.

Hermit Road (also called West Rim Road and Hermit Rim Road), which stretches from Grand Canyon Village to Hermits Rest, was built in 1912 by the

At the western end of Hermit Road, this arch welcomes visitors to Hermits Rest. When designing Hermits Rest, architect Mary Elizabeth Jane Colter departed from her plans for earlier structures at Grand Canyon National Park by evoking a simple lodge that might have been built by untrained mountain men, with an oversized fireplace, cobwebs, and soot accentuating its rustic quality. (Randy Moore)

Santa Fe Railway. Cars were initially banned from the road because they scared horses pulling open-air stagecoaches taking tourists to and from Hermits Rest. In 1935, NPS rebuilt and rerouted the road, which for many years was described as the best road in northern Arizona, "a boulevard in the wilderness," and "the most unique highway in the world" along the "dizzy edge of a stupendous chasm." Although Hermit Road ends at Hermits Rest, Grand Canyon extends another 180 rugged and largely inaccessible miles beyond the Rest. Hermits Rest is 35 miles from Desert View; elevations between these "bookends" of the developed areas of the South Rim range from 6,600 to 7,500 feet.

Hermit Trail, which starts just west of Hermits Rest, is a trail to the Colorado River that was a model for NPS trails in the 1920s and 1930s. This trail, which was promoted as being wide and comfortable, led to Hermit Camp, which was described by the Fred Harvey Company as "camping out deluxe" for its 11 tent-cabins, showers, and stables. Hermit Camp, which was near Hermit Creek on Tonto Platform, was operated by the Fred Harvey Company. Today, there are two designated campsites along the trail for backcountry permit holders.

Monument Creek Vista

Monument Creek Vista is an overlook along the South Rim's Hermit Road that is named for a 100-foot-high monument-like formation about a mile from the river along Monument Creek. (Monument Creek, which enters the Colorado River from river left near River Mile 93.5, forms Granite Rapid.) The overlook is also a shuttle bus stop that offers views of the headwaters of Monument Creek.

Monument Creek Vista is also the trailhead for part of the Hermit Road Greenway, which is a handicap-accessible trail that goes west to Hermits Rest. Parts of the 9-foot-wide, paved trail follow the old 1912 alignment of Hermit Road. The Hermit Road Greenway meets the U.S. Access Board's Accessibility Guidelines for Outdoor Developed Areas. Monument Creek Vista is 0.9 mile from Abyss and 1.7 miles from Pima Point.

Views at Hermits Rest are relatively limited by trees. Visitors to Hermits Rest can see Point Sublime (almost 800 feet higher than Hermits Rest), the Tonto Platform, and the Tower of Ra. A small store at Hermits Rest sells food, books, and souvenirs.

See also: Boucher, Louis; Colter, Mary Elizabeth Jane; Hermit Trail; Railroad

Further Reading

Anderson, Michael. 2001. *Along the Rim: A Guide to Grand Canyon's South Rim from Hermit's Rest to Desert View.* Grand Canyon, AZ: Grand Canyon Association.

HISTORIC SITES ACT OF 1935

The Historic Sites Act arose from the NPS's need for a stronger legal foundation for expanding its historic programs. The legislation was enacted primarily to organize the many federally owned monuments, historic sites, and parks in the National Park Service of the U.S. Department of the Interior. The act, which was signed into law on August 21, 1935, explicitly declared what the Antiquities Act of 1906 suggested, namely "that it is a national policy to preserve for public use historic sites, buildings, and objects of national significance for the inspiration and benefit of the people of the United States" (Historic Sites Act, 1935). The Historic Sites Act made preservation a governmental duty.

The first site protected by the Historic Sites Act was the Jefferson National Expansion Memorial in St. Louis, Missouri, on December 21, 1935. This National Memorial was created by President Franklin D. Roosevelt's Executive Order 7523 on December 21, 1935.

Several sites at Grand Canyon are protected by the Historic Sites Act of 1935. For example, Emery Kolb's residence/studio at the South Rim became an eyesore, but it was later renovated and today is a popular attraction in Grand Canyon Village. Emery's garage, where a skull and skeleton were found after Kolb's death, also still stands nearby.

See also: Grand Canyon Village

Further Reading

Historic Sites Act of 1935. (16 U.S.C. sec. 461–467). Retrieved June 1, 2017, at https://en.wikipedia.org/wiki/Historic_Sites_Act

HOGAN, DAN

Daniel Lorain Hogan (1866–1957) was a Flagstaff, Arizona, sheriff and part-time prospector who first visited Grand Canyon in 1890. The next year, he completed the first known rim-to-rim backpacking trip through Grand Canyon's central corridor. In 1893, Hogan—an orphan—discovered a vein of copper 1,100 feet below Maricopa Point; he and his business partner Henry Ward then filed a claim for the 20.64-acre Orphan Lode Copper Mine (with 4 rim-side acres), which is less than 2 miles west of today's Grand Canyon Village. (In some documents, Hogan referred to the mine as the Lost Orphan Mine.) There were more than 30 minable ores at Hogan's mine, which Hogan reached via his treacherous Hummingbird Trail (also called Hogan's Slide), which consisted of ladders, toeholds, ropes, and precarious steps across the cliff face.

After joining William Owen "Buckey" O'Neill's (1860–1898) regiment to fight with Teddy Roosevelt's Rough Riders in the Spanish-American War, Hogan returned to Grand Canyon. On March 23, 1906, President Roosevelt—Hogan's commander in the Spanish-American War—signed the certificate of patent giving the Orphan Lode Mine to Hogan. His mine was often inactive, so Hogan opened the 20-cabin Grand Canyon Trading Post on the rim above his mine. This inn, which by 1949 was called Rogers' Place (when it was leased to the son of Will Rogers Jr.), included a trading post, saloon, and swimming pool at rim's edge.

After rejecting NPS's offer of $5,555 for his undeveloped land, Hogan—who was wounded in one of the last fights with the Navajo—sold Lost Orphan Mine in 1947 to Madelaine Jacobs for $25,000. In 1953, after amateur prospectors found uranium at the site, Jacobs sold the mine and mineral rights to Western Gold and Uranium, which built a tramway 1,800 feet down the canyon's wall to access the expanded mine. By the mid-1950s, the mine was producing 1,000 tons of high-grade uranium ore (U_3O_8) per month that was processed in nearby Tuba City, Arizona. This uranium powered the country's nuclear power program. While this uranium was being mined, dust from the mine drifted through GCNP.

By the 1960s, however, scientists realized that uranium is abundant; indeed, it is more than 40 times more abundant than silver and more than 500-times more abundant than gold. This dropped the price of uranium. In 1963, President John F. Kennedy signed legislation authorizing NPS to acquire Orphan Mine on the condition that it could be mined for another 25 years and that the owners could operate Grand Canyon Inn through 1966. Western Gold and Uranium—after years of threatening to build a 400-room hotel on the rim—sold the land to NPS. By the time it closed on April 25, 1969, after 13 years of production, Orphan Mine had produced 495,107 tons of ore worth more than $40 million. Although tourists camped by the stockpiles of uranium for several years, today Hogan's abandoned mine—which can be seen from the path leading to the Powell Memorial—remains a low-level radiation area that is off-limits to tourists. The head frame at Orphan Mine, a landmark on the South Rim for more than 50 years, was removed in 2009.

Hogan died in 1957 in Veterans Hospital in Los Angeles, California.

See also: Maricopa Point

Further Reading

Lago, Don. 2015. *Grand Canyon: A History of a Natural World and National Park*. Reno: University of Nevada Press.

HOLMES, WILLIAM HENRY

William Henry Holmes (1846–1933) was an American artist, explorer, and archaeologist who began his career by producing beautiful, accurate, acclaimed drawings of shells and fossils at Smithsonian Institution. In 1872, Holmes replaced artist Thomas Moran with the Hayden Survey of the American West. Artist-cum-geologist Holmes later worked with Clarence Dutton, at which time Holmes illustrated the *Atlas* that accompanied Dutton's impressive *Tertiary History of the Grand Canyon, with Atlas* (Dutton, 1882).

Unlike Moran's images, which often combined scenes from different places and seldom included correct arrangements of Grand Canyon's rock formations, Holmes's illustrations were extremely detailed; his triptych *Panorama of Point Sublime* is one of the greatest images of Grand Canyon and a masterpiece of American scientific illustration. Holmes and Moran became the alter egos of Dutton and John Wesley Powell as they all described how people could see Grand Canyon while linking the canyon to bigger themes. Holmes was an unmatched geological artist; the images that Dutton famously described with words were what Holmes did with illustrations. Later in his life, Holmes succeeded John Wesley Powell as director of the Bureau of American Ethnology and later directed what became the National Museum of American Art.

Holmes later helped found modern archaeology, especially regarding pottery. He died on April 20, 1933, in Royal Oak, Michigan, and is buried beside his wife Kate in Rock Creek Cemetery in Washington, DC, beneath the epitaph, "They gave to art and science."

See also: Dutton, Clarence

Further Reading

Dutton, Clarence E. 1882. *Tertiary History of the Grand Canyon*. Washington, DC: U.S. Government Printing Office. (Reprinted 2001 by the University of Arizona Press, with an introduction by Wallace Stegner and a new foreword by Stephen J. Pyne.)

Kinsey, Joni L. 1998. *The Majesty of the Grand Canyon: 150 Years in Art*. Petaluma, CA: Pomegranate Communications.

HOLMSTROM, HALDANE "BUZZ"

Haldane "Buzz" Holmstrom (1909–1946) was born and raised in Oregon and worked on and off at a gas station most of his life. He was the first of his four siblings to finish high school. In his spare time, Holmstrom built boats. Buzz built his first boat in 1934 and, on October 25, 1937, began a historic 52-day, 1,100-mile trip from Green River, Wyoming, through Grand Canyon in a 15-foot-long, white, homemade Galloway-type boat made from "the right cedar tree" (Historical Marker Database, 2017). Holmstrom, who patterned his rowing after Nathan Galloway (i.e., rowing to slow his speed and to maneuver his boat), planned to go with a friend, but when his friend could not make it, Holmstrom decided to go alone. Holmstrom was guided by maps from Claude Birdseye's 1923 USGS expedition through Grand Canyon and Ellsworth Kolb's *Through the Grand Canyon from Wyoming to Mexico* (1914). He valued the solitude; as he wrote on the last night of his trip, "I know I have got more out of this trip by being alone than if I was with a party, as I have more time, especially at night, to listen and look and think and wonder about the grandeur that surrounds me."

When Holmstrom reached Lake Mead, motorboats would not tow him, so he spent the last four days of his trip rowing across the lake. He touched Hoover Dam on November 20, 1937, thereby becoming the first person to float through Grand Canyon by himself. On that trip, Holmstrom ran every rapid except Lava Falls. Holmstrom and an expedition of Carnegie–Caltech geologists, including Edwin McKee, were on the river at nearly the same time; their meeting during the trip was the first time that two river trips or two boatmen ever met in Grand Canyon. Holmstrom was also one of the last people to see Separation Rapid before it was submerged in Lake Mead. After his historic trip, Holmstrom went home to Oregon and resumed work at the gas station.

Holmstrom believed that he would be proud and happy after running Lava Cliff Rapid, the last of the "bad" rapids on his trip. Instead, Holmstrom wrote that:

> everything ahead seems kind of empty and I find I have already had my reward, in the doing of the thing. The stars, the cliffs and canyons, the roar of the rapids, the moon, the uncertainty and worry, the relief when through each one, the campfires at night, the real respect of the rivermen I met and others . . . This may be my last camp where the roar of a real rapid is echoed from the cliffs around and I can look at the stars and moon only through a narrow slit in the earth. The River and Canyons have been kind to me, I think my greatest danger is ahead, that I might get swellheaded over this thing. I am going to try to keep my mouth shut about it, go back to work in the old way, and have it only for a memory for myself. (Welch, Conley, and Dimock, 1998)

Fearing that someone would try to stop him if they knew his plans, Holmstrom had not told anyone about his trip. Nevertheless, the news of his historic feat got out and was later broadcast across the country. Needing money, Holmstrom did a few interviews but otherwise kept quiet. Holmstrom, who was uncomfortable with fame, was humble and deeply appreciative of the river; he viewed his accomplishment less with personal pride than with a deep appreciation and awe of the river's power and beauty. To Holmstrom, being immersed in that power and beauty was its own reward.

In 1938, Holmstrom began a second trip through Grand Canyon, this time accompanied by Amos Burg and Willis Johnson. During the trip, Burg rowed the 83-pound boat named *Charlie*; this boat, which was named for Charles Wheeler (who had donated $200 for the trip), was the first inflatable rubber raft to run Grand Canyon. Holmstrom rowed the same boat that he'd used the previous year for his solo trip through Grand Canyon. This time, however, he painted the boat red and named it *Julius F*, in honor of Julius F. Stone, who had sent $500 to support the trip. This was the first wooden boat to run all of the rapids in Grand Canyon. On this trip, Holmstrom ran Lava Falls.

During the 1938 expedition, Burg filmed a movie that was later released as *Conquering the Colorado* (1939). (Burg's film won an Academy Award in 1939.) Contrary to the movie's title, Holmstrom never spoke of "conquering" the river, instead emphasizing the river's beauty and how he "listened to the river." As Holmstrom noted, "Some people have said, 'I conquered the Colorado River.' I don't say so. It has never been conquered and never will. I think, anyone who it allows to go thru its canyons and see its wonders should feel thankful and privileged." Holmstrom got no money from the movie.

Holmstrom later enlisted in the Navy. After his discharge in 1945, he returned to Oregon and returned to work for the Bureau of Reclamation. The following year, he transferred to the U.S. Coast and Geodetic Survey to build and run boats. On

May 18, 1946, while on the second day of a surveying trip, Holmstrom died near Oregon's Grande Ronde River of a self-inflicted gunshot to the head. Holmstrom's death certificate lists Holmstrom's occupation as "Expert Boatman." He is buried at the Masonic Cemetery in Coquille, Oregon, beneath the epitaph, "Home is the sailor, home from the sea."

See also: Stone, Julius

Further Reading

Historical Marker Database. 2017. Green River, Wyoming Historical Markers. Retrieved July 1, 2017, at https://www.hmdb.org/results.asp?Town=Green%20River&State=Wyoming

Holmstrom, Buzz, and Dimock, Brad. 2003. *Every Rapid Speaks Plainly: The Salmon, Green, and Colorado River Journals of Buzz Holmstrom*. Flagstaff, AZ: Fretwater Press.

Welch, Vince, Conley, Cort, and Dimock, Brad. 1998. *The Doing of the Thing: The Brief Brilliant Whitewater Career of Buzz Holmstrom*. Flagstaff, AZ: Fretwater Press.

HOPI HOUSE

Just west of Verkamp's Visitor Center is the multistory, red sandstone Hopi House, which opened on January 1, 1905, to showcase Native American products and traditions. This building was the first of seven buildings at Grand Canyon designed by architect Mary Elizabeth Jane Colter between 1905 and 1937. Colter's other

Hopi House, which opened on January 1, 1905, was the first building at Grand Canyon designed by Mary Elizabeth Jane Colter. The building, which was patterned after traditional Hopi pueblos, continues to feature Hopi artists and their crafts. This photo of Hopi House and its Native American artists was taken ca. 1905. (Grand Canyon National Park Museum Collection)

iconic buildings in GCNP include Hermits Rest (1914), Lookout Studio (1914), Phantom Ranch (1922), Desert View Watchtower (1932), and Bright Angel Lodge (1935).

Colter patterned Hopi House after traditional Hopi pueblos at Third Mesa of Old Oraibi, an ancient village of the nearby Hopi Reservation. (The Hopi, whose small reservation north of Grand Canyon is surrounded by the Navajo Reservation, have lived in villages there atop three mesas for more than 900 years; this is the oldest community in the United States.) As in these pueblos, Hopi House has low, thatched ceilings and low doorways; most people must bow when they enter Hopi House. Colter repurposed several Western Union Telegraph poles when building Hopi House, so these wooden indoor beams in Hopi House are stamped "W.U.T." Low doors, low ceilings, and earth construction give Hopi House the feel of an adobe dwelling.

Hopi House was originally a market for arts and crafts produced by Native Americans, who lived and worked on-site. Although Native American artists no longer live in Hopi House, the building remains a showcase of handcrafted Native American rugs, kachinas, pottery, art, baskets, and jewelry. There are occasional presentations of Native American songs and dances on the patio at Hopi House.

See also: Colter, Mary Elizabeth Jane

Further Reading

Anderson, Michael. 2001. *Along the Rim: A Guide to Grand Canyon's South Rim from Hermit's Rest to Desert View.* Grand Canyon, AZ: Grand Canyon Association.

HOPI POINT

Hopi Point (elevation 7,043 feet), which is about 2 miles west of Grand Canyon Village, was originally named Rowes Point for Sanford Rowe (1853–1929), who opened a tourist facility south of there in 1892. Today, Hopi Point honors the Hopi Indians, who still live 100 miles east of Grand Canyon. The promontory, from which visitors can see 90 miles east to west, offers an excellent view of the Colorado River and, in the west, part of Granite Falls at Monument Creek. Hopi Point also features memorials to Claude Hale Birdseye, who led a pioneering USGS expedition through Grand Canyon in 1923, and to workers of the Civilian Conservation Corps, who made many improvements in GCNP in the 1930s.

Several mines were below Hopi Point. For example, in 1890, William Ashurst (the father of Senator Henry Ashurst, who cosponsored the legislation creating GCNP) was part of a "copper rush" there. Opposite Hopi Point on the north side of the river are Isis Temple, Horus Temple, and Osiris Temple, which were named for Egyptian deities by Clarence Dutton in the 1880s. A Fred Harvey brochure from the early 1900s described Hopi Point this way:

> [T]he panorama here, in broad daylight, seems to be the acme of scenic beauty. But when the afternoon sun . . . gradually sinks beneath the far mountain ranges, banked with clouds . . . the Canyon depths become a world of

mystery, with giant forms dimly outlined in the ghostly void. And then, miracle of miracles! The western sky becomes a riot of vivid colors, flaming across a field of turquoise, and fading to the afterglow.

Mohave Point

Mohave Point (elevation 6,975 feet) honors the Mohave Indians who lived along the Colorado River. Between Mohave Point and Hopi Point is The Inferno; below Mohave Point between Monument Creek and The Inferno is a jagged formation (capped with Supai Group) that Grand Canyon pioneer Emery Kolb named The Alligator. The Alligator points north-northwest to the Tower of Set, Horus Temple, and Osiris Temple; to the east of these three formations is Tower of Ra, and to the west is Isis Temple. Mohave Point also offers great views of the cliffs that border Abyss. In the east, the panorama seen from Mohave Point starts at Bright Angel Creek on the North Rim. Mohave Point, which is five miles east of Pima Point, offers expansive downstream views of the river that include Granite Rapid (at the end of Monument Creek), Hermit (Boucher) Rapid, and Salt Creek Rapid downstream.

Dana Butte, a mesa 2,000 feet below Hopi Point, dominates the vista at Hopi Point and points across the river to the flat-topped Shiva Temple, which is more than 600 feet higher than Hopi Point. This formation is bordered by Salt Creek and Monument Creek, which meet at the Colorado River and form Granite Rapids. Also visible are Trinity Canyon, Tonto Platform, and Cheops Pyramid; downstream are Point Sublime and Osiris Temple. Hopi Point—which extends farther into the canyon than any other viewpoint along Hermit Road—offers wide, unobstructed views and is a popular place for watching sunsets; views to the west featuring Cape Royal, Vishnu Temple, and Wotans Throne are often spectacular.

See also: Birdseye, Claude; Rowe, Sanford

Further Reading

Anderson, Michael. 2001. *Along the Rim: A Guide to Grand Canyon's South Rim from Hermit's Rest to Desert View.* Grand Canyon, AZ: Grand Canyon Association.

HYDE, GLEN AND BESSIE

In 1928, one day after divorcing husband Earl Helmick, Bessie Louise Haley married Glen Hyde in Twin Falls, Idaho. The couple decided to spend their honeymoon running the Colorado River through Grand Canyon on a 20-foot-long scow equipped with 20-foot-long oars at each end. Glen and Bessie went from Green River, Utah, to Bright Angel Creek in 26 days, at which time they hiked up to the South Rim to talk with river runner Emery Kolb. (The only guidebook that the Hydes took on their trip was Ellsworth Kolb's *Through the Grand Canyon from Wyoming to Mexico* [1914].) The Hydes had hoped to make money with their story, but Bessie apparently feared for her life; when she saw the new shoes on Emery Kolb's daughter, Bessie asked, "I wonder if I shall ever wear pretty shoes again."

Life jackets were common on river runners by this time, but Glen Hyde refused them, adding. "We don't need any artificial aids . . . I am going to do it without life jackets, or else." Kolb later recalled, "I did my best to get them to take some kind of life preservers, but they would not."

Just a few days later, with river flow at 8,000 cubic feet per second, the Hydes disappeared (probably around Mile 232) and presumably drowned after last being seen near Mile 95. Glen's 69-year-old father, Roland C. Hyde, who had already lost two other sons, sold almost everything he had to find his son. Requests for help with the search for the Hydes reached President Calvin Coolidge, whose Secretary of War (Dwight Davis) ordered an air search for the couple. While aboard an Army plane (the first plane flown within the Inner Gorge), Emery Kolb and GCNP ranger James Brooks spotted the Hydes' boat. Emery and his brother Ellsworth then repaired a boat they found at Diamond Creek and floated down the river to the Hydes' boat. The intact, upright boat included the Hydes' coats, Glen's rifle, and Bessie's camera and diary, with the last entry dated November 30, 1928. However, Glen and Bessie's bodies were never found. (Of the 10 life-jacketless drownings in Grand Canyon between 1880 and 1935, only three bodies were recovered.) Bessie Hyde, who came within 50 or so miles of being the first woman to complete a trip through Grand Canyon, was the first of a growing number of women to operate a boat in Grand Canyon.

The Hydes are commemorated by Honeymoon Rapids at River Mile 232, but the fate of the Hydes remains a mystery. In 1971, during a trip through Grand Canyon, Elizabeth "Liz" Cutler—a 61-year-old teacher from Pomeroy, Ohio—claimed that she was Bessie Hyde. Cutler's story appeared in *Outside Magazine* in 1985 and on NBC's *Unsolved Mysteries* in 1987 but was fiction. In 1992, a year after her death, friends settling river runner Georgie White Clark's estate found a copy of the marriage certificate of Glen and Bessie Hyde, prompting some to speculate that Georgie was Bessie Hyde. A subsequent investigation showed that she was not. The fate of the Hydes remains one of Grand Canyon's enduring mysteries.

See also: Kolb, Ellsworth and Emery

Further Reading

Dimock, Brad. 2007. *Sunk without a Sound: The Tragic Colorado River Honeymoon of Glen and Bessie Hyde*. Flagstaff, AZ: Fretwater Press.

I

IVES, JOSEPH CHRISTMAS

In 1857, the U.S. Army sent troops to Utah to quell rebellious Mormons. Leading the group was Army 1st Lt. Joseph Christmas Ives (December 25, 1829–1868), an 1852 U.S. Military Academy graduate. Because the overland route to Utah was arduous and difficult to supply, Ives's group tried to determine how far steamboats could navigate upstream on the Colorado River. Funds for Ives's exploration included $70,000 allocated by Secretary of War Jefferson Davis (later president of the Confederate States of America) in 1856 to determine supply routes for military operations in the West.

Treaty of Hidalgo

In 1846, the United States went to war against Mexico, primarily because of U.S. expansion and so-called Manifest Destiny. Two years later, on February 2, 1848, the Treaty of Guadalupe Hidalgo ended the Mexican-American War. This treaty, officially known as the Treaty of Peace, Friendship, Limits and Settlement between the United States of America and the Mexican Republic, was signed in Villa de Guadalupe Hidalgo, now a neighborhood in Mexico City.

Among other things, the Treaty of Hidalgo set the Rio Grande as the southern boundary of Texas and ceded much of northwest Mexico—today's American Southwest, including Grand Canyon—to the United States. In future decades, the United States wrested the area's land from its native residents. One of the first governmental surveys of the area was conducted by Lt. Joseph Christmas Ives, who led the Colorado Exploring Expedition in 1857–1858.

Ives and his 44-man crew explored the canyon by sailing up the Colorado River from its mouth aboard the 54-foot-long, iron-hulled, open-decked steamboat *Explorer*, which he built and tested in Philadelphia. The crew of 12 men included Smithsonian physician and geologist John Strong Newberry (1822–1892), the first geologist to see Grand Canyon. Ives and his crew often had to drag their boat across sandbars. All the while, they were watched by Native Americans, prompting Ives to report that he

> had a large crowd about me one day, and [I] exhibited several things that I supposed would interest them, among others a mariner's compass. They . . . thought we must be very stupid to be obliged to have recourse to artificial aid in order to find our way. (Ives, 1861)

Ives made it to Black Canyon, which is 350 miles above the mouth of the Colorado River and near the site of Hoover Dam. There, *Explorer* "brought up abruptly and instantaneously against a sunken rock . . . The concussion was so violent that the men near the bow were thrown overboard . . . and it was expected that the boat would fill and sink instantly" (Ives, 1861). This ended the navigation, and Ives concluded that Black Canyon was the farthest upstream point reachable by navigation of the Colorado River. (George Alonzo Johnson had steamed to near this point two months before Ives, but Johnson never published his report.)

Ives then split his crew; while some of his men returned downriver, Ives and others abandoned *Explorer* and hiked up to the rim of "Big Cañon." With help from conscripted Native American guides, Ives reached Diamond Creek. There, on April 5, 1858, Ives and his men became the first known non-Natives to reach the floor of Grand Canyon. The next people to see Grand Canyon would be members of John Wesley Powell's expedition in 1869.

Ives was the first to record his admiration and "wondering delight" for Grand Canyon's "stupendous formation" and scenery, noting that "the extent and magnitude of the system of canyons is astounding." Ives also complained about "the difficulty of getting out" (Ives, 1861) of the canyon. When he was boating up the river and surrounded by water, Ives noted the "majestic grandeur" of his surroundings. However, by the time he ended his exploration of Havasu Canyon and returned to the rim, the parched Ives declared "a more frightfully arid region probably does not exist up on the face of the earth" (Ives, 1861).

Ives's *Reports upon the Colorado River of the West* (1861) to the U.S. Senate, which was published by the War Department, was the first pictorial account of Grand Canyon. In that report, Ives became the first white person to praise the sublimity of Grand Canyon as "perhaps unparalleled in any part of the world." However, his report also famously claimed Grand Canyon was "a vast ruin" that was an "altogether valueless . . . profitless locality" resembling "the gates of hell" and would "forever be unvisited and undisturbed . . . After entering it there is nothing to do but leave" (Ives, 1861). Ives named Diamond Canyon, which is on the south side of Grand Canyon near River Mile 225.7.

Ives's expedition produced an early map of Grand Canyon that was drawn by topographer Baron Frederick W. von Egloffstein who, with Heinrich B. Möllhausen, also produced the first sketches of Grand Canyon. Those sketches exaggerated Grand Canyon's narrowness and depth. Ives subsequently helped build the Washington National Monument, after which he joined the Confederate Army in late 1861. Ives, who died in New York City on November 11, 1868, is honored at GCNP by Ives Point, which is just north of 120 Mile Creek.

See also: Newberry, John Strong

Further Reading

Ives, Joseph C. 1861. *Report upon the Colorado River of the West, Explored in 1857 and 1858.* Washington, DC: Government Printing Office. (Republished in 2005 by University of Michigan Library.)

J

JAMES, GEORGE WHARTON

George Wharton James (1858–1923) was a popular American writer, photographer, and speaker whose *In and around the Grand Canyon* (1900) was one of the earliest travel books about Grand Canyon. Later, James lived at Bass Camp while he wrote part of *The Grand Canyon of Arizona: How to See It* (1910). James, an ordained minister, was defrocked by the Methodist church after being charged with real estate fraud and sexual misconduct, but was later reinstated. Examples of sites in Grand Canyon named by James include:

Dana Butte directly north of Powell Memorial honors Yale University geologist James Wright Dana (1813–1895). Dana Butte is made mostly of Redwall Limestone.

Dutton Point on the east side of Powell Plateau was named for surveyor Clarence Edward Dutton (1841–1912), who began the tradition of naming sites in Grand Canyon for mythological deities. Dutton's *The Tertiary History of the Grand Canyon District* (1882) is a classic book about Grand Canyon.

Cheops Pyramid west of Phantom Canyon was named for its resemblance to the Great Pyramid of Cheops, the oldest (and only survivor) of the seven wonders of the ancient world. This peak is the easternmost of the sites in Grand Canyon named for Egyptian gods; other similarly named sites include Isis Temple, Horus Temple, and the Tower of Set.

Isis Temple, which is west of Phantom Ranch, honors Isis, the Egyptian goddess of nature. The top of Isis Temple consists of Coconino sandstone.

Tower of Set is a 6,012-foot red rock butte 1.5 miles north of the Colorado River that honors Set, the chief Egyptian god of disorder, violence, and war. This site featured prominently in Harriet Russell Strong's plan in 1918 for a trans-canyon tram.

Zoroaster Temple, near the head of Bright Angel Creek, is a butte at River Mile 84.5 honoring Zoroaster, a famous Persian religious leader (c.660–580 BCE). Zoroaster rejected deities and preached about perfection, good, and free will.

James was an early champion for the protection of Grand Canyon, noting in 1900, "The people ought to own, forever, this glorious inheritance. There should be no limitations placed upon their perfect enjoyment of it but such as are necessary for its preservation" (Kinsey, 1998). For many years, James's books were the standard guidebooks at Grand Canyon. Unlike most earlier books, James's books about Grand Canyon were written for tourists who were going to Grand Canyon. His books often emphasized the majesty of the canyon, which James believed "is one of those sights that words cannot exaggerate."

James later founded the Anti-Whispering Society, which was devoted to suppressing talking audiences, peanut fiends, and crying babies. James died on November 8, 1923, in St. Helena, California, and is buried in Mountain View Cemetery and Mausoleum in Altadena, California.

See also: Bass Trails and Bass Camp

Further Reading

James, George Wharton. 1900. *In and around the Grand Canyon*. Boston: Little, Brown, and Company.

James, George Wharton. 1910. *The Grand Canyon of Arizona: How to See It*. Boston: Little, Brown, and Company.

Kinsey, John L. 1998. *The Majesty of the Grand Canyon: 150 Years in Art*. Petaluma, CA: Pomegranate Communications.

K

KAIBAB PLATEAU AND KAIBAB NATIONAL FOREST

The Kaibab Plateau consists of 1,152 square miles of heavily forested land north of Grand Canyon. It is named for the Paiute word *Kai'-vav-wi*, which means "mountain lying down." (In English, this word became "Kaibab.") Mormon leaders had earlier named this area Buckskin Mountain and Buckskin Plateau for the abundance of mule deer here. The Colorado Plateau, of which Kaibab Plateau is a part, is the world's second highest plateau (the Tibetan Plateau is the world's highest). The North Rim is the southern border of the Kaibab Plateau.

Kaibab National Forest, whose 1.6 million acres border the North and South Rims of Grand Canyon, is part of the largest contiguous ponderosa pine forest in the United States. The logging industry along the South Rim helped produce Flagstaff and Williams, both of which were originally built from trees harvested near the South Rim. The forest was taken from the public domain in 1893 and later included Grand Canyon Forest Reserve. In 1906, President Theodore Roosevelt used part of the forest to create Grand Canyon Game Reserve. In 1907, the remaining Grand Canyon Forest Reserve became Grand Canyon National Forest, and the following year, the Grand Canyon National Forest became the Kaibab National Forest north of the Colorado River (it became part of the Coconino National Forest south of the river). In 1910, the western part of Coconino National Forest became Tusayan National Forest. In 1934, this forest became the Tusayan District of the Kaibab National Forest.

In 1919, GCNP was created from forest service land and was initially staffed by Forest Service rangers. In 1927, GCNP expanded by incorporating 45,000 acres of Kaibab National Forest on the North Rim and 2,000 acres on the South Rim. Today, Kaibab National Forest is one of six national forests in Arizona. All major roads to the North and South Rims go through Kaibab National Forest. There are an average of 100 fires per year in Kaibab National Forest, several of which are caused by humans.

Just north of Tusayan at the edge of Kaibab National Forest is a sculpture of Smokey Bear, an iconic advertising mascot of the U.S. Forest Service. Smokey Bear was created in 1944 in response to Japan's attempts to burn U.S. forests. The artist who drew Smokey was Albert Staehle (1899–1974), who was already famous for drawing other cute animals such as Borden's Elsie the Cow. Smokey Bear's most famous slogan—"Remember . . . Only YOU Can Prevent Forest Fires" followed three years later. His image, which is protected by federal law (Smokey Bear Act of 1952), is regulated by the U.S. Department of Agriculture. When the first living

Smokey Bear—a three-month-old American black bear cub rescued from the Capital Fire in Lincoln National Forest in New Mexico—lived at the National Zoo, it received more than 10,000 letters per week, prompting the U.S. Postal Service to give the bear its own ZIP Code (20252). The U.S. Forest Service's Smokey Bear Awards recognize service in preventing and increasing awareness of wildfires. In 1984, the U.S. Postal Service honored Smokey Bear by issuing a 20¢ stamp featuring Smokey and a cub in a tree.

When Smokey Bear was created, he wore a campaign hat of the NPS. This hat, in turn, was derived from those worn by cavalry who protected America's first national parks. State police still wear such hats; this is why truckers refer to state police as "bears" and "Smokey."

See also: North Rim; Squirrels

Further Reading

National Forest Foundation. 2017. Kaibab National Forest. Retrieved July 1, 2017, at https://www.nationalforests.org/our-forests/find-a-forest/kaibab-national-forest

Smokey Bear. 2017. Retrieved July 1, 2017, at https://smokeybear.com/en

KEN PATRICK TRAIL

This moderate, 9.8-mile (one-way) trail winds through cool Canadian Zone forests from Point Imperial to the North Kaibab Trailhead. The trail begins just west of the Point Imperial parking lot and hugs the canyon's rim for the first 3 miles; along the way, hikers enjoy beautiful views to the south and east and great views of Mount Hayden, Painted Desert, Little Colorado River Gorge, and Nankoweap Creek. The trail then crosses Cape Royal Road and goes west along a wooded wash and includes up-and-down sections that cross several washes. Along the way, hikers visit many spectacular viewpoints and meet the Uncle Jim Trail leading to Uncle Jim Point. About 4 miles from the North Kaibab Trailhead is the abandoned trailhead of the Old Bright Angel (i.e., Woolley) Trail. There is no water on the Ken Patrick Trail.

Ken Patrick Trail honors Kenneth Carmel Patrick (1933–1973), who was raised by his father Slim at Phantom Ranch at the bottom of Grand Canyon. Ken later worked as an NPS ranger at the North Rim before transferring to California's Point Reyes National Seashore. There, on August 5, 1973, he was murdered, thus becoming the first NPS ranger killed in the line of duty. Patrick is buried in GCNP's Pioneer Cemetery.

See also: Pioneer Cemetery and Other Gravesites

Further Reading

Zion National Park. 2017. Ken Patrick Trail from Point Imperial to North Kaibab Trailhead. Retrieved June 29, 2017, at http://www.zionnational-park.com/grand-canyon-ken-patrick.htm

KOLB, ELLSWORTH AND EMERY

Ellsworth Leonardson "Ed" Kolb (1876–1960) came to Grand Canyon from Pittsburgh, Pennsylvania, in October 1901, one month after Grand Canyon Railway arrived at the South Rim of Grand Canyon. Kolb, who arrived with $2, worked at several jobs, including woodchopper, construction worker, locomotive hostler, and bellhop at the Bright Angel Hotel on the South Rim. At Ellsworth's urging, and with reassurances to Ellsworth's mother from Ralph Cameron, younger brother Emery Clifford Kolb (1881–1976) joined his older brother at Grand Canyon the next year. Emery had served in World War II in the Army Signal Corps; Ellsworth had tried to serve, but his application was rejected.

The Kolbs bought a photography studio at Williams, Arizona, which Emery operated for a year. In 1903, the brothers dismantled the studio and gave the lumber to Ralph Cameron in exchange for space at the South Rim. There, the Kolbs opened a rim-side photography studio in a tent-cabin near Ralph Cameron's hotel; their first "darkroom" was a shallow mineshaft. The Kolbs' studio, which was made into a small wooden building in 1904, was the first permanent commercial photography studio at Grand Canyon. From their studio, the Kolbs catered to the growing number of tourists at the canyon. Some of the Kolbs' income involved photographing tourists on mules descending Bright Angel Trail. In 1903, when Ralph Cameron acquired ownership of Bright Angel Trail, the Kolbs collected $1 (of which the Kolbs kept only 30¢) from each tourist using the trail. During their careers, Kolbs photographed more than 50,000 mule strings descending into Grand Canyon.

In 1905, Emery married Blanche Bender; their first home was a tent-cabin on the South Rim. The Kolb brothers also owned an 894-square-foot cabin about 0.5 mile south of Kolb Studio. (This property, built in 1935, burned down on February 19, 1979.) Blanche worked as the Kolbs' clerk and bookkeeper until her death in 1960, but Emery's marriage strained his relationship with Ellsworth. In 1906, the Kolb brothers built a wooden darkroom at Indian Garden (3,000 feet below the rim), where they used the site's water for photofinishing. This building was destroyed by a flashflood in 1910, after which the brothers built a replacement. For years thereafter, Emery ran the 9-mile round-trip from their rim-side studio as many as three times per day to use Indian Garden's clean water to develop the brothers' pictures. Kolb stopped his daily round-trips when water became available at the South Rim around 1930. Edith Kolb (1907–1979), the only child of Emery and Blanche Kolb, became the first Anglo child born (and the only such child to live and grow up) at Grand Canyon Village. Edith was named for Emery and Ellsworth's sister Edith, who died a month after birth.

In 1912, using a newly invented Pathé-Frères motion picture camera that they bought for $250, Emery and Ellsworth Kolb filmed a 45-minute-long movie titled *The Grand Canyon Film Show* documenting their boat trip through Grand Canyon. For the trip, the Kolbs—who were the 26th and 27th people to complete a boat ride through Grand Canyon—sought boating advice not from Robert Stanton but instead from Nathaniel Galloway. Prospector and newspaperman Charles Russell,

Although the Kolb brothers became famous for being the first to take moving pictures on a river trip through Grand Canyon in 1911–1912, they posed for their most iconic photo at the canyon's South Rim. The brothers titled this photo, which shows Ellsworth lowering Emery into a crevice, "The View Hunters at Work in the Canyon," but it has since become known colloquially as "Kolb on a Rope." This photo was taken along the Trail of Time between markers for 3,220 and 3,330 Million Years. (Library of Congress)

who had run some of the Colorado River in Grand Canyon, confidently offered "to furnish boats, men, and provisions [to the Kolbs] for the trip for $10,000 and take the party through." However, Ellsworth declined Russell's offer, noting that Russell had lost three boats before getting through half of Grand Canyon's rapids and that he recovered his last boat by chance. In 1913, Galloway was offering to take motion picture companies through Grand Canyon for $3,000 but had no takers.

The Kolbs' daredevil expedition was only the eighth successful trip through Grand Canyon, and their movie was the first movie of a river trip through Grand Canyon. The movie included a few admirations of the scenery but focused mainly on the brothers' adventure, including flipping boats, getting thrown overboard, crashing against rocks, and smashing a hole in one boat. The Kolbs—who spent just over 100 days on the river between Green River, Wyoming, and Needles, California—used 16-foot-long, Galloway-type boats (rowed Galloway-style) and Frederick Dellenbaugh's *A Canyon Voyage* (1908) as their guide for the trip. Ernest Kolb (1891–1987), the youngest brother of the Ellsworth and Emery, accompanied his brothers for about 20 miles; the Kolbs were also assisted on part of their trip by Hubert "Bert" Lauzon (1886–1951), who became a Grand Canyon legend after working for William Bass and as a Grand Canyon constable and park ranger. The Kolbs made no pretense about the purpose of their trip; as Ellsworth noted: "We could not hope to add anything of importance to the scientific and topographic knowledge of the canyons, and merely to come out alive at the other end did not make a strong appeal to our vanity. We were there as scenic photographers in love with their work" (Lago, 2015). After their trip, the Kolbs advertised themselves as "Kolb Brothers—Photographic Explorers."

The U.S. Forest Service banned the Kolbs from showing the film at the canyon, so the brothers (or, in most cases, only Emery) showed it as they toured various cities across the United States. Starting on April 15, 1915, when they were finally allowed to show the movie at Grand Canyon, the Kolbs showed their movie—later supplemented with footage from a 1921 USGS expedition (that included Ellsworth as a hired boatman)—in their studio every day until Emery's death in 1976, as well as during lectures they gave throughout the country. Today's visitors to Kolb Studio can see the projector the Kolbs used to show the movie.

For the expedition, Emery named his boat *Edith* for his 4-year-old daughter, and Ellsworth named his boat *Defiance* "because nobody loves me." Both of these boats are part of GCNP's collection of historic boats. Instead of revealing Grand Canyon, the Kolbs used their movie and Grand Canyon to promote themselves and serve tourists. To promote their movie, the Kolbs hired kids to distribute cards claiming that "you will be amazed" by the "authentic motion pictures and thrilling story" of the Kolbs' trip through Grand Canyon.

The first account of the Kolbs' adventure appeared in *The American* magazine in 1913. During this time, Emery Kolb was giving well-attended presentations across the country about his trip through Grand Canyon. In a standing-room-only talk in Chicago, Alexander Graham Bell introduced Kolb to Gilbert Grosvenor (1875–1966), the president of the National Geographic Society. Grosvenor was impressed with Kolb, later telling him, "The fact that over 200 people were content to stand throughout your entire entertainment is a great tribute to your power as a speaker and the nature of your subject" (National Geographic, 2000). Grosvenor, the first full-time editor of *National Geographic* and a pioneer of photojournalism, was interested in the Kolbs' story, and, the following year, *National Geographic* dedicated 85 pages of its August issue to the Kolb brothers and Grand Canyon. That same year, Macmillan publishers published Ellsworth's photo-laden *Through the Grand Canyon from Wyoming to Mexico*, a matter-of-fact account of his and his brother's do-it-yourself voyage through Grand Canyon on the Colorado River. For $3.50, purchasers received a signed copy of the book, perhaps with a signed photograph of both brothers. (By the 1960s, the hardcover price rose to $7.) Ellsworth wrote the book, and both Ellsworth and brother Emery shared photographic credits. The foreword to Kolb's book was written by celebrated author Owen Wister, whose *The Virginian* (1902) invented the cowboy Western genre. Wister criticized the railroad's attempts to evict the Kolbs from Grand Canyon:

> But is the Santa Fe wise in its persecution of the brothers Kolb? . . . Why, because the brothers Kolb are independent, crush their little studio, stifle their trade, push these genius artists and lovers of nature from the Canyon that nobody has photographed or can photograph as well?

Grand Canyon later became the backdrop for a variety of Harlequin romances written by authors such as Anne Marie Duquette, Ann Collins, and Patricia Chandler.

In 1915, the success of the Kolbs' movie prompted the brothers to add a small showroom (now part of the auditorium) to their studio that allowed them to show their movie to more customers. This movie was the first film of river running in Grand Canyon. Although a dispute with the Fred Harvey Company prevented the Kolbs from showing their movie at Grand Canyon for several years, Emery showed the film daily at Kolb Studio from 1915 until his death in 1976; this was the longest run in the world for any movie. Most of the photos sold by the Kolbs featured tourists aboard mules. The Kolbs also intentionally photographed some of the same views photographed on the Powell expedition, to which the Kolbs added their copyright. This was the first such deliberate rephotography in the American West. The Kolbs' photographs shaped views of what GCNP and other national parks should look like.

In 1922, barnstorming stunt pilot Royal V. Thomas, with Ellsworth Kolb as his cameraman, landed the first airplane—a custom-built Thomas Special—in Grand Canyon. Thomas and Kolb's stunt, which was approved by GCNP Superintendent Walter Crosby, was assisted by GCNP rangers. Thomas landed on, and took off from, a 90-foot-wide, 400-foot-long runway prepared by NPS rangers at Plateau Point on the Tonto Platform. One commentator declared the feat to be the most dangerous stunt ever done. Two weeks after the publicity stunt, the Fred Harvey Company paid Thomas to repeat the flight. This time the trip's photographer was from Fox News Company, and the film from this trip (i.e., not Kolb's film of the original landing) was shown across the country. Emery Kolb secured permits from the U.S. Forest Service to build airstrips on both sides of GCNP, but he never built the strips.

In 1923, Emery Kolb was part of the Birdseye USGS expedition through Grand Canyon to identify potential sites of dams. During this expedition, Edith Kolb (Emery's daughter) met her father at Hance Rapid and, while there, became one of the first women to run a major Grand Canyon rapid. The Flagstaff, Arizona, newspaper reported that Edith had inherited her adventurousness from her father and uncle.

By 1924, the business relationship between the Kolb brothers dissolved (Emery later claimed that Ellsworth had a nervous breakdown, and the Kolbs flipped a coin to decide which brother would leave Grand Canyon). Emery won two of the three flips and rights to Kolb Studio, and Ellsworth moved to Los Angeles, where he worked for an aeronautics firm. (Some researchers believe the coin toss story is apocryphal.) Ellsworth, who returned to Grand Canyon occasionally and helped Emery, received a monthly stipend of $150 from Emery until Ellsworth died in his sleep in Los Angeles in 1960.

Emery Kolb completed the last major addition to Kolb Studio in 1925. The Fred Harvey Company later tried to run Kolb out of Grand Canyon by building Lookout Studio, which blocked views of and access to Kolb Studio. Many people mistook Lookout Studio for Kolb Studio. The enraged Emery Kolb retaliated by printing postcards describing the Lookout Studio as "an obstruction."

Although Emery owned several expensive cameras, he took many of his most famous pictures with a Brownie camera given to him by writer Zane Grey. Grey also gave Kolb the oars that hang in the darkroom in Kolb Studio. Elsewhere in the home are three original paintings by North Rim artist Gunnar Widforss, to whom Emery Kolb had loaned money. NPS occasionally offers tours of the Kolb residence.

In 1930, Emery renewed his 10-year contract to do business at Grand Canyon. Kolb paid the government $250 plus 4 percent of his gross profits (excluding his fees for lectures, which he delivered across the United States). In 1940, Emery and Barry Goldwater were part of a *Ripley's Believe It or Not!* that was broadcast live from the banks of the Colorado, which was described as the most dangerous and exciting feat a human had ever tried. Emery, by this time a celebrity at Grand Canyon, once accompanied Herbert Hoover on a hunting trip on the North Rim. By 1962, Emery had sold his studio, adjacent garage, and village residence to NPS for $65,000.

In 1974, the 93-year-old Emery Kolb was flown by helicopter to the confluence of the Little Colorado and Colorado River, after which he ran the Colorado River for the last time (he went as far as Crystal Rapid). Two years later, the legendary Kolb—the last living pioneer of Grand Canyon—died in a Flagstaff hospital at age 95 and was buried near his brother Ellsworth and beside his wife Blanche in Pioneer Cemetery on the South Rim. (Blanche had died in 1960 of a heart attack in Kolb Studio.) That same year, Ellsworth Kolb's *Through the Grand Canyon from Wyoming to Mexico* went out of print after 27 pressruns. There are now 29 variants of that book, with most having 76 photographic plates (and some having as few as 29). When the University of Arizona Press reprinted Ellsworth's book, river runner Barry Goldwater added an introduction.

Emery Kolb's 72 years at the South Rim are the record for longest residence at Grand Canyon Village. During his remarkable career, Kolb photographed more than 3 million visitors at Grand Canyon. Kolb, who established a scholarship for valedictorians of Grand Canyon High School, planned to leave his business to his daughter Edith, but she was indifferent about the business, and GCNP eventually took over Kolb's five-story, 23-room building. Emery Kolb left behind thousands of photos and hundreds of pieces of camera equipment.

Shiva Temple

Shiva Temple is a wooded butte named for Shiva, the Hindu destroyer of the ego and universe. Shiva Temple, the largest of the North Rim buttes, rises 4,000 feet from its base.

In 1937, scientists from American Museum of Natural History examined Shiva Temple and Wotans Throne to see whether plants and animals growing atop these buttes—which had been isolated from the North Rim for 100,000 years—had evolved new species. They had not; the species there were like those of adjacent areas. However, the scientists found Anasazi ruins, deer antlers, boxes of film, and tissues left earlier by Emery Kolb and his friends. Undeterred, GCNP superintendent Miner Tillotson urged the scientists to study other such "sky islands" in the canyon, especially Fishtail Plateau, which is about 25 miles downstream from Shiva Temple. Today, Shiva Temple—whose top spans 6 square miles—is a popular place for climbers.

After Emery died in 1976, the Grand Canyon Natural History Association briefly continued photographing mule rides descending Bright Angel Trail. Within a few months, however, this ended, and Kolb Studio was boarded up and closed. The building remained vacant for more than a decade, at which time NPS considered it an eyesore and urged that it be razed. However, it was saved when Congress passed a law prohibiting the removal of any structure in NPS more than 50 years old (the property was also protected by its inclusion in the Grand Canyon Village Historical District). On April 13, 1990, the Grand Canyon Natural History Association (now the Grand Canyon Association) opened their store at Kolb Studio. Profits were used to restore the rest of the building.

The Kolbs are commemorated in GCNP by Kolb Arch on river right near River Mile 52.5. This 147-foot-long arch, which is the longest arch in Grand Canyon, was found by river runner Barry Goldwater. It is also known as Kolb Bridge and Kolb Natural Bridge.

See also: Cameron, Ralph; Grand Canyon Natural History Association; Grand Canyon Village; Pioneer Cemetery and Other Gravesites

Further Reading

Kolb, Ellsworth. 1914. *Through the Grand Canyon from Wyoming to Mexico.* New York: Macmillan Company.

Lago, Don. 2015. *Grand Canyon: A History of a Natural World and National Park.* Reno: University of Nevada Press.

National Geographic. 2000. Rock of Ages. Retrieved July 1, 2017, at http://ngm.nationalgeographic.com/flashback/2000#/2000-07-FB.jpg

Naylor, Roger. 2017. *The Amazing Kolb Brothers of Grand Canyon.* Grand Canyon, AZ: Grand Canyon Association.

Suran, William C. 2003. *The Brave Ones: The Journals and Letters of the 1911–1912 Expedition down the Green and Colorado Rivers by Ellsworth L. Kolb and Emery C. Kolb.* Flagstaff, AZ: Fretwater Press.

Suran, William C. 1991. *The Kolb Brothers of Grand Canyon.* Grand Canyon, AZ: Grand Canyon Natural History Association.

KOLB STUDIO

Ellsworth Kolb arrived at Grand Canyon from Pittsburgh, Pennsylvania, in 1901, and his younger brother Emery joined him the next year. The brothers opened Kolb Studio on a 20-by-55-foot ledge on Ralph Cameron's Cape Horn Mining Claim at the trailhead of Bright Angel Trail. The Kolbs had obtained this place in 1903 in exchange for lumber the Kolbs gathered when they tore down a photography studio in Williams that they had bought for $425. In 1904, Kolb Studio was the first permanent commercial photography studio (and one of the first commercial businesses) at Grand Canyon. The Kolbs collected $1 (but kept only 30¢) from each person who ventured down Cameron's Bright Angel Toll Road (now Bright Angel Trail).

The Kolbs later added to the house, including places for their living quarters and a small showroom in 1915. The Kolbs' five-story, 8,000-square-foot building that hangs on the South Rim eventually served as a home, studio, and theater for the Kolbs. In 1905, Emery married Blanche Bender. Emery and Blanche had one daughter, Edith Blanche Kolb.

The Kolb brothers' primary source of income initially involved selling postcards (including those of bargirls from Williams), but later the Kolbs began photographing hikers and mule caravans descending Bright Angel Trail. (After visiting Kolb Studio, Walt Disney also began selling photos of tourists at Disneyland after it opened in 1955.) One of the brothers (usually Emery) would then run to Indian Garden (3,000 feet below), where he would use the water there to develop the film, and then race back up the trail to sell the pictures for 25¢ each at the end of the mule ride. The Kolbs also got water from Rowe Well, a watering hole a few miles south of Grand Canyon.

The Kolb brothers were not the first people to run the Colorado River through Grand Canyon, but they were the first to film such a journey. For their 101-day, 1,167-mile trip in 1911–1912, Emery named his boat *Edith* for his 4-year-old daughter, and Ellsworth named his boat *Defiance* "because nobody loves me." The Kolbs' 45-minute movie titled *Grand Canyon Film Show* was shown daily at Kolb Studio until 1976, making it the longest-running movie in history. In 1932, the Kolbs added recorded narration to the movie, but Emery continued to introduce the movie to guests. The Kolbs also showed the movie across the country, including in Pittsburgh, New York (at Carnegie Hall), and Washington, DC. The showing in Chicago was attended by Gilbert Grosvenor, the publisher of *National Geographic*. When Grosvenor included an article about the Kolbs and Grand Canyon in the August 1914 issue of his magazine, interest in Grand Canyon increased dramatically.

The brown building just a few yards west of Kolb Studio is Emery Kolb's garage. When Emery died, a canoe (given to Emery by North Rim pioneer David Rust) was found in the garage containing parts of a human skeleton. (Although some people suspected that the skeleton belonged to Glen Hyde, they were wrong; Emery had found the skeleton in 1931 below Yavapai Point.) Today, the five-story, 23-room Kolb Studio is operated by the Grand Canyon Association and houses a gallery, store, and exhibits. From the studio, visitors have excellent views of the switchbacks of Bright Angel Trail, Indian Garden, Battleship, and Cheops Pyramid.

See also: Grand Canyon Village; Kolb, Ellsworth and Emery

Further Reading

Anderson, Michael. 2001. *Along the Rim: A Guide to Grand Canyon's South Rim from Hermit's Rest to Desert View.* Grand Canyon, AZ: Grand Canyon Association.

L

LAKE MEAD AND HOOVER DAM

Near the southwest edge of Grand Canyon, on the Arizona–Nevada border, the Colorado River flows into Lake Mead, which began filling behind Boulder Dam (now known as Hoover Dam) in 1935. At the time, this dam—which was built in less than five years—was the largest hydroelectric dam in the world, and it remains the tallest concrete-arch dam in the United States. At its crest, Hoover Dam is 45 feet thick, and at its base it is 660 feet thick. When Lake Mead is full, the 726-foot-high, 1,244-foot-long dam backs the Colorado River up to the southwestern end of GCNP. The 3.25 million cubic yards of concrete used to build the dam (not counting the additional 1.5 million cubic yards of concrete for supporting structures) would build a two-lane highway from New York to San Francisco. Boulder Dam provided work (during the Great Depression) for more than 4,000 men, most of them heads of families; many thousands more earned a living by making materials and machinery for the dam.

Hoover Dam and Lake Mead, which changed the economy, landscape, and fate of the American Southwest, helped transform large sections of a barren desert into fertile farmland. However, contrary to popular belief, Hoover Dam was not the first dam on the Colorado River; that distinction goes to Laguna Dam, which was built near Yuma, Arizona, in 1909.

Lake Mead, which is about 30 miles from Las Vegas, is a National Historic Landmark named for Elwood Mead (1858–1936), head of the Bureau of Reclamation from 1924 to 1936. In 1936, NPS and the Bureau of Reclamation created Boulder Dam Recreation Area, the first national recreation area in the United States. In 1947, the name of this site was changed to Lake Mead National Recreation Area. This recreation area was expanded in 1950 with the addition of Davis Dam (near Bullhead City, Arizona), which created Lake Mohave. Lake Mead National Recreation Area, which includes three of the four American desert ecosystems (Mohave, Great Basin, and Sonoran), ranges from 500 feet to 7,000 feet above sea level.

Boulder Dam was first presented to Congress in 1922. When it was approved six years later by President Calvin Coolidge, the site for the dam was Boulder Canyon. The dam's site was later moved several miles downstream to Black Canyon, where topographer Roland Burchard—a member of Birdseye's 1923 USGS expedition through Grand Canyon—identified and mapped the best site for the dam. Despite this change in location, most people continued to refer to the dam by its original name: Boulder Dam. After being elected president in 1920, Warren G. Harding appointed California engineer Herbert Hoover as Secretary of Commerce. In that position, Hoover led much of the work on the Colorado River Compact and the

Boulder Dam Project. When the dam was officially announced on September 17, 1930, Secretary of the Interior Ray Lyman Wilbur announced:

> I have the honor and privilege of giving a name to this structure. In Black Canyon, under the Boulder Canyon Project Act, it shall be called Hoover Dam . . . [because Hoover was] the great engineer whose vision and persistence, first as chairman of the Colorado River Commission in 1922, and on so many other occasions since, has done so much to make this dam possible. (Turner, 2016)

(By that time, Hoover was Wilbur's boss and the 31st president of the United States.) Many people opposed the name because they blamed Hoover for the Great Depression. Although some people referred to the dam as Boulder Dam, Congress and others referred to it as Hoover Dam, and in 1935 the U.S. Postal Service even issued a 3¢ stamp honoring the name "Boulder Dam." However, Hoover was soundly defeated in the election of 1932 by Franklin D. Roosevelt, who replaced Wilbur with Harold Ickes as Secretary of the Interior. Soon after his appointment, Ickes announced the dam would be named Boulder Dam because "the men who pioneered this project knew it by this name" (Turner, 2016). President Roosevelt was on hand to dedicate the dam—as Boulder Dam, almost two years ahead of schedule—on September 30, 1935. (After the ceremony, Roosevelt became the first U.S. president to visit Las Vegas.) Hoover was not invited to that ceremony, and Roosevelt did not mention Hoover's name.

The controversy about the dam's name was not officially resolved until April 30, 1947, when President Harry Truman signed House Resolution 140 (passed unanimously by Congress) naming the dam Hoover Dam:

> As President, Herbert Hoover took an active part in settling the engineering problems and location of the dam in Black Canyon . . . the construction contracts were signed under his administration, and when he left office, construction had been pushed to a point where it was more than a year ahead of schedule. (Turner, 2016)

Hoover Dam was built from 1931 to 1936 at a cost of $49 million ($700 million in 2017) and 112 lives. The primary purposes of Hoover Dam were to harness the water of the mercurial Colorado River, produce a reservoir for storing water, and generate hydroelectric power for growing population centers in the Southwest.

Lake Mead (elevation 1,221 feet) is the largest human-made reservoir (as measured by volume) in the world. When full, Lake Mead is 112 miles long, has a maximum depth of 590 feet, has 759 miles of shoreline, covers 247 square miles, and contains more than 26 million acre-feet of water. Lake Mead stores water for Nevada, Arizona, California, and Mexico. However, in early 2018, Lake Mead was more than 130 feet below capacity because of drought and increased usage of the lake's water; this deficit of water is visible as a "bathtub ring" of mineral deposits on cliffs surrounding the lake. In 2017, the scheduled depletions of Lake Mead were not sustainable.

The filling of Lake Mead submerged several communities in Nevada, most notably the small town of St. Thomas. St. Thomas, which was founded in 1865 by Mormon pioneers in the Moapa Valley, became a ghost town as it slowly went under Lake Mead early in the summer of 1938. The town's residents were given a five-year notice to leave town; by the time the town's last 11 inhabitants left, materials worth salvaging had been removed from old buildings, and the local cemetery had been moved 7 miles away to near Overton. On June 11, its last day of business, Postmaster Rox Whitmore of the St. Thomas post office canceled more than 4,000 letters from stamp collectors across the United States who wanted the last postmark from what would be the "submerged city." Hugh Lord, the last resident of St. Thomas, left in 1938 when he woke up and water from the rising lake was lapping around his bed.

Pearce Ferry

Pearce (often misspelled Pierce) Ferry, which is at River Mile 280 of the Colorado River at the southern end of Pearce Bay, marks the boundary between Grand Canyon and Lake Mead. The site was discovered in 1862 by Mormon explorer Jacob Hamblin during the first recorded trip around Grand Canyon. The ferry is named for Harrison Pearce (1818–1889), who was sent to northern Arizona by Brigham Young around 1876. Pearce lived with his son in a stone building (Fort Pearce) that is now beneath Lake Mead. Pearce Ferry is at the end of 52-mile Pearce Ferry Road.

When both Lake Mead and Lake Powell are full, they store more than 50 million acre-feet of water, an amount equivalent to more than two years of the Colorado River's average flow. However, in early 2018, neither lake was even 60% full. When Lake Mead was last full in 1998, St. Thomas was under more than 100 feet of water. As the lake shrank, it revealed areas that had been submerged for decades. The drop in the lake's surface has also enabled divers to see a B-29 Superfortress, a military relic that crashed into the lake in 1948 on a secret mission. In 2016, the popular World War II airplane was accessible to divers.

A variety of concessionaires in Las Vegas (24 miles away) offer tours of Hoover Dam for more than 800,000 people per year. Lake Mead National Recreation Area receives more visitors than either Grand Canyon or Yellowstone National Park.

See also: Thematic Essays section: Whose Water Is It?

Further Reading

National Park Service. 2017. Lake Mead. Retrieved July 5, 2017, at https://www.nps.gov/lake/index.htm

Turner, Jim. 2016. *The Mighty Colorado River: From the Glaciers to the Gulf*. Tucson, AZ: Rio Nuevo Publishers.

LAKE POWELL AND GLEN CANYON DAM

Just northeast of Grand Canyon, along the border between Utah and Arizona, is Lake Powell, the second-largest (behind Lake Mead) human-made reservoir in the United States. Lake Powell is held back by Glen Canyon Dam. A dam in Glen Canyon was studied as early as 1924, but those plans were abandoned in favor of what became Hoover Dam, which was completed in 1936. Although USGS hydrologist Eugene La Rue—a respected authority regarding the Colorado River and a member of Claude Birdseye's 1923 USGS expedition—opposed the Hoover Dam, the construction of Glen Canyon Dam did include some of La Rue's ideas.

Construction of the 1,560-foot-long, 710-foot-high concrete arch-gravity dam across the deep, sandstone gorges of Glen Canyon began in 1956, when President Dwight D. Eisenhower pressed a button on his White House desk that triggered the first blast to reroute the Colorado River. By 1960, workers were pouring concrete at the dam around the clock; it took 400,000 loads, each consisting of 24 tons of concrete, to build the dam.

Construction of Glen Canyon Dam was substantially complete by 1963, but the dam was not dedicated until September 22, 1966, by Claudia "Lady Bird" Johnson. The dam and hydroelectric plant are separated by a beautiful, 86,000-square-foot lawn of Kentucky bluegrass. At its crest, the dam is 25 feet wide, and at its base it is 300 feet wide. Glen Canyon Dam, which is operated by the Bureau of Reclamation, is 15.5 miles upstream on the Colorado River from Lees Ferry. In 1969, the Secretary of the Interior and others recommended that the dam's name be changed to Dwight D. Eisenhower Dam to acknowledge Eisenhower's support of the Colorado River Storage Project. (The name was not changed.) GCNP lies between Lake Powell and Lake Mead, which is held back by Hoover Dam.

In 1983, a high snow melt filled Lake Powell to record levels and threatened structural problems for Glen Canyon Dam. In response, dam operators opened the dam's eight turbines, four bypass tubes, and two spillways, thereby increasing flow of the Colorado River in GCNP to more than 90,000 cubic feet per second (cfs). At Crystal Rapid, this flow flipped several boats, and 90 people were thrown overboard (one of whom died). Soon thereafter, GCNP temporarily closed part of the river and began extracting more than 130 river runners from the canyon.

Lake Powell, which is in a high desert amid porous land, loses huge amounts of water—approximately 800,000 acre-feet, or 6 percent of the Colorado River's flow—per year to evaporation and seepage. Despite these loses, Lake Powell and Lake Mead provide 90 percent of the stored surface water in the Colorado River Basin. Most of this water is used for agriculture.

Glen Canyon Dam has dramatically changed the Colorado River and its various ecosystems in Grand Canyon. The dam has stabilized the flow of the Colorado River, helped boaters to plan for future years (because they knew there would be an adequate flow of water), and increased the length of the river running season in Grand Canyon. However, it has also produced a variety of changes and problems:

Before Glen Canyon Dam, the average flow during late summer was less than 4,000 cfs and up to 38,000 cfs in the spring. Today, however, the flow of water through Grand Canyon is no longer determined by nature but instead is regulated by Glen Canyon Dam in response to power needs in the Southwest. The flow is relatively constant; the peak release of water from Lake Powell into GCNP occurs near midday when demands for electricity are highest.

Before Glen Canyon Dam, approximately 66 million tons of sediment passed Lees Ferry in Grand Canyon each year. Another 20 million tons entered the canyon from tributaries (e.g., Paria River and Little Colorado River). This silt-laden water was often described as "too thick to drink, too thin to plow." Today, however, Glen Canyon traps more than 70 percent of the river's silt, an amount equal to that loaded in 30,000 dump trucks per day. Before Glen Canyon Dam, the Paria River carried only 0.16 percent of the Colorado River's sediment, but today it contributes more than 4 percent; this makes the Paria one of the largest contributors of silt in GCNP and an important source for beach replacement. Unless the Colorado River is stained by silt from tributaries such as the Paria and Little Colorado Rivers, the river's water is clear.

Before Glen Canyon Dam, water temperatures in the Colorado River in Grand Canyon varied with the season. Water was often above 80°F in summer and near freezing in winter. Because water comes from about 200 feet below the surface of Lake Powell, the water entering Grand Canyon is near 48°F year-round.

As a result of this lack of silt and the water's cold temperature, four of the eight species of native fish in Grand Canyon—for example, Colorado pikeminnow (*Ptychocheilus lucius*) and bonytail chub (*Gila elegans*)—that had evolved in the untamed river's warm, muddy water are gone. They have been replaced by exotic species such as rainbow trout and brown trout that thrive in the clear, colder, postdam water. Native frogs, lizards, insects, and other animals are increasingly rare.

Glen Canyon Dam ended the natural cycle of flooding and the replenishment of sediments in Grand Canyon. Because most of the Colorado River's silt is trapped behind Glen Canyon Dam, beaches throughout Grand Canyon have eroded significantly—and, in many cases, disappeared—since the construction of Glen Canyon Dam. This, in turn, has allowed invasive species such as tamarisk to crowd out native species.

Before Glen Canyon Dam, seasonal flows of water through Grand Canyon fluctuated between 3,000 and 90,000 cfs, and often exceeded 100,000 cfs. These natural floods carried away some rockslides, mitigated risk from fires, and scoured pockets of encroaching vegetation from beaches. Today, however, water flow through Glen Canyon Dam is relatively steady and seldom exceeds 25,000 cfs. As a result, vegetation that was once regularly scoured from beaches has now taken over those beaches, thereby changing ecosystems in the canyon.

Invasive species such as tamarisk (a Middle Eastern plant used to create windbreaks in California's Imperial Valley in the early 1900s) now dominate many areas of Grand Canyon. Although the northern tamarisk beetle (*Diorhabda carinulata*, which eats tamarisk) migrated into Grand Canyon in 2009, it has not (as of 2018) had a major impact. One of the most recent invaders in Grand Canyon is the coin-sized, filter-feeding quagga mussel (*Dreissena rostriformis bugensis*), which is indigenous to Ukraine. After first being detected in the United States in 1989, the mussel was detected in Lake Mead in 2007, in Lake Powell in 2012, and at Lees Ferry in 2014. Quagga are prolific breeders; one female quagga produces up to a million eggs per year over its 5-year lifetime. With no natural predators in GCNP, populations skyrocket. Left to grow, these groups of quagga can clog water pipes (much like plaque slowly clogs heart arteries) and damage ecosystems.

Other changes to Grand Canyon produced by Glen Canyon Dam include less driftwood, more riverside marshes, and more waterbirds. Although river runners continue to encounter churning rapids throughout Grand Canyon, Glen Canyon Dam has tamed the Colorado River in Grand Canyon. Today's rapids, wildlife, scenery, and navigational challenges in Grand Canyon are far different from those faced by John Wesley Powell and other Grand Canyon river running pioneers. Fewer than 400 people boated through Grand Canyon before Glen Canyon Dam was finished.

GCNP began using controlled floods (high-flow releases) having water flows up to 45,000 cfs for 3–4 days to try to restore sensitive habitats and beaches in Grand Canyon. The first of these controlled-floods occurred in March 1996 and involved the release of 117 billion gallons of water from Lake Powell at a rate that would have filled an Olympic-size swimming pool every 2.5 seconds. The initial results of this flood, which lowered Lake Powell by 2.5 feet, seemed positive; many beaches reappeared after the floods, and some vegetation was scoured from the riverbanks. However, these positive results were short-lived; most of the new beaches created by the controlled floods disappeared within a year, and the problematic vegetation along the river soon reappeared. GCNP repeated the high-flow releases in 2004, 2008, 2013, 2014, and 2016. These floods do not change the annual amount of water released from Lake Powell to Lake Mead; subsequent releases are adjusted to compensate for the high volumes released during the high-flow floods.

When full, Lake Powell has an elevation of 3,700 feet, a surface area of 161,390 acres, and a potential volume of 24 million acre-feet of water. The dam cost $135 million ($986 million in 2016) and 18 lives. The lake's average depth is 132 feet, and its maximum depth is 583 feet. The closing of Glen Canyon Dam's gates on March 13, 1963, backed up water 30 miles into Cataract Canyon and into more than 90 side-canyons. It also reduced the flow of water in Grand Canyon to 900 cfs. Only seven people ran the Colorado River through Grand Canyon that year.

It took more than a decade to fill Lake Powell. In early 2018, however, drought and increased usage of the lake's water resulted in the lake being only 60 percent full. The lake's shoreline—normally 1,960 miles, which exceeds that of the entire West Coast of the United States (1,293 miles)—had shrunk by more than 30 percent.

Glen Canyon National Recreation Area was established by the U.S. Congress on October 27, 1972 to "provide for public outdoor recreation use and enjoyment of

Lake Powell and lands adjacent thereto in the States of Arizona and Utah and to preserve the scenic, scientific, and historic features contributing to the public enjoyment of the area." The site covers more than 1.25 million acres of land between Lees Ferry and Glen Canyon Dam, of which Lake Powell covers only 13 percent. Glen Canyon National Recreation Area, which straddles the Arizona–Utah border, entertained more than 3.2 million visitors in 2017. The site includes the Carl B. Hayden Visitor's Center, named for Carl B. Hayden (1877–1972), an Arizona senator who served for over 50 years. When Arizona became a state in 1912, Hayden was its first representative.

See also: Thematic Essays section: Whose Water Is It?

Further Reading

Brower, David Ross. 1988. *The Place No One Knew: Glen Canyon on the Colorado.* Layton, UT: Peregrine Smith.

Martin, Russell. 1999. *A Story That Stands like a Dam: Glen Canyon and the Struggle for the Soul of the West.* Salt Lake City: University of Utah Press.

National Park Service. 2017. Glen Canyon. Retrieved July 7, 2017, at https://www.nps.gov/glca/index.htm

LAVA FALLS TRAIL

Lava Falls Trail is the shortest trail from the rim to the river in GCNP. This difficult, steep, 1.8-mile trail descends 2,540 feet from the North Rim just west of an extinct volcano called Vulcans Throne to the Colorado River. The trailhead is at the end of a rough, 2-mile road that branches, about 3 miles from the rim, from the road leading to Toroweap Overlook. The trail, of which only a small part is visible from the trailhead, begins with a relatively easy 5-minute hike that leads to a ledge and the first indication of the challenges ahead. The trail includes several switchbacks covered by cinders, and the second half of the descent is even steeper. The trail is sometimes difficult to follow; be on the lookout for cairns for guidance.

Lava Falls Trail includes no shade, water, or shelter; the hike is often steep, and the footing is often unstable. Loose footing and boulders make falls common on this trail. In the summer, some people may not be able to carry enough water on this trail. Lava Falls Trail, which is easily underestimated because of its short length, is for experienced hikers.

Lava Falls Trail ends 0.3 mile upstream from Lava Falls Rapid, which hikers can hear during most of their descent to the river. Lava occurs on both sides of the river in this area and extends 60 miles downstream. River runners at Lava Falls drop 14 feet in 100 yards.

Lava first appears about 3 miles upstream from Lava Falls. When John Wesley Powell noted that some of this lava is hundreds of feet above the river, he commented, "What a conflict of water and fire there must have been here! Just imagine a river of molten rock running down into a river of melted snow. What a seething and boiling of the waters; what clouds of steam rolled into the heavens!" (Arizona State University, 2017). The rapid here was formed from debris flows from

Prospect Canyon, which joins the river at the top of the rapid. Powell and Robert Stanton, who led the first two expeditions to see Lava Falls, refused to run the rapid. Lava Falls, which has the fastest water in GCNP (up to 30 miles per hour, depending on conditions), was first run successfully by Flavell and Montéz in 1896 (George F. Flavell and Ramón Montéz ran all the rapids except Soap Creek). Lava Falls, with Crystal Rapid, is one of the most challenging rapids in GCNP.

See also: Rapids

Further Reading

Arizona State University. 2017. Lava Falls. Retrieved July 1, 2017, at http://grandcanyonhistory.clas.asu.edu/sites_coloradorivercorridor_lavafalls.html

LEES FERRY

Lees Ferry is a calm, relatively shallow site near the confluence of the Paria and Colorado Rivers that marks River Mile 0 on maps of the Colorado River through Grand Canyon. The ferry, which is between where water flows out of Glen

Lees Ferry, which designates River Mile 0 on maps of Grand Canyon, is a relatively shallow part of the Colorado River that is accessible from both sides of the river. The area is named for ferry operator John D. Lee, who was executed for his role in the Mountain Meadows Massacre of 1857. Cars, horses, and wagons were ferried across the river here from the early 1870s until 1928, after which vehicles and people crossed the river at the newly completed Navajo Bridge. Today, Lees Ferry is where most river runners begin their trips through Grand Canyon. (Randy Moore)

Canyon and into Grand Canyon, is the only site within hundreds of miles in either direction where Grand Canyon's walls go from steep to sloping and where the river can be accessed from both sides. The original site of Lees Ferry, then named Pahreah Crossing, was located in 1858 by Jacob Hamblin and a Paiute Indian named Naraguts. Hamblin was the first known person to cross the river here in 1864. At Lees Ferry, the upstream dip of the Echo Cliffs Monocline brings the erodible Moenkopi and Chinle Formations to river level; people reach the river by following the bedding planes within these formations. When John Wesley Powell's expedition passed Lees Ferry in 1869, they saw only Native American ruins, prompting boatman Jack Sumner to describe the scene as being "desolate enough to suit a lovesick poet" (Ghiglieri, 2003).

Lees Ferry is named for John Doyle Lee (1812–1877), a Mormon adopted by Brigham Young after Mormon leader Joseph Smith was assassinated. In 1871, Young told Lee to move the Mormons to what is now Utah; Lee, who urged Young to establish a permanent settlement at the ferry, realized that what became Lees Ferry was key to the Mormons' expansion from Utah into northern Arizona. In 1874, the Mormons fortified the area with Lees Ferry Fort, the remains of which are still there; this building was meant to help keep the peace between the Mormons of Utah and the Navajo Indians of Arizona. There were no attacks at the ferry, and the building was subsequently used as a trading post, dining hall, residence, and school. The inscription "J. Hislop 1889" beside the front door of the fort commemorates John Hislop, a surveyor and civil engineer who was here for a Christmas dinner with the Robert Stanton railroad survey. The small rock cabin nearby was a post office that was open from 1879 to 1912 (ferry operator Warren Johnson was the area's first postmaster) and again from 1922 to 1923.

On December 25, 1871, Lee arrived at the crossing (with two of his 19 wives, and several of his more than 50 children) with hopes of establishing a permanent ferry service for Mormon settlers heading to Arizona. By late 1872, the ferry was operating. Lee's first customers were 15 Navajos going north, and the first large group of Mormon settlers—70 wagons plus horses and cattle—crossed the river 5 months later. Earlier, however (on September 7–11, 1857), Lee had been part of the Utah Militia that attacked and killed more than 120 people in an emigrant wagon train at Mountain Meadows, Utah (this massacre is part of the background of the Sherlock Holmes novel *The Sign of Four*). Only 18 children were spared in the attack, and 17 of these children were adopted by Lee (one child), Jacob Hamblin (three), and others. The next summer, a report accused Lee "and a few other white men" (Berger, 2007) of the massacre. Brigham Young sent Lee, who was hoping to evade the law, to what became Lees Ferry; there, Lee used a boat named *Cañon Maid* left by John Wesley Powell to ferry people across the river. Lee charged 75¢ for each horse and $3 for each wagon to cross the river. (This ferry, whose remote site had protected Lee from authorities, operated until 1928; the next year, it was superseded by the first Navajo Bridge.) Lee often looked for approaching officers from Lees Lookout, a hill atop the Chine Formation that is northwest of Lees Ferry. Today, visitors to Lees Ferry can hike to this hilltop for excellent views of the ferry, Paria River, and Vermilion Cliffs.

Lee, a field commander who became a scapegoat for a wider guilt associated with the Mountain Meadows Massacre, was excommunicated in 1870 by Young when the church was accused of harboring a fugitive. On November 7, 1874, Lee was arrested for the killings while he was in Panguitch, Utah, visiting one of his wives. After his first trial ended with a hung jury, Lee was convicted of first-degree murder at his second trial by a jury of 12 Mormons for his complicity in the Mountain Meadows Massacre. Only seven witnesses, including Jacob "The Buckskin Apostle" Hamblin, were called by the prosecution. On March 23, 1877, after shaking hands with attendees and posing for a photo by James Fennemore (who had lived with Lee at Lonely Dell) while sitting atop his coffin, Lee was executed by a firing squad at the site of his alleged crimes. Lee, who denied killing anyone, was the only person to be executed for the massacre; he is also the only person blamed for the massacre ("one of the most lamentable tragedies in the annals of the West") on a plaque mounted near Mountain Meadows. Lee's last words included another proclamation of innocence, "I have been sacrificed in a cowardly, dastardly manner" (Legends of America, 2017). When Lee was shot, his body fell into his coffin, after which the coffin was closed. After two of his sons opened the coffin briefly to lay Lee's temple robes atop the body, Lee was buried beneath a simple monument in Panguitch, Utah. On April 20 and May 8–9, 1961, Lee was posthumously reinstated "to membership and former blessings" by the Mormon Church. Lees Ferry (also known as Compact Point, which divides the Upper and Lower Basins), is a mile downriver from Lees Ferry and is named for Lee.

Even before Lee's execution, people were searching for Lee's alleged "hidden treasure" of several buckets of gold. For example, Isaac Haight—another indicted leader of the Mountain Meadows Massacre—searched for Lee's gold under the alias "Brown." Haight was guided by Grand Canyon pioneer William Bass, who said that he had been given a map to the treasure by Lee's 17th wife, Emma Louise Batchelor Lee (1836–1897). Bass and Haight found the trail to be impassible, and Lee's alleged treasure was never found.

After Lee's execution in 1877, the ferry was operated by Lee's widow, Emma, who named the isolated area Lonely Dell. This was the name that Powell had printed on government maps. Emma then sold her interest to the Mormon Church. Subsequent operators of the ferry included Joseph L. Foutz (1879), Warren Marshall Johnson (1879–1896, with his two wives—Permelia Jane Smith and Samantha Nelson, and their 20 children), James S. Emett (1896–1909), Grand Canyon Cattle Company (1909–1910), and Coconino County (1910–1928). The last trace of John and Emma Lee disappeared when Warren Johnson tore down their cabin and used it for firewood. Emma later married Franklin H. French, a prospector who helped improve Tanner Trail from the South Rim to the Colorado River. There are two monuments to Johnson (1838–1902) and his family near Marble Canyon Lodge, but the largest monument there is to John Lee. That 10-foot-high monument, erected In 1961 near Navajo Bridge, describes Lee as a "frontiersman, trailblazer, a man of great faith, sound judgment, and indomitable courage . . . who with superhuman effort and in the face of almost insurmountable obstacles, maintained this ferry which made possible the colonization of Arizona."

In 1880, Lees Ferry—which was then operated by ferryman Warren Johnson—was first included in the U.S. Census, which reported 27 "permanent" residents. Within a few years, following the discovery of gold in nearby Glen Canyon, gold miners flocked to Lees Ferry. The most extensive gold mining operation was led by Charles H. Spencer, the head of American Placer Company; Spencer hoped to mine gold from the Chinle Shale cliffs behind Lees Ferry. Spencer brought a large steamer and tons of equipment to the ferry, but the venture failed, and Spencer left broke in 1912. (His equipment and buildings still litter the area around the original crossing.) By 1900, the census listed only 16 people living at Lees Ferry. Spencer returned briefly to Lees Ferry in 1932–1933 but failed again, this time leaving for good.

Lees Ferry, which served as a prominent gateway to Arizona for 54 years, was not reliable; it was often inoperable, and nine people drowned there. Nevertheless, the ferry continued to operate until June 7, 1928, when it capsized and dumped a Model T and three men into the river. All of the people aboard the ferry drowned—including ferry operator Adolpha Johnson; this was the largest known number of drowning deaths by people not trying to go downstream. The ferry never reopened, for the 1929 opening of Navajo Bridge—already under construction just 4 miles downstream—allowed faster, safer, and more efficient travel between Utah and Arizona. That bridge is now used only for foot traffic, while another one nearby carries vehicles.

Today, Lees Ferry—which is about 15 miles downstream from Glen Canyon Dam and less than a mile upstream from GCNP—is at the northeast end of GCNP, is part of the Glen Canyon National Recreation Area, and is a famous site for trout fishing and launching boats. The area includes hiking trails to places of geologic and historic interest:

American Placer Corporation Office, built in 1911, was the first of eight stone buildings built by Spencer at the site. This was the headquarters of Spencer's gold mining operation.

Pioneer Ferry Site, which is where most people crossed the Colorado River in the 1870s and 1880s. The first large group, consisting of 70 wagons and associated horses and cattle, crossed here in May 1873.

Lees Ferry Fort, which was never attacked, was built in 1874 and later became the home of Warren Johnson, the ferryman who replaced John Lee. (Johnson was accompanied by his four children and two wives.) In the 1920s, USGS employees lived in this building.

The wrecked *Charles H. Spencer* paddlewheel steamboat, which launched in 1912. This 92-foot-long steamer was meant to carry coal from Warm Creek, Utah (now in Lake Powell), to Lees Ferry. However, the underpowered boat, which was designed for the relatively calm Sacramento River, made only three trips before being abandoned when Spencer left the area. In the 1920s, the boat sank, and it remains visible in the river.

Spencer Bunkhouse, built in 1911, is the only remaining bunkhouse (of three) built here by Spencer. In 1921, the building was remodeled and became a schoolhouse and by 1964 was a USGS laboratory. (*Note:* Grand Canyon's

> Spencer Canyon on river left near River Mile 246 is named for a different Charles Spencer, who was a friend and scout of the Hualapai.)

Nearby are Lonely Dell Ranch, an orchard (now maintained by NPS), a cemetery on the Paria River floodplain, a USGS guesthouse (built in 1950 with stones from two of Spencer's bunkhouses), and other ruins of Spencer's mining operation. The cemetery, which includes 20 graves of the 21 people buried here between 1874 and 1933, includes four of Warren and Permelia Johnson's children who died from diphtheria (brought to Lees Ferry by a visiting family) within seven weeks of one another in 1891. This cemetery was the first European cemetery at Grand Canyon, and the first person buried here was James Webb Jackson (1837–1874). The cemetery was patrolled by GCNP rangers as early as 1962. The fence surrounding the cemetery was installed by Joseph S. Johnson in 1963 at a family reunion.

Lees Backbone, which is directly across the river from Lees Ferry, includes remnants of a road called the Long Dugway. This hand-built, one-lane, wagon road, which is visible from River Mile 0.1–0.5 in the Navajo Nation on river left, was used from 1898 to 1928 to access Lees Ferry from the south and to allow wagons and cars to bypass the gullies and boulders at the ferry. A bell on river left was rung by travelers trying to get the attention of ferry operators on the other side of the river. Lees Ferry is a unique place; for example, Barry Goldwater considered Lees Ferry one of his favorite spots because it was beautiful, historic, and restful.

Since 1922, Lees Ferry has also been the dividing point between the Upper and Lower Colorado River for water rights in this section of the Colorado River. The Upper Basin (Colorado, New Mexico, Utah, Wyoming, and northern Arizona) and the Lower Basin (Arizona, California, and Nevada) share (with two states in Mexico) equally the water divided by an imaginary line near Lees Ferry. This is why Lees Ferry has been referred to as "both the physical and spiritual heart of water history in the arid West" (Kupel, 2017). Plaques commemorating Lees Ferry are near the ferry. Another plaque commemorating "Lee Ferry" is in front of Marble Canyon Lodge near Navajo Bridge.

The USGS's narrow, concrete, gaging station on river left at Lees Ferry was built and began operations in 1921. For the first few years, the station was comanaged by Southern California Edison Company. This and other such stations downriver (e.g., near the Kaibab Suspension Bridge at River Mile 88.1) were automated in 1977, but people continue to periodically sample the water in GCNP. The first NPS ranger station opened at Lees Ferry in 1958. On June 11, 1974, NPS bought Lonely Dell for preservation as a historic site. The Lees Ferry Historic District was placed on the National Registry of Historic Places that same year.

See also: Rapids

Further Reading

Berger, Todd R. 2007. *It Happened at Grand Canyon*. Guilford, CT: TwoDot Publishers.

Brooks, Juanita. 1992 (1962). *John Doyle Lee: Zealot, Pioneer Builder, Scapegoat*. Logan: Utah State University Press.

Ghiglieri, Michael, ed. 2003. *First through Grand Canyon: The Secret Journals and Letters of the 1869 Crew Who Explored the Green and Colorado Rivers*. Flagstaff, AZ: Puma Press.

Kupel, Doug. 2017. Lee's Ferry, Revisited. Retrieved July 1, 2017, at http://www.gcrg.org/docs/gtslib/kupel-02.pdf

Legends of America. 2017. Mountain Meadows Massacre assassins. Retrieved July 1, 2017, at http://www.legendsofamerica.com/ut-mountainmeadowsassassins4.html

Reilly, P. T. 1999. *Lee's Ferry: From Mormon Crossing to National Park*. Logan: Utah State University Press.

LIPAN POINT

Lipan Point (elevation 7,360 feet), which is 0.5 mile north of East Rim Drive, is accessible along a paved road that climbs more than 70 feet to a parking lot atop a summit. Lipan Point was originally named Lincoln Point (for Abraham Lincoln) but was renamed in 1902 to honor the Lipan Tribe of Apache Indians of Texas, who were allies of the Comanches.

Lipan Point offers one of the best views of the Colorado River in GCNP. The viewpoint also features extensive views of the tilted, red-and-white layered Grand Canyon Supergroup, which are rare sedimentary layers that have eroded away farther downstream. At Lipan Point, which is on river left near River Mile 68.5, the layers of Supergroup are near the bottom of the canyon and dip 12° to 15° toward the northeast. Views to the northwest include Zoroaster Temple, Wotans Throne, Vishnu Temple, Cape Royal, Grandview Point, Yaki Point, Isis Temple, Shiva Temple, and Chuar Butte (Chuar Butte and Temple Butte hide the confluence of the Colorado and Little Colorado Rivers). The northeastern horizon features Echo and Vermilion Cliffs. Directly below Lipan Point is Cardenas Butte, 75-Mile Creek, and its tributaries, which join the Colorado River between Nevills Rapid and Hance Rapid (which are visible approximately 3.8 miles to the west). Just beyond Hance Rapid, where the river begins cutting into the Vishnu Schist, the Inner Gorge begins. The trailhead of the rugged, 8-mile Tanner Trail is near the Lipan Point parking lot.

To the east of Lipan Point (and at the end of Unkar Creek) is Unkar Delta, which was the site of the largest known Anasazi community found in Grand Canyon. These Native people grew corn, beans, and squash. After wintering at the bottom of Grand Canyon, they moved to the rim in the spring to hunt game and plant crops. The Colorado River at Lipan Point is less than 2 miles from the South Rim but almost 8 miles from the North Rim.

From Lipan Point, which offers the most spacious views of Grand Canyon from the South Rim, visitors can see parts of six climatic zones, ranging from the arctic zone near the top of San Francisco Peaks to the warmest climate at the bottom of Grand Canyon. Lipan Point is also a popular place to watch sunset and sunrise. These views are so popular that in 1999, the U.S. Postal Service issued a 60¢ stamp featuring sunrise at Lipan Point.

See also: Thematic Essays section: How the Colorado River Carved Grand Canyon

Further Reading

Anderson, Michael. 2001. *Along the Rim: A Guide to Grand Canyon's South Rim from Hermit's Rest to Desert View.* Grand Canyon, AZ: Grand Canyon Association.

LITTLE COLORADO RIVER

The Little Colorado River is a tributary of the Colorado River that originates in east-central Arizona's White Mountains (near Mount Baldy at an elevation of almost 10,000 feet) and drains the Painted Desert region of western New Mexico and eastern Arizona. The Little Colorado, which joins the Colorado River at River Mile 61.7, is more than 300 miles long, but only the headwaters and parts close to the Colorado River flow year-round. (Much of the water above Cameron is diverted for irrigation.) The rest of the river contains water only after flash floods or snow melt. As it nears the Colorado River, the Little Colorado River carves a gorge up to 3,000 feet deep; this steep-walled gorge begins about 9 miles below Cameron and 41 miles above the river's confluence with the Colorado River. At Cameron, Arizona—a gateway to GCNP—the Little Colorado River is crossed by U.S. Highway 89.

In 1869, as he explored Grand Canyon, John Wesley Powell called the Little Colorado River the Flax River for the flax that grows along the river. (The Navajos called it Red Stream; the Havasupai called it Salty River.) Powell's expedition saw the river soon after a flash flood, and the water was laden with silt. George Bradley and Jack Sumner, who were with Powell, described the river as:

> a lo[a]thesome little stream, so filthy and muddy that it fairly stinks. It is only 30 to 50 [yards] wide now and in many places a man can cross it on the rocks without going on to his knees . . . [The Little Colorado was] as disgusting a stream as there is on the continent . . . half of its volume and 2/3 of its weight is mud and silt. [It was little but] slime and salt . . . a miserably lonely place indeed, with no signs of life but lizards, bats and scorpions. It seemed like the first gates of hell. (Ghiglieri, 2003)

In fact, when the Little Colorado River is not stained by floodwater, its water resembles the powdery blue water of Havasu Creek because of its dissolved travertine and other forms of limestone (from Blue Springs, which comes out of Redwall Limestone about 13 miles up the canyon). At the confluence, the water of the Little Colorado River is easily distinguished from the artificially (i.e., dam-fed) clear, greenish-brown water of the larger Colorado River. During the summer, water of the Little Colorado is also 20° warmer than that of the Colorado River.

The confluence of the Little Colorado and Colorado Rivers, which is best viewed from Cape Solitude (3,400 feet above the confluence) ends Marble Canyon, which is sacred to Hopis, Havasupais, Zunis, Navajos, and other Native Americans. The confluence is also an excellent habitat for the humpback chub, a native species of GCNP that is endangered.

Beamer Cabin

Beamer Cabin, which is just upstream from the confluence of the Colorado and Little Colorado Rivers, is part of a small cliff of Tapeats Sandstone along the south bank of the Little Colorado River. The "cabin" was built (i.e., chiseled out of rock) by Ben Beamer, a prospector who lived there from 1890 to 1892 and built a trail (Beamer Trail) from the cabin to the Tanner Trail. Beamer, who tried unsuccessfully to grow crops here, built his cabin from rocks he gathered from a nearby Puebloan ruin that John Wesley Powell had noted in 1869 and that had been occupied as early as the 4th century BCE. Beamer was the first person to describe the now endangered humpback chub (an endangered fish in Grand Canyon), which he described in the July 17, 1892, issue of the *Denver Republican* as sometimes being "so thick that you can lean over the water's edge and pull them out by the tail two at a time." Soon after the passage of the Endangered Species Act in 1973, humpback chub was one of the first animals added to the list.

Hopi oral histories refer to Beamer Cabin as "the white man's store." Today, Beamer Trail, which spanned from Palisades Creek to his cabin, receives relatively few visitors.

Also near the confluence of the Little Colorado and Colorado Rivers are the Hopi salt mines (which were deposited by the Tapeats Sea 550 million years ago) and the Hopis' sacred *Sipapu* (*Sipapuni*), which is a spring atop a travertine dome the Hopis believe is the opening through which their ancestors emerged (one Hopi word for Grand Canyon is *Öng'tupqua*, meaning "a place of salt"). The midair collision of TWA and United airplanes in 1956 also occurred near this confluence. At the mouth of the Little Colorado River, the rim is 3,760 feet above the river.

See also: Thematic Essays section: The Colorado River's Historic Sites; Gateways to Grand Canyon; Sipapu

Further Reading

Ghiglieri, Michael, ed. 2003. *First through Grand Canyon: The Secret Journals and Letters of the 1869 Crew Who Explored the Green and Colorado Rivers.* Flagstaff, AZ: Puma Press.

LITTON, MARTIN

Clyde Martin Litton (1917–2014) was a vocal conservationist and longtime Grand Canyon river runner who led opposition to the construction of dams on the Colorado River, including in GCNP. After first floating through Grand Canyon in 1955 with P. T. Reilly (as one of the first 300 people to do so after John Wesley Powell), Litton returned to Grand Canyon many times and in 1971 founded Grand Canyon Dories to run commercial trips through Grand Canyon. Litton traversed Grand Canyon many times in dories and believed that rubber rafts, which later became the primary way that most people go through Grand Canyon, are as unnatural in Grand Canyon as dams.

Litton, who adapted dories to the Colorado River, began the tradition of naming dories after lost or compromised places. In 1962, the first two dories ran through Grand Canyon; those dories were *Portola* and *Susie Too*; Litton later changed the name of *Susie Too* to *Music Temple*. This name commemorated Music Temple, an acoustic grotto in Glen Canyon (discovered by John Wesley Powell) that is now submerged in Lake Powell. In 1976, Litton donated *Music Temple* to GCNP, where it is now part the park's collection of historic boats.

In 1967, Litton—whose conservation work was featured in several documentaries—published *The Life and Death of Lake Mead*. In 1999, the 82-year-old Litton became the oldest person to ever run a dory through Grand Canyon. Five years later, Litton broke his own record by again becoming the oldest person to row through Grand Canyon at age 87. In his conservation work, Litton was uncompromising, for he realized that once areas were developed, their natural traits were lost forever. Litton's philosophy about places such as Grand Canyon was simple: He believed that we have no right to change these places.

Litton sold Grand Canyon Dories in 1989. He died on November 30, 2014, at age 97 in Portola Valley, California. In 2016, the 24-minute documentary *Martin's Boat* honored Litton's life.

See also: Sierra Club

Further Reading

Brower, Kenneth. 2014. Appreciation: Lessons from the man who stopped Grand Canyon dams. *National Geographic*, December 2, 2014. Retrieved July 1, 2017, at http://news.nationalgeographic.com/news/2014/12/141202-grand-canyon-dams-colorado-river-martin-litton-conservation/

LOPER, BERT

Albert A. "Bert" Loper (1869–1949) was a river running pioneer—especially of the Colorado River—in the American West who was born in Bowling Green, Missouri, on the day that John Wesley Powell discovered the confluence of the Colorado and San Juan Rivers. By the age of 13, Loper was on his own, and he supported himself as a mule skinner, ditch digger, miner, farmer, and soldier in the Spanish-American War.

In 1917, Loper and the South Rim's Ellsworth Kolb became the first people to boat through Utah's Westwater Canyon. In 1920, Loper was the lead boatman on a USGS survey looking for a suitable site for Black Canyon Dam (now called Hoover Dam). Three years later, however, Loper was overlooked (in favor of Emery Kolb) as boatman for the USGS Birdseye Expedition through Grand Canyon.

In 1939, Don Harris—who later owned a commercial river running company—needed advice about running the Colorado River through Grand Canyon. When he contacted the 69-year-old Loper, the two boaters decided to make the trip together. This was Loper's first run through Grand Canyon and made him the 69th person through Grand Canyon. Along the way, Loper became the oldest person ever to pilot a boat through Lava Falls and one of the first to run every rapid in Grand Canyon.

Ten years later, the 79-year-old Loper tried another run (with six other people) through the Canyon aboard *Grand Canyon*, which Loper built himself. The group

On July 8, 1949, famed river runner Bert Loper died in 24.5 Mile Rapid during his second trip through Grand Canyon. Loper's boat, *Grand Canyon*, was found downstream and pulled ashore at River Mile 41.6, where it soon became a monument to Loper, "The Grand Old Man of the Colorado." In this photo taken in 2014, author Randy Moore sits among the crumbling remains of Loper's boat. (Randy Moore)

launched on July 7 (three weeks shy of Loper's 80th birthday). The next day, in 24.5 Mile Rapid—not far from where Peter Hansbrough drowned on July 15, 1889—Loper died of a heart attack; his motionless body was last seen floating downstream. That evening, Loper's boat was found at River Mile 41.6, where it was dragged ashore; soon thereafter, pioneering Grand Canyon river runner Norman Nevills painted an epitaph on the boat's bow memorializing Loper as "The Grand Old Man of the Colorado." Although only fragments now remain of Loper's boat, a plaque honoring Bert now stands nearby. In 1975, 19-year-old Curtis Verploegh found Bert's remains on river left just past Cardenas Creek (near River Mile 70.1), some 46 miles from where Loper was last seen. Although Bert wanted to be buried along the Colorado River, he was buried in Sandy, Utah, beside his wife Rachel, who had died 52 days earlier. In 1950, on the day before what would have been Loper's 81st birthday, a 6-foot-tall memorial to the legendary Loper was dedicated in a park in Green River, Utah.

See also: Thematic Essays section: The Colorado River's Historic Sites

Further Reading

Dimock, Brad. 2007. *The Very Hard Way: Bert Loper and the Colorado River.* Flagstaff, AZ: Fretwater Press.

"LOST WORLDS"

In the summer of 1937, in an effort to complement Charles Darwin's findings in Galápagos with those possible in Grand Canyon, biologist Harold E. Anthony (1890–1970) of New York's American Museum of Natural History organized an expedition to search for new species atop Shiva Temple and Wotans Throne. Anthony reasoned that because these wooded "islands in the sky" mesas had been isolated from the canyon's rims for thousands of years, they could house new species unknown to humans. The expedition was hyped by museum director and famed dinosaur hunter Roy Chapman Andrews (1884–1960) as an expedition to "lost worlds," which had been made popular by Arthur Conan Doyle's 1912 novel *The Lost World* and the 1933 movie *King Kong*.

Grand Canyon pioneer Emery Kolb offered to guide Anthony's group to the top of Shiva Temple, but his offer was rebuffed by Anthony and NPS. To extract his revenge, Kolb and a group of his friends—including Gordon Berge, Ralph White, Ruth Stephens Baker (Kolb's darkroom assistant who was also the babysitter for Kolb's grandson), and Kolb's daughter Edith Kolb Lehnert—made plans for their own expedition to Shiva Temple. A few weeks before Anthony's expedition was to start, Kolb and his crew flew to the North Rim. After camping at Point Sublime, they climbed up the north side of Shiva Temple so that they would not be seen by people looking through telescopes at the South Rim's Kolb Studio and The Lookout. To make sure that Anthony realized that he had been beaten to the top of the formation, Kolb and his group left behind two Kodak film boxes and three empty tomato cans. Stephens, who had lived for 8 years in a three-room tent-house at Grand Canyon, was the first woman to set foot atop Shiva Temple; she left several lipstick-stained tissues so that Anthony would know that. The group also planted a flag that would be visible from Kolb Studio. Stephens, who feared that Kolb would lose his job if NPS learned of the hike, kept their adventure a secret for years.

When Anthony and his group (which included GCNP Superintendent Miner Tillotson and GCNP naturalist Edwin McKee) began their expedition, they had supplies delivered from planes by parachutes. Anthony stayed atop Shiva Temple for 10 days but found no new species; all of the plants and animals living there were the same as those at Grand Canyon's rims. Anthony's report did not mention the earlier visit by Kolb.

Wotans Throne was scaled by a different group, which included George B. Andrews (the son of Roy Chapman Andrews). They found some ancient ruins but no unusual plants or animals.

See also: Kolb, Ellsworth and Emery

Further Reading

Leavengood, Betty. 2007. *Grand Canyon Women: Lives Shaped by Landscape*. Grand Canyon, AZ: Grand Canyon Association.

M

MARBLE CANYON

Marble Canyon, the northernmost section of Grand Canyon, is a relatively straight, 60-mile section of Grand Canyon between the mouth of the Paria River (just south of Lees Ferry) and the confluence of the Little Colorado and Colorado Rivers. John Wesley Powell described the area this way:

> We have cut through the sandstones and limestones met in the upper part of the canyon, and through one great bed of marble a thousand feet in thickness. In this, great numbers of caves are hollowed out, and carvings are seen which suggest architectural form, though on a scale and so grand that architectural terms belittle them. As this great bed forms a distinctive feature of the canyon, we call it Marble Canyon. (Brian, 2015)

In fact, Powell was seeing the thick Redwall Limestone, which resembles marble when polished by the Colorado River.

Vista Encantada

Vista Encantada (Spanish for "enchanting view"), which is 7 miles south of Point Imperial, is an 8,480-foot-high overlook along the North Rim that offers a picnic area and excellent views into Marble Canyon, parts of Nankoweap Canyon, and Brady Peak (elevation 8,107 feet). This peak is made primarily of Hermit Shale with a Coconino Sandstone top.

The area was originally named Marble Gorge, but that name was changed to Marble Canyon in 1951. The Kaibab Formation becomes visible within the first mile after river runners enter Marble Canyon, and the Toroweap Formation becomes visible just past River Mile 2. Marble Canyon ends at Mile 61.7, which is where the Colorado River is joined by the Little Colorado River. The head of Marble Canyon is the eastern boundary of GCNP, and the left bank of Marble Canyon—from Lees Ferry to the Little Colorado River—belongs to the Navajo. Ironically, no marble is visible in Marble Canyon.

See also: Little Colorado River

Further Reading

Brian, Nancy. 2015. *River to Rim: A Guide to Place Names along the Colorado River in Grand Canyon from Lake Powell to Lake Mead*. Second edition. Flagstaff, AZ: Earthquest Press.

MARBLE CANYON DAM

The 1,900-foot-drop in elevation in Grand Canyon between Lees Ferry and Lake Mead is one of the biggest untapped hydroelectric potentials in the United States. In the 1920s, engineers began planning dams in Grand Canyon to exploit this potential, and USGS envisioned these dams in Grand Canyon as part of "an unbroken string of hydroelectric dams from the mountains of Colorado and Wyoming to the last canyons of the Colorado River along the California/Arizona border" (Rink and Dimock, 1997). However, these dams in Grand Canyon were postponed in the 1930s in favor of Hoover Dam, which could store more water for irrigation and flood control.

Marble Canyon Dam, like Bridge Canyon Dam downstream, was meant to be a "cash register" dam, meaning that it would sell electricity to pay its cost and fund future projects. Some of the power generated by the dam would be used to lift water from the Colorado River to Phoenix, Tucson, and central Arizona. Interestingly, the proposed site of the Marble Canyon Dam was not among the more than 20 such sites identified by the USGS Birdseye Expedition in 1923.

Most of the planning for the 310-foot-high, $238-million Marble Canyon Dam focused on two sites—one at River Mile 32.8 and another at River Mile 39.4—in Grand Canyon. Both of these sites would have produced reservoirs in Arizona, thereby enabling Arizona to store water, and states in the Upper Basin to guarantee their obligation of water (detailed in the Colorado River Compact) to other states. But Marble Canyon Dam was expensive, for it required a 25-mile road across the Navajo Reservation to connect the nearest highway to a worksite 2,500 feet below the canyon's rim, which was one of the most inaccessible dam sites ever explored by the Bureau of Reclamation. All of the dam's materials, workers, and supporting supplies were to be delivered to the River Mile 32.8 site by helicopter and down a 3,400-foot cable. The dam would have created Marble Canyon Reservoir, which could hold 363,000 acre-feet of water and generate 2.31 billion kilowatts of power. Keeping Marble Canyon Reservoir free of silt would have required silt-retention dams in tributary canyons (e.g., a 380-foot-high dam in Paria Canyon). Even with these dams, Marble Canyon Reservoir was forecast to fill with silt in only 104 years.

Plans for Marble Canyon Dam began when the U.S. Bureau of Reclamation included the dam in its Pacific Southwest Water Plan in the 1940s. Exploratory drilling began in 1951. The site at River Mile 39.4 was the preferred site because of its additional 30 feet of drop and 45 percent greater capacity for the reservoir produced by the dam. At this site, river runners can still see two 6-foot-high drifts of talus from drilling into cliffs of Redwall Limestone on both sides of the river. There are more than 30 drill holes near river level here. This dam, had it been built, would have flooded 40 miles of Grand Canyon, including Redwall Cavern and Vasey's Paradise. Nearby are the remains of a camp where workers worked, relaxed, and slept.

An accompanying 38-mile-long, 36-foot-wide tunnel, which would have diverted 90 percent of the Colorado River's water to power generators, would have eliminated the need for more dams between Marble Canyon Dam and Bridge Canyon Dam, which was located at the other end of Grand Canyon. Government officials claimed that there would always be a "scenic trickle" of 1,000 cubic feet per second from Marble Canyon Dam for aesthetic and recreational purposes, but this diversion of water would have drastically reduced the flow of the Colorado River in GCNP.

In 1963, public opposition to the dam began to grow, but Marble Canyon Dam was endorsed by Floyd Dominy, the powerful director of the Bureau of Reclamation. In 1969, Marble Canyon Dam became moot when President Lyndon B. Johnson, on his last day in the White House, created Marble Canyon National Monument. On January 4, 1975, this monument was absorbed into GCNP when President Gerald R. Ford signed the Grand Canyon National Park Enlargement Act (Public Law 93–620). This legislation had been sponsored by Arizona Senator Barry Goldwater who, in 1940 at the age of 31, became the 72nd person to boat through Grand Canyon. The Grand Canyon National Park Enlargement Act created today's boundaries of GCNP.

See also: Thematic Essays section: Whose Water Is It?; Litton, Martin; Sierra Club

Further Reading

Boyer, Diane E., and Webb, Robert H. 2007. *Damming Grand Canyon: The 1923 USGS Colorado River Expedition*. Logan: Utah State University Press.

Rink, Glenn, and Dimock, Brad. 1997. Damming Grand Canyon. *Boatman's Quarterly Review*, Spring. Retrieved July 9, 2017, at http://www.gcrg.org/bqr/10-2/bcd.html

MARICOPA POINT

Maricopa Point (elevation 6,988 feet), which was earlier called Hollenbeck Point, is named after the Maricopa Indians of south-central Arizona. The point is 1 mile east of Hopi Point and 2 miles west of Grand Canyon Village. An area just west of here (i.e., between Powell Point and Maricopa Point) was once the site of the most intense mining work ever to occur in GCNP. In 1890, Daniel Lorain Hogan (1866–1957)—a Flagstaff sheriff, orphan, Havasupai guide, and part-time prospector—found a vein of copper in the Coconino sandstone 1,100 feet below the rim. Three years later, Hogan filed a 20-acre claim here for Lost Orphan Mine that included land from the site of his discovery up to (and including four acres atop) the rim. (President Theodore Roosevelt, who was Hogan's commanding officer in the Spanish-American War, signed the papers in 1906 to give this property to Hogan.) Although Hogan exported little copper, he converted the site to a tourist facility consisting of 20 cabins, a trading post, and a saloon.

In 1947, after rejecting the NPS offer to buy his land for $5,555, Hogan sold his property for $25,000 to Madeleine Jacobs, who discovered in 1951 that the site contained large amounts of uranium ore. Jacobs sold the property to Western Gold and Uranium, Inc., whose uranium from 1956 to 1969 fueled the nation's atomic energy program (at one time, Orphan Mine employed 45 men working two daily shifts and was the largest producer of uranium in the United States). While the mine was open, tourists continued to stay on the rim at the nearby Grand Canyon Inn.

In 1962, after Western Gold announced plans to build a 600-room, 18-story hotel at the rim above the mine, President John F. Kennedy signed Orphan Mine Bill, which allowed Western Gold to continue to mine the site as negotiations continued to close the site. The nation's need for uranium ore later declined, and in 1969 the last shipment of uranium left the South Rim. By 1987, the site was part of GCNP, which closed the mine and all mining operations there. Today, Orphan Mine and the surrounding area are designated Rehabilitated Areas that are off-limits to tourists.

Grand Canyon's Only Endangered Plant

Sentry milk-vetch (*Astragalus cremnophylax* var. *cremnophylax*) is a tiny, lavender-flowered plant that was discovered at Grand Canyon in 1903 by botanist Marcus E. Jones. *A. cremnophylax* is endemic to Grand Canyon, meaning that it grows nowhere else in the world except Grand Canyon. In 1903, Jones described the plant as "common" at Grand Canyon, but only 40 years later it was rare. Today, factors such as habitat loss, drought, and tourists' trampling have made sentry milk-vetch the only federally registered endangered plant in GCNP.

Sentry milk-vetch is a perennial herb that grows in cracks and shallow soil pockets in Kaibab limestone. It grows only within 25 feet of the rim (*cremnophylax* means "gorge watchman"). The plant grows to about 1 inch tall and 1–6 inches in diameter, and its short stems have compound leaves each having 5–9 leaflets. Sentry milk-vetch, which is a member of the pea family (Fabaceae), is pollinated by Mason bees, flies, and ants. It blooms in the spring and after summer rains.

There are now only four populations of *Astragalus cremnophylax* at Grand Canyon. The largest of these populations resides in a fenced area near Maricopa Point.

Maricopa Point offers a 180° view of several miles of the Tonto Trail around Horn Creek and Dana Butte, but the Colorado River is hidden as it flows in the Upper Gorge. Visitors can see parts of Bright Angel Trail along Garden Creek, but most of the trail is obscured by a large red rock formation west of Bright Angel Trail and north of Maricopa Point (and between Maricopa Point and the Plateau Point Trail) called the Battleship (originally named for the battleship *Iowa*), which is in the Supai Group. Farther east is Wotans Throne and Vishnu Temple. Visitors looking north will see a panorama that features (left to right) Shiva Temple, Isis Temple, Cheops Pyramid, Bright Angel Point, Sumner Butte, the pyramidal Brahma Temple, and Zoroaster Temple.

Visitors walking from Maricopa Point to Powell Memorial will see the Hopi Fire Tower. In the late 1960s, there were often more firefighters in GCNP than NPS rangers. Today, however, NPS no longer staffs the park's four fire towers and instead uses aerial monitoring to search for fires (the towers at Hopi Point and elsewhere in GCNP are now used primarily to monitor air quality). Writer Edward Abbey's work as a lookout at a fire tower for four seasons inspired his *Black Sun* (1971).

See also: Thematic Essays section: John Wesley Powell; Hogan, Dan; Trail of Time

Further Reading

Anderson, Michael. 2001. *Along the Rim: A Guide to Grand Canyon's South Rim from Hermit's Rest to Desert View.* Grand Canyon, AZ: Grand Canyon Association.

MARSTON, OTIS "DOCK"

Otis Reed "Dock" Marston (1894–1979) was an early river runner and the 83rd person through Grand Canyon. In 1942, Marston and his son Garth were with Norm Nevills on the 19th trip through Grand Canyon. Marston later became a boatman for Nevills, but their relationship soured. Marston often tried to destroy Nevills's reputation, at one point even claiming that Nevills's fatal 1949 plane crash was a suicide.

Marston is associated with a variety of "firsts" during his 22 trips through Grand Canyon. For example, Marston was the first person, or part of the first group:

- To intentionally swim a Grand Canyon rapid without any kind of flotation device. Marston swam through Deubendorff Rapid in 1947.
- To complete a motorized run of the canyon, aboard *Esmeralda II* in June 1949. At the end of this trip, the crew—Edward "Ed" Hudson (who built and piloted the boat), Marston, Oakland attorney Bestor Robinson, Willie Taylor, and Ed Hudson—were greeted by NPS officials offering their congratulations for the accomplishment. This trip began the era of hard-hulled motorboats in Grand Canyon, which lasted into the 1960s. *Esmeralda II* is in GCNP's collection of historic boats.
- To be associated with a helicopter rescue of river runners in 1950.
- To successfully down-run through Grand Canyon with outboard motors in 1951.
- To run Colorado River through Grand Canyon on the highest flow ever navigated, in 1957 at 126,000 cubic feet per second.
- To successfully run upriver through Grand Canyon aboard a jet boat, in 1960.
- To successfully run Grand Canyon with a stock motorboat, in 1950.
- To run Grand Canyon on the dammed Colorado River after the bypass tubes of Glen Canyon Dam were closed, in 1963, aboard a SportYak II. This boat was the smallest boat to traverse Grand Canyon. Like *Esmeralda II*, it is part of GCNP's collection of historic boats.

In 1953, Marston mailed Christmas cards listing the first 100 Grand Canyon river runners. The names included many historic figures at Grand Canyon, including John Wesley Powell, Robert Brewster Stanton, Nathaniel Galloway, the Kolb brothers, Claude Birdseye, Haldane "Buzz" Holmstrom, and Barry Goldwater. The following year, Marston corrected the list, for he had not included Leigh Lint, who accompanied the 1923 USGS expedition led by Claude Birdseye. Marston's Christmas card in 1954 also listed river runners numbers 101–206 (stopping at 200 would have omitted six of the people on Bus Hatch's first commercial trip through Grand Canyon).

In 1947, Marston began writing a manuscript about the first 100 river runners through Grand Canyon. He assumed that the work would take him six months, but he worked on the ever growing manuscript until he died in San Francisco, California, on August 30, 1979. Part of Marston's manuscript was published in 2014 as *Powell to Power: A Recounting of the First One-Hundred River Runners through the Grand Canyon*. Despite his many accomplishments, Marston is not commemorated by a feature at Grand Canyon. Marston's ashes were spread at Lava Falls by Bill Belknap. Marston's historical records at The Huntington Library in San Marino, California, include 492 boxes, 60 albums, 163 reels of film, and 38 video cassettes.

See also: Grand Canyon Museum

Further Reading

Marston, Otis Reed "Dock." 2014. *From Powell to Power: A Recounting of the First One Hundred River Runners through the Grand Canyon*. (Edited by Tom Martin.) Flagstaff, AZ: Vishnu Temple Press.

MATHER, STEPHEN

Stephen Tyng Mather (1867–1930) was an American industrialist and conservationist whose first career involved being president and owner of the Thorkildsen-Mather Borax Company. (Mather invented the popular phrase "Twenty Mule Team," which helped spur sales of his Death Valley borax soap.) After becoming wealthy, Mather shifted his focus to conserving public lands. In 1917, after lobbying for the establishment of an agency to operate national parks, Mather was appointed the Assistant Secretary of the Interior and Director of NPS. Like conservationist John Muir, whom he admired, Mather viewed nature as a refuge for the human spirit, yet understood the need to build a political backing for the national parks. This meant getting lots of people to the parks and making sure they enjoyed their visits. So began NPS's formal and extensive partnerships with railroads and hotels in the American West. The success of this approach later produced problems.

As director of NPS, Mather helped increase the number of national parks and monuments, introduced concessions to NPS, and produced criteria for adding properties to the NPS. Mather, who believed that the primary criterion for the establishment of a national park was its scenery, organized 57 units of NPS (there are now more than 400). Mather served as NPS director until January 1929, when a stroke forced him to retire. He died a year later, on January 22, 1930, and is buried in Mather Cemetery in Darien, Connecticut.

The most visited rim-side viewpoint in GCNP is the South Rim's Mather Point. There, Mather is honored by a large, bronze plaque that was dedicated in 1932 on July 4 (Mather's birthday). The plaque reads as follows:

Stephen Mather, who was the first director of the National Park Service, often used personal funds to finance park improvements. This photo from 1921 shows Mather at the Powell Memorial at the South Rim's Powell Point. (Grand Canyon National Park Museum Collection)

> He laid the foundation of the National Park Service, defining and establishing the policies under which its areas shall be developed and conserved unimpaired for future generations. There will never come an end to the good that he has done.

This quote came from a speech given by Michigan Congressman Louis Cramton to the U.S. House of Representatives when Mather resigned as director of NPS. The plaque honoring Mather was originally on a stone pedestal just east of Yavapai Observation Station. Later, the plaque was mounted on a boulder by the sidewalk at Yavapai, after which that boulder and plaque were moved to their current position near Mather Point. An identical plaque is just outside the Administrative Building at the North Rim and elsewhere in NPS. Indeed, 67 such plaques honor Mather in national parks, national monuments, and state parks scattered from Alaska to the Virgin Islands and from Maine to Hawaii. (Another plaque was mounted, at the request of Mather's assistant Horace Albright, outside the NPS Director's Office in Washington, DC). Seven of the 67 plaques are in Arizona; only California, with 12, has more.

Mather often used personal funds to support NPS; for example, in 1923 Mather and his friends bought what became Pipe Spring National Monument on the

Arizona Strip. At that monument, you'll find a plaque honoring Mather. Ironically, Mather opposed plaques and other memorials in NPS, claiming that national parks should showcase nature, not plaques and transient heroism, and that parks should not be cluttered with monuments "like courthouse squares."

See also: Mather Point; National Park Service

Further Reading

Shankland, Robert. 1951. *Steve Mather of the National Parks*. New York: Alfred Knopf.

MATHER POINT

Mather Point (elevation 7,120 feet), which is 9.8 miles (as the condor flies) from the North Rim, is the most visited viewpoint in GCNP; virtually everyone who visits the South Rim enjoys their first view of Grand Canyon at Mather Point. On the walk leading to Mather Point, visitors see the Landmark and Tribal Medallion, which lists the eight tribes—Southern Paiute, Hopi, Zuni, Navajo, Apache, Yavapai, Hualapai, and Havasupai—that call Grand Canyon home. These tribes also consider Grand Canyon essential to their development and identity. The nearby Mather Amphitheater hosts programs and talks by GCNP rangers.

Grand Canyon Greenway

Grand Canyon Greenway is a paved set of multiuse, accessible trails throughout GCNP for nonmotorized travel, including people on foot, in wheelchairs, on bicycles, and (in some areas) on horseback. The trails, which link several major points of interest in GCNP, were designed to disperse people from crowded sites to other places where they could see the park from different views and under their own power. Popular examples of the Grand Canyon Greenway include the trails from Mather Point to Pipe Creek Overlook (South Rim), from Monument Creek Vista to Hermits Rest (South Rim), and from Bright Angel Point to the North Kaibab Trailhead (North Rim).

Construction of the trails, which were envisioned in GCNP's General Management Plan in 1995, was started in 1999. Today, there are 73 miles of Grand Canyon Greenway on the North and South Rims.

Mather Point, which was developed in 1953, is 4,600 feet above the Colorado River. The view from Mather Point, which is somewhat restricted by Yaki and Yavapai Points, extends from the lower end of Garden Creek (including part of Bright Angel Trail), across Pipe Creek and east to Cedar Ridge (at the top of the Supai) and O'Neill Butte, which are prominent features along South Kaibab Trail. In the east is Cape Royal, the easternmost overlook on the North Rim. Also visible are Vishnu Temple, Isis Temple, and, in the foreground, the relatively flat Tonto Platform, along which the Tonto Trail links Bright Angel Trail to the west and South Kaibab Trail to the east. Just across the river is a green oasis called Phantom Ranch,

The most popular overlook in Grand Canyon National Park (GCNP) is the South Rim's Mather Point, which is near the Landmark and Tribal Medallion. Mather Point is where most people visiting the park get their first glimpse of Grand Canyon. (Max Richard/National Park Service)

which offers lodging for people who hike or ride mules to the river (tourists cannot fly in a helicopter to Phantom Ranch). On clear days, visitors can see almost one-fourth of GCNP from Mather Point.

Mather Point honors Stephen Tyng Mather, a Borax Corporation executive who in 1914 complained in a 26-page letter to the Department of the Interior about the poor condition and management of U.S. national parks. Interior Secretary Franklin K. Lane (a former classmate of Mather) responded with a single sentence: "Dear Steve, if you don't like the way the national parks are being run, come on down to Washington and run them yourself" (Saber, 2017). Mather accepted the challenge and, three years later, became the first director of NPS. Mather helped create NPS, including formulating its mission and goals, hiring its managers, promoting its value to Congress and the public, and protecting and expanding its sites. The wealthy, manic Mather, who often paid NPS bills with personal funds, died on January 22, 1930, one year after retiring from NPS. Mather's house in Darien, Connecticut is a National Historic Landmark.

Mather was succeeded at NPS by his pragmatic long-time assistant, Horace Albright (1890–1987). Together, Mather and Albright defined much of what people expect of national parks today. Albright is honored at GCNP by the Albright Training Center, which was established in 1963. In 1980, President Jimmy Carter awarded Albright—who was the second director of NPS—the Presidential Medal of Freedom, the nation's highest civilian honor. In 2013, the Albright Training Center was listed on the National Register of Historic Places.

See also: Mather, Stephen

Further Reading

National Park Service. 2017. Stephen Tyng Mather. Retrieved June 4, 2017, at https://www.nps.gov/bestideapeople/Mather.html

Saber, Kate. 2017. The visionaries. National Parks Conservation Association. Retrieved July 5, 2017, at https://www.npca.org/articles/1017-the-visionaries#sm.000imk8bne0ndg611k210pebvec31

MATTHES, FRANÇOIS

François Emile Matthes (1874–1948) was a geographer whose maps of the upper Grand Canyon and other sites were among the first topographic maps of areas that became national parks. Matthes, a naturalized American who was born in Holland, began his 51-year career with USGS in 1896. His work often involved mapping remote areas of the American West.

When Matthes began mapping Grand Canyon in the spring of 1902, the only maps of the area were the charts produced more than three decades earlier by John Wesley Powell. The work was arduous and required Matthes to create several trails. Because there was no bridge across the Colorado River, it took Matthes six days to cross the river and reach the other rim. Parts of the trails built by Matthes and his colleagues in Bright Angel Canyon from the North Rim were later improved by David Rust and are now part of North Kaibab Trail. Matthes did all of his work with a plane table, of which Grand Canyon provided "an extreme test of the efficiency of that instrument" (Matthes, 1962).

Matthes produced the first topographic maps of Grand Canyon. His first maps at Grand Canyon featured Zoroaster Temple and parts of Bright Angel Creek and took nearly two years to finish. Matthes's maps were the best maps available for several decades. In 1933, a group of geologists attending the 16th International Geological Congress sent Matthes a signed, panoramic photo of Grand Canyon, noting, "We have been using your maps and have marveled at them. There are no other such topographic maps in all the world" (Matthes, 1962).

Matthes continued Clarence Dutton's use of "heroic nomenclature," adding the names Krishna Shrine and Solomon Temple to the canyon's geography. Here are other examples of sites in Grand Canyon named by Matthes:

Lipan Point, which is just west of Desert View, is the trailhead of Tanner Trail and one of the widest points in GCNP. It was earlier named Lincoln Point (for Abraham Lincoln) but was renamed in 1902 by Matthes to honor the Lipan tribe of Apache Indians of Texas. From Lipan Point, visitors can see parts of six climatic zones, ranging from the arctic zone near the top of the San Francisco Peaks to the warmest climate at the bottom of Grand Canyon.

Visitors also have an excellent view of Hance Rapid, which is a 30-foot-drop near River Mile 77 of the Colorado River. Lipan Point is an excellent place to see the Colorado River, Grand Canyon Supergroup, and sunsets.

Wotans Throne is a high, flat-topped butte southwest of Cape Royal that is visible from many parts of the South Rim. Matthes named Wotans Throne for Wotan, the Norse god of war.

Walhalla Plateau and Glade on the North Rim near River Mile 84 is named for the great Scandinavian hall of gods in Asgard to which the souls of slain Viking warriors and other heroes went. "Glade" refers to an opening in a forest. This area, which had been referred to by Mormon settlers as Greenland Plateau, was named by Matthes in 1902.

Matthes's contributions to Grand Canyon and other national parks earned him a Citation for Distinguished Service, the highest commendation awarded by the U.S. Department of the Interior.

Matthes died in El Cerrito, California, at age 74 on June 21, 1948. His ashes were scattered at Yosemite Valley. Matthes is honored by François Matthes Point (on river right near River Mile 84; elevation 8,020 feet) and the North Rim's unmarked François Matthes Trail, a 10.6-mile (round trip), former logging road through pine and aspen trees that begins near Vista Encantada Overlook along Cape Royal Road.

See also: North Kaibab Trail

Further Reading

Matthes, François E. 1962. *François Matthes and the Marks of Time: Yosemite and the High Sierra*. Retrieved March 3, 2017, at http://www.yosemite.ca.us/library/matthes/francois_matthes.html

MCKEE, EDWIN "EDDIE"

Edwin Dinwiddie "Eddie" McKee (1906–1984) was introduced to Grand Canyon when he was in the Boy Scouts in Washington, DC, where he grew up. McKee was a member of Boy Scouts, Troop 1, whose scoutmaster was Grand Canyon pioneer François Emile Matthes. McKee enrolled in the U.S. Naval Academy preparing for a career in the Navy but left school after three years. In 1927, Matthes arranged McKee's first visit to Grand Canyon, where McKee worked as a summer intern with paleontologist Charles W. Gilmore and ecologist John C. Merriam. Merriam was developing GCNP's first interpretive programs, and McKee spent much of his time working on displays at the Yavapai Point Observation Station and Trailside Museum on the South Rim. McKee appreciated the impact of Grand Canyon on visitors ("within its walls, one is in another world, both scientifically and geologically") and spent many years trying to help people appreciate the canyon's significance. McKee, who was not related to North Rim pioneers Thomas and Elizabeth McKee, influenced GCNP's public programs for decades to come; some of his exhibits were displayed there for more than 50 years.

McKee was inspired by his experiences at Grand Canyon, and soon enrolled at Cornell University to study geology. Two years later, following the drowning of GCNP naturalist Glen E. Sturdevant at Horn Creek Rapid, the 22-year-old McKee became GCNP's second park naturalist and, later, the GCNP's chief park ranger. McKee continued to publish *Grand Canyon Nature Notes*, which Sturdevant had started in 1925. (The publication ended in 1935.) While at Grand Canyon, McKee began dating biologist Barbara Hastings, who was studying mammals on the North Rim; their weekends together required McKee to often hike rim-to-rim-to-rim. After Eddie and Barb married on December 31, 1929, they jointly began a bird-banding center and became close friends with many Havasupai; these friendships enabled the McKees to amass a collection of 184 Havasupai baskets (from 32 basket makers), which they described in their book titled *Havasupai Baskets and Their Makers* (1975). McKee acknowledged the Havasupai by, for the first time, using their names (e.g., Manakaja, Wescogame) to denote layers of the Supai Formation. Barb shared in all of Eddie's work.

McKee published seven books, four monographs, and many other publications about Grand Canyon. To McKee, "the most remarkable and unique feature of Grand Canyon is the clarity and simplicity of the Earth's history as recorded by the layers in its walls, comparable to the pages and chapters of a book." McKee compiled the first checklists of Grand Canyon's reptiles, amphibians, birds, and mammals, and in 1929, while hiking down the Tanner Trail, he discovered (and later described) the endemic Grand Canyon rattlesnake (*Crotalus oreganus abyssus*), a subspecies of the Western Diamondback rattlesnake that lives only at Grand Canyon. In 1932, McKee, Emery Kolb, and 130 others founded the Grand Canyon Natural History Association, which later became today's Grand Canyon Association. This organization's goal was to promote scientific studies of geology, biology, ethnology, archaeology, and related subjects at Grand Canyon. McKee's self-published *Ancient Landscapes of the Grand Canyon Region* (1931), which was originally sold at GCNP for 50¢, enjoyed 30 revised printings before it went out of print in 1985 (the year following McKee's death).

In 1938, because of NPS's policy preventing rangers from staying at one park for more than 10 years, McKee rejected the NPS offer to transfer to Yosemite and instead became the assistant director of research of the Museum of Northern Arizona in Flagstaff. From 1942 to 1952, McKee (with only a bachelor's degree) became a geology professor and administrator at the University of Arizona at Tucson. From 1953 until his retirement in 1977, McKee was a USGS research geologist in Denver, Colorado, where he directed the Paleotectonic Maps Section. McKee was awarded an honorary doctorate from Northern Arizona University in 1957, and in 1964 he organized the first symposium about the formation of Grand Canyon. After retiring, McKee continued to study sedimentology in his lab at USGS. McKee published his last research paper about Grand Canyon in 1984.

McKee died in Denver on July 23, 1984, and Barb died in 1998. Their ashes are interred at the South Rim's Pioneer Cemetery beneath a small tombstone made of water-sculpted Tapeats Sandstone from Pumpkin Spring; the tombstone tells visitors, "They shared a love of natural history and the Grand Canyon." Nearby are the graves of paleontologist David White (with whom McKee first worked as a summer intern in 1927) and geologist John H. Maxson. When the McKees' son came to Grand Canyon to replace the plaque at his parents' grave, he brought cement but not a brush. He finished his job by using a brush left at the nearby grave of Grand Canyon artist Gunnar Widforss.

In GCNP, McKee is commemorated by McKee Point (near River Mile 177) and McKee Amphitheater, which is between GCNP Headquarters and the South Rim Trail.

See also: Pioneer Cemetery and Other Gravesites; Yavapai Point

Further Reading

Berger, Todd R., ed. 2008. *Reflections of Grand Canyon Historians: Ideas, Arguments, and First-Person Accounts*. Grand Canyon, AZ: Grand Canyon Association.

MISSION 66

Mission 66 was a 10-year program initiated in 1956 by NPS director Conrad L. Wirth (with the support of President Dwight D. Eisenhower) to modernize, expand, and generally reinvigorate the visitor services, employee residences, interpretive centers, staffing, and resource management at NPS properties in preparation of the 50th anniversary of NPS in 1966. Mission 66 produced 575 campgrounds, 1,197 miles of roads, 1,502 parking lots, and 100 visitor centers, the first of which was in 1957 at GCNP (at a cost of almost $500,000). Mission 66 updated the visual identity (i.e., brand) of NPS and most national parks. Nearby was the so-called Mather Business Zone, which included services such as a bank, post office, and gas station.

Mission 66 was meant to help NPS deal with growing numbers of visitors. The end of World War II produced a surge in visitation at national parks, and many parks were jammed with tourists. This was especially true at Grand Canyon; on June 19, 1946 alone, 4,262 people came to Grand Canyon in more than 1,000 cars (during all of June that year, visitation totaled 98,545 people). On April 11, 1947, the millionth car arrived at Grand Canyon. GCNP, like many other national parks, needed updating.

The $1 billion (over the program's 10 years) Mission 66 project improved facilities such as roads, utilities, and employee housing. One of the biggest Mission 66 projects at GCNP was the 12.4-mile-long trans-canyon water pipeline, on which construction began in 1965. In addition to bringing water to new sites, this project pioneered the use of helicopters to move people and materials to remote locations in GCNP. Mission 66 also developed the concept of "visitor centers" at many parks. The Visitor Center at the South Rim, which was completed in 1957, was one such center that resulted from Mission 66. Other improvements at GCNP produced by Mission 66 included Thunderbird Lodge (built in 1968) and the 10-acre Horace M. Albright Training Center (built in 1963–1965). The training center, which trains 700 NPS employees per year, was added to the National Register of Historic Places in 2013.

An often overlooked part of Mission 66 involved NPS's decision to hire a New York graphics design firm to redesign the NPS's iconic "arrowhead" logo, which dated from 1951. The new and uninspiring "triangles and cannonballs" logo, which appeared in 1966, was quickly abandoned and might have been quickly forgotten, had it not appeared on more than 110 million 5¢ postage stamps.

See also: Grand Canyon Village

Further Reading

Allaback, Sarah. 2000. *Mission 66 Visitor Centers: The History of a Building Type*. Washington, DC: National Park Service.

MORAN, THOMAS

Thomas Moran (1837–1926) was an American painter who became famous for his paintings of the American West. Although no photographs were taken during John Wesley Powell's 1869 expedition through Grand Canyon, Powell's exploits (associated with Thomas Moran's panoramas of the canyon) made him a national hero. Moran's paintings, which were often sponsored (and featured in advertisements) by the Atchison, Topeka and Santa Fe Railway, captured the public's imagination, and his gigantic 7-by-12-foot painting titled *The Grand Canyon of the Yellowstone* (purchased in 1874 by Congress for the then gigantic sum of $10,000) from the Hayden Survey to Yellowstone inspired Congress to make Yellowstone the first national park in 1872. This painting was also the first landscape by an American artist to hang in the U.S. Capitol. Thereafter, Moran sometimes signed his name "T-Y-M" for Thomas "Yellowstone" Moran. (In 1913, Elliott Daingerfield's *The Genius of the Grand Canyon* sold for $15,000.) Although the railroad once owned several of Moran's paintings, today its collection includes only one: *Grand Canyon of Arizona, from Hermit Rim Road*, which Moran painted in 1912–1913.

Thomas Moran was an artist whose famous paintings of Grand Canyon and other sites in the Southwest were sponsored by the Santa Fe Railway to promote tourism. One of these paintings of Grand Canyon, *The Chasm of the Colorado*, hangs in the Smithsonian American Art Museum. For several decades, more people saw this painting of Grand Canyon than saw Grand Canyon itself. (Library of Congress)

John Wesley Powell, appreciating the political power of Moran's work, convinced Moran to accompany him (and Jack Hillers) to the North Rim of Grand Canyon in 1873. During this trip, Moran visited and sketched views of Toroweap, where he began painting his first oil painting of Grand Canyon, the influential 7-by-12-foot *The Chasm of the Colorado*. This ominous painting, which includes features from several North Rim overlooks, shows shadows of rain falling in the inner canyon while sun bathes other parts of the canyon. The Colorado River is near the center of the painting. Like many of Moran's paintings, there is no simple focus, muted colors create a seemingly endless view, and contrived foregrounds lead viewers' eyes to the center of the painting. The painting, which was used by the Atchison, Topeka and Santa Fe Railway for its calendars for several years, shows what Moran considered "by far the most awfully grand and impressive scene that I have ever yet seen" (Lago, 2015). Congress bought Moran's painting and hung it across from his earlier painting of Yellowstone.

The Chasm of the Colorado introduced the American public to Grand Canyon's stunning beauty; indeed, for several decades, more people saw Moran's painting than saw Grand Canyon. *The Chasm of the Colorado*, which is owned by the Department of the Interior, now hangs beside *The Grand Canyon of the Yellowstone* in the second-floor lobby of the Smithsonian Institution's Museum of American Art.

Moran never did return to Yellowstone, but he returned to Grand Canyon often because "of all places on earth, [it] is the most inspiring in pictorial possibilities." Moran was invited to the South Rim in 1892 as a guest of the Atchison, Topeka and Santa Fe Railway, which paid his expenses in exchange for the copyright to one of Moran's paintings that could be copied and hung in railroad stations and hotels. From 1899 to 1924, Moran spent most winters painting at Grand Canyon as a guest of the railroad. Much of Moran's popularity and the railroad's success depended on Charles A. Higgins (1858–1900), who was in charge of the railroad's advertising and marketing programs aimed at luring people to Grand Canyon and other sites in the American West. Visitors at El Tovar Hotel, where Higgins often stayed, can see a silent tribute to Higgins; when they face the north porch and look at the first-floor fascia, visitors see the last sentence of Higgins's *Titan of Chasms: The Grand Canyon of Arizona* (1906): "Dreams of mountains, as in their sleep, they brood on things external."

Like Charles Dutton before him, Moran began his work with photographs and sketches but then attempted to produce something even more inspiring than reality. Moran had little interest in replicating what he saw in Grand Canyon; as he often noted, "Art is not nature" (Candlin, 1986), and he believed that technical accuracy could ruin paintings. This is why most of his paintings emphasize golden-yellow schemes (like those of European landscapes) and have no simple focus. This approach provided a sharp contrast to the accurate and detailed work of William Henry Holmes, whose paintings often included foregrounds that lead to detailed horizons. Many artists consider Holmes's three-part *Panorama from Point Sublime*, which appeared in Dutton's *Tertiary History of the Grand Canyon District* (1882, and was probably inspired by looking east from Toroweap), to be the most magnificent image of Grand Canyon ever.

Moran didn't value technical accuracy. Instead, he tried to capture the canyon's idealized landscape and overall impact on the human spirit for, according to Moran, an artist's goal is

> to produce for the spectator of his pictures the impression produced by nature on himself . . . My personal scope is not realistic. All my tendencies are toward idealization . . . Topography in art is valueless. The motive or incentive . . . was the gorgeous display of color that impressed itself upon me . . . and while I desired to tell truly of nature, I did not wish to realize the scene literally to preserve and convey its true impression. Every form introduced into the picture is within view from a given point, but the relation of the separate points is not always preserved. (Johns, 1996)

Moran often darkened and narrowed the canyon's gorges, barely acknowledged geological strata, and rearranged the canyon's buttes, mesas, and other features to convey the feelings of the expedition he was describing. After the arrival in 1901 of the Santa Fe Railway at the South Rim increased tourism, Moran's paintings became more accurate.

When John Wesley Powell wrote about his expeditions through Grand Canyon for *Scribner's Monthly* in 1875, he included Moran's illustrations, even though Moran had never been down the river. Moran's art was also included in Clarence Dutton's influential *Tertiary History of the Grand Cañon District* (1882). Several of Moran's other paintings, one of which (*Cliffs of Green River*, 1887) hangs near the top of the stairway in El Tovar Hotel, helped convince President Theodore Roosevelt to create the Grand Canyon National Game Preserve (without the approval of Congress) in 1906. As one critic noted, Moran "awakened the American consciousness to the permanent value of those wide, measureless expanses of wilderness, of sky and mountain and extravagances of Nature, as natural resources of beauty, to be prized and conserved and held as great national parks" (Arizona State University, 2017). Moran named Grand Canyon's Tower of Set (Moran spelled it "Sett"), a pinnacle of Supai sandstone that dominates the skyline from River Mile 91.

In 1917, just one year after the creation of NPS, the National Gallery of Art staged an exhibition of art about national parks; 17 of the portraits and 4 of the paintings were by Moran. Moran died on August 25, 1926, at age 89 in Santa Barbara, California. In 2014, the U.S. Postal Service issued a Forever stamp featuring one of Moran's paintings of Grand Canyon. Moran, who is commemorated at GCNP by Moran Point, is one of only two artists (with Gunnar Widforss) to have overlooks named after them at GCNP.

See also: Thematic Essays section: John Wesley Powell; Dutton, Charles; Moran Point

Further Reading

Arizona State University. 2017. Landscape art. Retrieved June 3, 2017, at http://grandcanyonhistory.clas.asu.edu/history_visualarts_landscapeart.html

Candlin, Enid Saunders. 1986. Art is not nature. *The Christian Science Monitor*, March 6, 1986. Retrieved July 4, 2017, at https://www.csmonitor.com/1986/0306/umoran.html

Higgins, Charles A. 1906. *Titan of Chasms: The Grand Canyon of Arizona*. Chicago: Santa Fe Railway.

Johns, Joshua. 1996. Thomas Moran and the American landscape: The lure of the West. Retrieved April 28, 2017, at http://xroads.virginia.edu/~cap/nature/cap3.html

Kinsey, Joni Louise. 1992. *Thomas Moran and the Surveying of the American West*. Washington and London: Smithsonian Institution Press.

Lago, Don. 2015. *Grand Canyon: A History of a Natural World and National Park*. Reno: University of Nevada Press.

Wilkins, Thurman. 1998. *Thomas Moran: Artist of the Mountains*. Norman: University of Oklahoma Press.

MORAN POINT

Moran Point (elevation 7,160 feet), which is 8 miles south (as the condor flies) of the North Rim's Cape Royal, features the layered, eastward-dipping Grand Canyon Supergroup; this, with nearby Lipan Point, is one of the few places along the South Rim where these formations are visible. Here, visitors can also see the Colorado River, Coronado Butte, Red Canyon (where Grand Canyon pioneer John Hance had several mines), Grandview Point, Horseshoe Mesa, and the Grand Canyon Metamorphic Suite.

Many people believe that Moran Point is named for famed artist Thomas Moran, who first came to the South Rim in 1892 (as a guest of the Santa Fe Railway) and spent nearly every winter at Grand Canyon from 1899 to 1929. Moran's paintings of Grand Canyon, which were often painted in golden-yellow hues, were misleading, for they usually rearranged geological strata and other features. Moran had little interest in accuracy but instead tried to capture the impressions the canyon made on people who witnessed it. (After the Santa Fe Railway arrived at the South Rim in 1901, Moran's paintings became more realistic.) Moran's painting titled *Grand Canyon of the Yellowstone* (1872) helped galvanize support for Yellowstone, the first national park. (That painting hangs in the Smithsonian Museum of American Art in Washington, DC) Today, artists continue to be inspired by the shadows and light at Moran Point.

Some historians, however, claim that Moran Point is named for artist Peter Moran (1841–1914), Thomas's brother. Peter came to the South Rim in 1881 (more than a decade before Thomas), possibly with explorer and Army Capt. John Gregory Bourke (1846–1896), who some claim named the area in Peter's honor.

See also: Moran, Thomas

Further Reading

Anderson, Michael. 2001. *Along the Rim: A Guide to Grand Canyon's South Rim from Hermit's Rest to Desert View*. Grand Canyon, AZ: Grand Canyon Association.

MOTORBOATS IN GRAND CANYON NATIONAL PARK

For the first few decades of commercial river running in Grand Canyon, customers and their guides floated down the Colorado River on small, oar-powered boats. By the 1950s, however, several GCNP outfitters began using outboard motors to push larger, pontoon-like rafts carrying larger numbers of customers—sometimes, as many as 40 per boat—on faster, shorter, cheaper, and more profitable trips through Grand Canyon. Indeed, between 1969 and 1971, the number of outfitters increased from 11 to 20, and the number of people who paid those outfitters to float through Grand Canyon rose from 3,500 to 10,500. The Colorado River in Grand Canyon was becoming crowded.

In 1964, Congress passed the Wilderness Act, which directed NPS to determine which of its lands could be classified as wilderness. "Wilderness," which usually bans motorized vehicles, is the highest level of conservation protection; only Congress can designate land as "wilderness" (National Park Service, 2017). In 1966, GCNP drafted its Wilderness Planning Procedures Statement, declaring "boating except with motorboats and airboats is an acceptable use of park wilderness" (Ingram, 2003). Although this policy—which was issued when river running in GCNP was starting to grow—could have ended the use of motorized boats in GCNP, the ban on such boats was not implemented. In 1971, NPS explicitly advocated a ban on motorboats in GCNP, despite the fact that motorboats cut the time in half for a trip through GCNP. Five years later, in 1976, NPS again banned the use of motorboats with this statement:

> The Colorado River totaling about 17,009 acres is now recommended as wilderness so as to perpetuate the primitive qualities of the canyon with increased opportunities for solitude and enjoyment of the beauty and natural significance of the Grand Canyon. To achieve this, all visitor use of the river will be without motors, and more nearly like the experience of earlier explorers. A three-year study of the river with public participation has shown that visitor appreciation, understanding, and enjoyment of the Grand Canyon will be enhanced by this type of use. The objective for visitor use is to provide the opportunity for an intimate association with the river environment with good opportunities to see and gain understanding of the natural and historic features of the inner canyon. (Ingram, 2003)

Politicians tried to play both sides of the controversy. For example, in 1978, Arizona Senator, former Republican presidential nominee (1964), and Colorado River runner Barry M. Goldwater (1909–1998) promised to ban motorboats in GCNP because he believed them to be "incompatible and unnecessary." After claiming to speak as an environmentalist and noting that he had been "up and down that river so many times," Goldwater proclaimed, "I, too, regret the commotion and noise made by the motorized vehicles on the Colorado River in the Grand Canyon, and will work for the plan to discontinue their use." Goldwater added that rowing the

Grand Canyon "is a delightful way to go through, because it is quiet, you do not have the smell of gasoline, you do not have the roar of outboard motors." Goldwater wanted the government to stay out:

> Government got their hands on it and started this environmental business. If there is a faster way to destroy a beautiful place of scenery than to make a wildlife preservation out of it or to get an environmental group into it, I do not know what it could be. (Ingram, 2003)

Commercial river running companies pushed back. When outfitters protested that the ban on motorboats was elitist, Goldwater said that motorized boats were acceptable in the park and complained that eliminating them would "deprive a great many people the opportunity" to run the river. In 1980, Goldwater—with Utah Senators Jake Garn and Orrin Hatch—sponsored legislation that maintained the use of motorboats in GCNP (the legislation called it "commercial motorized watercraft excursions"; Hayes, 1999). After noting that motorized boats "left no environmental impact," Goldwater announced, "I'm not convinced of the need to ban" the boats. Goldwater then changed his position again, claiming that he opposed motorized boats in GCNP "because I think it's a special place which I don't think God made for everyone to see" (Ingram, 2003). Throughout the Congressional testimony associated with the legislation, Hatch did not mention the fact that his father, Jesse Hatch, was a cousin of Bus Hatch, the founder of Hatch River Expeditions, a commercial outfitter that would benefit financially from the use of motorized boats in Grand Canyon (Hayes, 1999).

Advocates of motorized boats in Grand Canyon gained another ally in 1981 when James Watt became Secretary of the Interior. Watt, who had helped organizations sue NPS to continue the use of motorized boats in Grand Canyon, believed that environmentalists were "the greatest threat to the ecology of the West" (Hayes, 1999). Watt wanted to increase the input and impact of concessionaires in NPS, and he emphatically wanted motorized boats in GCNP. As he noted, "I don't like to paddle and I don't like to walk" (Prochnau and Thomas, 1981).

Despite many claims and much political posturing, motorized boats were not banned from GCNP, and the proposal to designate more than 90 percent of Grand Canyon "wilderness" awaits action. Today, motorized boats remain the most popular way for more than 25,000 people per year to "run the Grand Canyon." As of 2018, there was no congressionally designated wilderness anywhere in GCNP.

See also: Goldwater, Barry

Further Reading

Hayes, Byron. 1999. Motorized rafting: An illegitimate industry? The 35 year wilderness war for Grand Canyon. *The Waiting List*, 3 (2), May. Retrieved May 7, 2017, at https://rrfw.org/sites/default/files/documents/Motorized_Rafting_-_An_Illegitimate_Industry.pdf

Ingram, Jeff. 2003. *Hijacking a River: A Political History of the Colorado River in the Grand Canyon*. Flagstaff, AZ: Vishnu Temple Press.

National Park Service. 2017. Wilderness. Retrieved July 1, 2017, at https://wilderness.nps.gov/faqnew.cfm

Prochnau, Bill, and Thomas, Valerie. 1981. The Watt controversy. *The Washington Post*, June 30. Retrieved July 1, 2017, at https://www.washingtonpost.com/archive/politics/1981/06/30/the-watt-controversy/d591699b-3bc2-46d2-9059-fb5d2513c3da/?utm_term=.58bd16ba66e4

MULES

Mules result from mating a male donkey with a female horse. Mules are hardier, more useful, cheaper, longer-lived, more patient, and less obstinate than donkeys, which are a type of domesticated horse. Mules have long been important at Grand Canyon; prospectors and early settlers—as well as today's visitors—have used them to haul supplies and themselves into and around Grand Canyon. The value of Grand Canyon's mules has even been recognized by politicians in Washington, DC. For example, in 1939, Congress and President Franklin D. Roosevelt agreed that the Fred Harvey Company should be paid $125 for a mule named "Pima" who died while packing supplies to a CCC camp in the canyon.

Mules live 30–35 years and at Grand Canyon are ridden when they are between the ages of 6–25. They can also carry more weight than donkeys. Mule rides, which have been used to see Grand Canyon since 1886, are available at the North and South Rims of Grand Canyon. Each of GCNP's 150–180 mules, most of which come from Tennessee, walk about 800 miles per year into and out of the canyon; they usually "work" 3–4 days per week. Mules at the South Rim go into Grand Canyon

Mules continue to do important work for people who live and work in and around Grand Canyon. This work is honored by the U.S. Postal System's postmark at Supai, Arizona, a small community of Havasupai in Grand Canyon. The post office at Supai is the last such office in the United States that transports mail by horses or mules. (Randy Moore)

along Bright Angel Trail and come out of Grand Canyon on the South Kaibab Trail. Mules stop at least 30 times on their way out of the canyon and at every other switchback. About 60 of the mules at the South Rim (each of which weigh 1,000 to 1,200 pounds) are housed at the barn across from Bright Angel Trailhead, while the others stay near Yaki Point. In 2017, the Bright Angel History Room displayed the saddle of Ross Knox, who rode mules more than 40,000 miles between the South Rim and Phantom Ranch in Grand Canyon.

When they arrive at Grand Canyon, most mules are used to pack supplies to, and mail from, Phantom Ranch; later, they are promoted to carrying humans. The six quarts of urine and 44 pounds of manure produced each day by each mule are conspicuous reminders to hikers that they are sharing the trails with mules. Although mules have been replaced by tractors and trucks elsewhere, they remain valuable at Grand Canyon. (So do horses; in 2017, GCNP had three patrol horses—Rio, Squirt, and Maverick.)

GCNP buys approximately 20 new mules each year. About every 45 days, farriers replace the mules' shoes, a process that takes about 45 minutes. Grand Canyon's Mule Barn, which includes the longest continually operational blacksmith shop in the United States, is near the head of Bright Angel Trail on the South Rim. In 2017, a 1-hour ride along the North Rim cost $45 per person, a three-hour trip to Supai Tunnel cost $90 per person, and round-trips (including accommodations and meals) to Phantom Ranch cost $515 per person. Guides take up to 10 mule riders at a time. There have been no tourist fatalities on mule rides at Grand Century in the last century.

See also: Bright Angel Trail; North Kaibab Trail; South Kaibab Trail

Further Reading

Brownold, Tom, and Dimock, Brad. 2010. *The Grandest Ride*. Tucson, AZ: Rio Nuevo Publishers.

N

NANKOWEAP CANYON

Nankoweap Canyon at Mile 52.5 was named by John Wesley Powell after his second expedition through Grand Canyon. The canyon is seemingly close to the North Rim but is designated a Primitive Use Area. Construction of the 14-mile-long Nankoweap Trail, which enabled pack animals, equipment, and supplies to be taken to Grand Canyon's remote Marble Gorge, was supervised by Charles Walcott; after spending 72 days along the trail, Walcott described the area as "a very lonely, quiet place . . . nothing but cliffs and cañons. Rocks, rocks, rocks . . . To me, [Grand Canyon] is intensely interesting and worth a year of exclusion" (Annerino, 2017). (Walcott, who discovered some of the oldest fossils in Grand Canyon and helped establish the canyon's age, later became the third director of USGS.) Parts of the unmaintained trail, such as the narrow parts of the Supai Traverse beginning near River Mile 6 on the trail, are exceedingly difficult and should be visited only by experienced hikers.

Nankoweap Canyon includes more than 40 archaeological sites, most of which were built between 900 and 1150 CE by prehistoric Pueblo people (e.g., the Nankoweap granaries visible from the river are one of several ancestral sites of Hopis). The meaning of the canyon's name is unclear; it could refer to the canyon's deep echoes, or to a Navajo raid that killed several Paiute Indians.

At the head of Nankoweap Rapid (Mile 52.3), the rim is 3,200 feet above the river. When hikers look up Nankoweap Canyon from the river, they see the North Rim's Point Imperial.

See also: North Rim; Point Imperial

Further Reading

Annerino, John. 2017. *Hiking Grand Canyon: A Detailed Guide to More Than 100 Trails*. New York: Skyhorse Publishing.

Hefley, Flood. 2013. *Grand Canyon Nankoweap: Hiking the Notorious Horsethief Trail*. Boulder, CO: Johnson Books.

NATIONAL PARK SERVICE

Grand Canyon National Park is part of the U.S. National Park Service (NPS), a federal agency that manages all of the national parks, national monuments, and other historical and conservation properties. NPS is part of the Department of the Interior, which was established on March 3, 1849, the last day of the 30th Congress. Since 1872, NPS has grown from a single park to an organization that includes more than 415 historic, recreational, natural, and cultural sites in every state, the District of Columbia, American Samoa, Guam, Puerto Rico, and Virgin Islands. These properties comprise 4.5 million acres of oceans, lakes, and reservoirs, more than 68,500 archaeological sites, 27,000 historic structures, 21,000 buildings, 12,250 miles of trails, 5,500 miles of roads, and more than 121 million objects in museum collections.

NPS preserves the national and cultural resources and values of the NPS for the enjoyment, education, and inspiration of this and future generations. NPS is responsible for 16 of the 19 World Heritage Sites in the United States (one of which is GCNP). The NPS emblem includes a sequoia (representing vegetation), a bison (representing animals), and mountains and water, all arranged as part of an arrowhead (representing historical and archaeological sites and values).

On March 1, 1872, President Ulysses S. Grant signed "An Act to set apart a certain Tract of Land lying near the Headwaters of the Yellowstone River as a public Park." This created Yellowstone National Park, the first national park in the United States. Because there was no state governmental agency to manage the park, the federal government assumed control of the park. Initially, each national park was managed independently. Starting in 1911, there were annual legislative attempts to create NPS, but each failed. However, to remedy managerial inconsistences among the parks, NPS was created on August 25, 1916, when President Woodrow Wilson signed the National Park Service Organic Act requiring NPS to "conserve the scenery and the natural and historic objects and wildlife therein, and to provide for the enjoyment of the same in such manner and by such means as will leave them unimpaired for the enjoyment of future generations" (National Park Service, 2017). This act created NPS and protected the 40 national parks and monuments then in existence. On June 10, 1933, President Franklin D. Roosevelt's Executive Order 6166 reorganized NPS by consolidating all of the national parks, national monuments, national military parks, national cemeteries, national memorials, and national capital parks (e.g., those controlled by agencies such as the Departments of War and Agriculture) into the NPS.

Today, NPS's more than 415 properties span 84.4 million acres; this is more land than is covered by New Mexico, the fifth largest state. Of this land, 4.3 million acres are owned privately. The largest national park—Wrangell-St. Elias National Park in Alaska—covers 13.2 million acres (which is more than 16 percent of the entire NPS). The smallest property of NPS is the Thaddeus Kosciuszko National Monument, which covers 0.02 acres (approximately 100 square yards).

NPS, which employs more than 20,000 workers, includes properties that attract more than 330 million visitors per year. The 10 most visited units of NPS handle more than 25 percent of these visitors. The most visited unit of NPS is Blue Ridge Parkway with more than 15 million visitors per year; GCNP ranks 10th, with nearly 6 million visitors per year. Among national parks, only Great Smoky Mountains National Park has attendance (10.7 million visitors) that exceeds that of GCNP. The NPS budget is approximately $3.4 billion per year. Visitation to NPS properties adds $34.9 billion to the economy per year, supports more than 310,000 jobs, and provides more than $12 billion in labor income. The top expenses of visitors at NPS properties are for lodging (31 percent), food and beverages (27 percent), gasoline and oil (12 percent), admission and fees (19 percent), and souvenirs (10 percent).

In early 2018, NPS also had a backlog of more than $12 billion in deferred maintenance. At GCNP, deferred maintenance totaled $370 million, most of which was due to utilities such as the aging trans-canyon pipeline. For example, all of the freshwater used at the North Rim arrives in a single aluminum pipe from Roaring Springs that was built more than 50 years ago. In 2017, a rockslide destroyed more than 300 feet of this pipeline, thereby closing North Kaibab Trail.

See also: Mather, Stephen

Further Reading

National Park Service. 2017. Organic Act of 1916. Retrieved June 1, 2017, at https://www.nps.gov/grba/learn/management/organic-act-of-1916.htm

NAVAJO BRIDGE

In the 1870s, people in Utah began moving to Arizona but soon encountered a formidable barrier to their journeys: miles and miles of Grand Canyon. One of the

River runners see this downstream view of Navajo Bridge near River Mile 4. The original (upstream) bridge, which opened to much fanfare in 1929, ended commercial river crossings at Lees Ferry. When the downstream bridge opened in 1995, the older bridge became a pedestrian crossing. Navajo Bridge, which is approximately 470 feet above the river, is the last place for cars to cross the Colorado River until Hoover Dam, which is more than 340 miles downstream. (Randy Moore)

few places that wagons could get to the Colorado River was at what came to be known as Lees Ferry, which had been established by (and named after) John D. Lee in 1873. However, this ferry was unreliable, dangerous, and sometimes closed by floods.

Construction of Navajo Bridge began in June 1927, about 4 miles downstream from Lees Ferry. Some of the materials used to build the bridge were ferried across the Colorado River at Lees Ferry, which continued to operate after construction of the bridge began. On June 7, 1928, three people were killed at Lees Ferry when the ferry sank; because the bridge was nearing completion, the ferry was not replaced. Navajo Bridge was funded by what is now the Arizona Department of Transportation and the U.S. Department of the Interior's Bureau of Indian Affairs (the eastern end of the bridge is on Navajo Nation land). During construction of the spandrel arch bridge, one worker (Lane McDaniel) died when he fell from the bridge.

On January 12, 1929, the Navajo Steel Arch Highway Bridge—now commonly known as Navajo Bridge—opened to traffic and carried U.S. Route 89A 800 feet across Colorado River's Marble Canyon. The opening was described by the *Coconino Sun* (a Flagstaff, Arizona, newspaper) as "the biggest news in southwest history." Although the bridge was in a remote location, its dedication on June 14–15, 1929, attracted more than 6,000 visitors who arrived in 1,217 cars to hear speeches by the president of the Mormon Church and the governors of Arizona, New Mexico, and Utah. At the time, this bridge was the highest steel arch bridge in the world. Because the dedication occurred during Prohibition, the bridge was christened with a bottle of ginger ale.

Navajo Bridge was originally known as Grand Canyon Bridge, but the Arizona legislature changed its name in 1934 to Navajo Bridge. Navajo Bridge, which is near River Mile 4.2, is the last bit of "civilization" river runners will encounter on the Colorado River until they get to the suspension bridges that lead to Phantom Ranch, which is 88 miles downstream.

Navajo Bridge—one of just seven land crossings of the Colorado River for more than 700 miles—joined the National Register of Historic Places in 1981. However, by 1990, the original bridge—which was only 16 feet wide and had a 40-ton capacity—could no longer meet the demands of increased traffic in the area. Its narrowness had also resulted in 72 accidents and eight deaths.

After rejecting plans to expand the original bridge, highway officials began building a new bridge adjacent to the original Navajo Bridge. During construction of this bridge, a 1.625-inch error in the positioning of anchor bolts produced a $42,000 delay. The new bridge, which opened on May 2, 1995 (and was dedicated on September 14 of that same year), resembles the original bridge. Whereas the first bridge had been christened with a bottle of ginger ale, the second bridge was christened with a bucket of Colorado River water.

The original Navajo Bridge is now used by pedestrians. The east side of the bridge, which is on Navajo land, includes booths at which vendors sell jewelry and Native American crafts. The nearby Navajo Bridge Interpretation Center, which was dedicated on June 17, 1997, on the west side of the bridge, describes the area's history and features self-guided walks, exhibits, memorials, and a bookstore and gift shop. California condors often perch in the ironworks of the bridges.

Feature	Original Bridge	New Bridge
Cost	$390,000	$14,700,000
Length, ft	834	909
Width of road, ft	18	44
Length of steel arch, ft	616	726
Height above river, ft	467	470
Arch rise, ft	90	90
Amount of steel, lb	2,400,000	3,900,000
Amount of concrete, yd^3	500	1,790
Amount of steel reinforcement, lb	82,000	434,000

See also: Thematic Essays section: The Colorado River's Historic Sites; Marble Canyon

Further Reading

National Park Service. 2017. Navajo Bridge. Retrieved June 2, 2016, at https://www.nps.gov/glca/learn/historyculture/navajobridge.htm

NAVAJO POINT

This South Rim viewpoint, which is 0.5 mile west of Desert View, offers views of the Colorado River sweeping west (almost to its confluence with the Little Colorado River) and several miles of the Tanner Trail. The elevation here is 7,498 feet, making Navajo Point the highest point on the South Rim (however, people atop Desert View Watchtower are about 10 feet higher). Navajo Point honors the Navajo Nation, which is the largest reservation of Native Americans in the United States. In the Grand Canyon area, Navajo call themselves *Diné*, meaning "the people."

Navajo Point is the general area where Spanish explorers saw Grand Canyon in 1540. Directly below Navajo Point are Escalante and Cardenas Buttes along Tanner Canyon. To the west are Shiva Temple (20 miles away) and Cape Royal, and to the east are the Marble Platform, Desert View Watchtower, and Palisades of the Desert, which is a line of cliffs that form the east wall of the canyon beyond Desert View. To the northwest are Zoroaster Temple, Wotans Throne, Vishnu Temple, Cape Royal, and Chuar Butte. Eight miles (as the condor flies) across from Navajo Point is Point Imperial, which is 8,803 feet above sea level and 6,000 feet above the Colorado River. The most expansive views from Navajo Point and Desert View are to the north and west, so the panoramas here are often photographed in the morning.

See also: Desert View

Further Reading

Anderson, Michael. 2001. *Along the Rim: A Guide to Grand Canyon's South Rim from Hermit's Rest to Desert View.* Grand Canyon, AZ: Grand Canyon Association.

NEVILLS, NORMAN

In 1938, Norman Davies Nevills (1908–1949) became the first commercial river runner in Grand Canyon (and its first river outfitter) when he took botanists Elzada Clover (1897–1980) and Lois Jotter (1914–2013) from the University of Michigan on a 666-mile trip from Green River, Utah, to Lake Mead. Clover and Jotter were the first women to run the Colorado River through Grand Canyon. Their 43-day expedition to catalog Grand Canyon's plants made Nevills famous. Nevills and Clover planned to write a book about their famous trip, but they never did.

Nevills, who named his river running operation Nevills Expedition, got the idea for commercial river running in 1936 when he was paid to row Ernest "Husky" Hunt to Lees Ferry. Nevills invented the profession of river running in Grand Canyon; he ran seven trips through Grand Canyon (at the time, the previous record was two).

In 1940, Nevills took Barry Goldwater through Grand Canyon. Footage from Nevills's fourth trip through Grand Canyon in 1942, which included Otis Marston and Ed Hudson, was made into a short film titled *Facing Your Danger*, which won an Academy Award for Short Subjects in 1946. Nevills advertised his expensive ($1,500–$2,500) trips through Grand Canyon in magazines such as *Desert Magazine*, and other magazines such as *The Atlantic*, *Arizona Highways*, *National Geographic*, and *Saturday Evening Post* praised Nevills's skills as a boatman. Nevills, a publicity-conscious showman who accounted for more than one-third of the first 100 people to boat through Grand Canyon, never lost a customer or flipped a boat while working as an oarsman.

Just as Nathan Galloway had earlier redesigned boats for running the Colorado River in Grand Canyon, so too did Nevills modify Galloway-type boats for his expeditions. Nevills's boats were made of marine plywood and were therefore easier to build than Galloway boats. Nevills's boats were also wider than Galloway boats. Nevills and his friend Don Harris built several boats, two of which—*Mexican Hat* and *WEN* (Nevills's favorite boat, which was named for his father, W. E. Nevills)—are now in GCNP's collection of historic boats.

Norm Nevills, an airplane pilot, ran his last trip through Grand Canyon in the summer of 1949. By this time, Nevills had made river running an economically viable operation, and he had plans to expand his fleet of riverboats, make a movie for Walt Disney about Grand Canyon, and take a group of Sierra Club members through Grand Canyon. However, on September 19, 1949, Norm and his wife, Doris, were killed when their Piper PA-12 Super Cruiser plane *Cherry II* crashed soon after takeoff at Mexican Hat, Utah. In December 1949, Jim and Bob Rigg spread their mingled ashes over Mexican Hat, the lower Colorado River in Grand Canyon, and elsewhere.

After the Nevills' deaths, their company was taken over by boatmen Frank Wright and Jim Rigg, who renamed it Mexican Hat Expeditions. This company, which ran trips until 1969, later became Canyoneers, Inc. Thanks to the efforts of Frank Masland, who was a passenger with Nevills in 1948 and 1949, Norm and Dorris are commemorated by Nevills Rapid (River Mile 75.5), Doris Rapid (River Mile 138.4, just upstream from where Doris fell overboard during a 1940 trip), and a bronze cenotaph at Navajo Bridge. (Dock Marston strongly opposed these and other honors for the Nevills.) That cenotaph was designed by artist Mary Ogden Abbott, who was one of Nevills's passengers through Grand Canyon in 1949. Masland also tried to get officials to name the reservoir behind Glen Canyon Dam Lake Nevills, but they chose the name Lake Powell. The Nevills' cenotaph, which was dedicated at a service at Navajo Bridge in July 1952 led by Barry Goldwater and featuring Arizona governor Howard Pyle, tells readers that Norm and Doris "sought and ran and mastered the wild and secret waters . . . by the river they loved so well in the desert that was their home . . . They run the rivers of eternity."

See also: Grand Canyon Museum

Further Reading

Reilly, P. T. 1987. *Norman Nevills: Whitewater Man of the West*. Salt Lake City: Utah Historical Society.

Staveley, Gaylord. 2006. Norman Nevills. Grand Canyon River Guides. Retrieved April 4, 2017, at http://www.gcrg.org/bqr/17-1/nevills.html

Webb, Roy, ed. 2005. *High, Wide, and Handsome: The River Journals of Norman D. Nevills*. Logan: Utah State University Press.

NEWBERRY, JOHN STRONG

John Strong Newberry (1822–1892) was an American physician, geologist, and explorer who, in 1857–1858, explored Grand Canyon with a U.S. War Department expedition led by Lt. Joseph Christmas Ives. Ives's expedition occurred just 10 years after Mexico had ceded its northwestern territories (which included Grand Canyon) to the United States. Newberry's report, which comprised half of Ives's *Report on the Colorado River of the West, Explored in 1857 and 1858* (1861), claimed that Grand Canyon was carved by the Colorado River:

> [These canyons] belong to a vast system of erosion, and are wholly due to the action of water. Probably nowhere in the world, has the action of [water] produced results so surprising . . . [Nowhere], so far as we know, are the secrets of its structure so fully as revealed here . . . [Grand Canyon is] the most splendid exposure of stratified rock that there is in the world. (Ives, 1861)

Newberry was a graduate of Cleveland Medical School; because the PhD degree had not yet come to the United States, a medical degree was the only postgraduate degree for students interested in science.

Newberry, an invertebrate paleontologist, was the first person to describe Grand Canyon's geology, and the first geologist to convincingly attribute the formation of Grand Canyon to water and the uplift of the continent. Newberry, who recognized Grand Canyon as a geological "paradise," also named several rock layers in Grand Canyon. Newberry's most detailed work was at the mouth of Diamond Creek, where Egloffstein's dark sketches exaggerated the canyon's walls.

Newberry, who was originally interested in sanitation and public health, worked for 24 years as a geologist and paleontologist at what is now Columbia University.

Pioneering geologist John Strong Newberry did much of his work in Grand Canyon near Diamond Point, which dominates this downstream view from the Colorado River from near River Mile 224. This landmark is about as high as Navajo Bridge, which river runners pass under near River Mile 4.5. Near Diamond Point is Diamond Creek Takeout, which is the first place since Lees Ferry (River Mile 0) where a road—stretching 21 miles from Route 66—reaches the Colorado River. (Randy Moore)

Newberry served as president of the American Association for the Advancement of Science, was one of the 50 founding members of the National Academy of Science, and helped found the Geological Society of America. Newberry, with Grove Karl Gilbert and Clarence Dutton, made Grand Canyon a focus of geology in America. He was also a strong advocate of what became known as "intelligent design," which linked geology with biology and theology.

Newberry died on December 7, 1892, in New Haven, Connecticut, and is buried in Lake View Cemetery in Cleveland, Ohio. He is honored at Grand Canyon by Newberry Butte near River Mile 81 and by Newberry Point near River Mile 125.

Newberry's student and protégé Grove Karl Gilbert (1843–1918) visited Grand Canyon in 1871–1872 with a Joseph Ives–inspired expedition led by Lt. George Wheeler (1842–1905). Gilbert, who was one of John Wesley Powell's closest advisors, tested and refined generalizations made earlier by Powell (e.g., about river flow and erosion) and named the Colorado Plateau ("Within this region of great disturbance is a restricted area of comparative calm . . . It is called the Colorado River Plateau Province"). Gilbert, who pioneered efforts to unify geology with mechanics and mathematics, was among a group of 20th-century geologists named "America's Greatest."

See also: Ives, Joseph Christmas

Further Reading

Ives, Joseph C. 1861. *Report on the Colorado River of the West, Explored in 1857 and 1858.* Washington, DC: Government Printing Office. (Reprinted in 2005 by the University of Michigan Library.)

NORTH KAIBAB TRAIL

North Kaibab Trail, which is the only maintained trail from the North Rim to the Colorado River, is the least used and most challenging of the Corridor Trails. It was originally called Bright Angel Trail because it was an extension of the Bright Angel Trail from the South Rim, but in 1926 it was renamed North Kaibab Trail when the upper parts of the trail to Roaring Springs were rebuilt by NPS. (The trailhead of the Old Bright Angel Trail, now abandoned, is 4 miles along the Ken Patrick Trail from the North Kaibab Trailhead, and the abandoned original trail goes from there to Roaring Springs.) Like the South Rim's Bright Angel Trail, the North Kaibab Trail follows Bright Angel Fault, which creates natural breaks in Grand Canyon's cliffs. Today's North Kaibab Trail replaced an older trail to the river that crossed Bright Angel Creek more than 90 times (today's trail crosses the creek only six times).

North Kaibab Trail stretches 14.2 miles through Bright Angel Fault from the North Rim to Phantom Ranch and the Colorado River. This photo was taken north of Roaring Springs, which is about 4.6 miles down the trail and 3,000 feet below the trailhead atop the North Rim. (Randy Moore)

Services at the North Rim operate from mid-May to mid-October, and the road to North Rim (Arizona 67) is usually open another few weeks before it is closed by snow. North Kaibab Trail is open year-round but can become dangerous in winter when snow and ice accumulate. Along its 14.2 miles from the North Rim to the Colorado River, the North Kaibab Trail passes through every ecosystem found between Canada and Mexico, ranging from the Hudsonian Zone atop the rim to the Lower Sonoran Zone at the river. During the descent, the air becomes warmer and drier. When it was completed in 1928, the North Kaibab Trail completed the only NPS-maintained rim-to-rim trail in GCNP.

The North Kaibab Trail's trailhead (elevation 8,250 feet), which is almost 1,000 feet higher than South Rim trailheads, is just off the North Rim Entrance Road, 1.5 miles north of Grand Canyon Lodge. This area includes a parking lot and mule coral; the trailheads for the Ken Patrick Trail and Uncle Jim Trail are also here. At the trailhead, the most common vegetation is ponderosa pine (*Pinus ponderosa*), white fir (*Abies concolor*), and quaking aspen (*Populus tremuloides*). The top 4.7 miles of the trail (i.e., from the North Rim to Roaring Springs) are the steepest and most challenging parts of the trail; almost half of the trail's descent occurs in this part of the trail.

North Kaibab Trail's first descent is through Kaibab Limestone for 0.7 mile to the Coconino Overlook, which is 500 feet below the rim and near the bottom of the Toroweap Formation. From this scenic overlook, hikers look down on Roaring Springs Canyon toward its intersection with Bright Angel Canyon. Switchbacks then twist and turn 1.3 miles to Supai Tunnel (elevation 6,800 feet), which was blasted from the Supai formation by CCC workers in the 1930s. This section of the trail, which hikers share with mules, replaced the old Bright Angel Trail to the northeast; traces of that abandoned trail can be seen on the far slope of the drainage. At Supai Tunnel, there are an outhouse and rest area; water is available here when it's warm. However, there are no sweeping views because hikers are in a canyon.

About 2.7 miles from the trailhead is Bridge in the Redwall, which was built after a flood (triggered by 14 inches of rain in 36 hours on the North Rim) in 1966 destroyed much of the trail. Here, the dominant vegetation includes pinyon pine, cliffrose (*Cowania mexicana*), and broadleaf yucca (*Yucca baccata*). However, careful observers will see two ponderosa pines in the drainage below; these trees are nearly 1,000 feet below their normal range. Bridge in the Redwall crosses Roaring Springs Canyon. Beyond this bridge, the trail continues to follow Roaring Springs Fault to a tall spire called The Needle. Here, rock layers on the eastern side of the fault are about 189 feet lower than those on the western side.

Approximately 2.3 miles beyond Bridge in the Redwall (and 4.7 miles from the trailhead) is Roaring Springs (elevation 5,200 feet). At this point, hikers have descended 3,040 feet below the rim. Here, water from rain and snowfall at the North Rim percolating through rock layers meets the impervious Muav Limestone. This creates a water table above the limestone that, after draining through caverns, emerges as a 400-foot-high waterfall. Water from Roaring Springs has provided water for the North Rim (more than 3,600 feet above) since 1928 and for the rest of GCNP since 1979. The pump house near Roaring Springs (built in 1979; the power lines for these pumps are prominent) pumps water to the North Rim and Indian Garden, from which another pump pushes water to the South Rim. Pump house operator and famed artist Bruce A. Aiken (b. 1950) lived, worked, and raised a family for 33 years (1973–2006) at the house at Roaring Springs, which is now a NPS ranger station. Many of Aiken's more than 350 images of Grand Canyon are featured in *Bruce Aiken's Grand Canyon: An Intimate Affair* (2007).

At Roaring Springs, hikers can see the Tonto Group, which includes the brown Tapeats Sandstone, the greenish-gray Bright Angel Shale, and the buff-colored Muav Limestone. Trees prominent at higher elevations are almost gone (except in washes and shaded slopes), and have been replaced by plants such as Mormon tea (*Ephedra viridis*). The hike from Roaring Springs to the river is long (9.2 miles) but not overly difficult. The first descent is into the alluvium-littered Tapeats Narrows, beyond which hikers cross the Dox Sandstone. This red sediment is the uppermost layer of the tilted rocks comprising the Grand Canyon Supergroup. In many other areas of GCNP, the Supergroup is gone, leaving the Tapeats Sandstone atop the Vishnu Schist at the Great Unconformity. Vegetation here includes desert plants, with mesquite (*Prosopis glandulosa*) being abundant.

Cottonwood Camp, which is 6.8 miles from the trailhead, is approximately halfway between the river and trailhead. At the camp, hikers find seasonal water and restrooms; the vegetation here includes cattails, horsetails, willows, and tall cottonwoods. NPS built this camp in the 1920s as a layover for mules coming from and going to the North Rim, and the area was redeveloped in the 1930s by the CCC. Just beyond Cottonwood Camp is a rise that provides an excellent view of the South Rim and, to the northwest, Oza Butte. All of the rock layers above this point are sedimentary, meaning that they originated by the accumulation and solidification of sediments.

About a mile beyond Cottonwood Camp is a 0.4-mile-long side trail that takes hikers across a bridge to an amphitheater carved by the 140-foot-high Ribbon Falls. This waterfall was named by François Matthes, who believed the water resembled ribbons flowing in the wind. The water in Ribbon Creek goes through limestone and, in the process, dissolves calcium carbonate that precipitates into deposits called travertine near these falls. Vegetation here includes mosses and maidenhair ferns. Some Zuni tribes believe their ancestors emerged at Ribbon Falls.

Soon after Cottonwood Camp, hikers enter The Box, a narrow slot-canyon having steep walls. Portions of the trail through The Box were blasted away during renovations in 1926–27 so hikers could avoid crossing Bright Angel Creek dozens of times. At the north end of The Box is the Bass Formation, a limestone formation that is the oldest (1.25 billion years old) layer of sedimentary rock in GCNP. Bass Formation sits atop Precambrian schist, a metamorphic rock (i.e., a rock made of sedimentary rocks exposed to heat and pressure). The steep walls of The Box are made of Vishnu Schist 1.7 billion years old.

The North Kaibab Trails then leads to Phantom Ranch, which is just upstream from where Bright Angel Creek enters the Colorado River. The area around modern-day Phantom Ranch was opened as a tourist camp in 1906 by E. D. Woolley and

David Rust. In 1913, following a stay by former president Theodore Roosevelt on a hunting trip, Rust's Spartan tent-camp became known as Roosevelt Camp. When the Fred Harvey Company renovated the camp in 1922, it changed the name of the camp to Phantom Ranch (for Phantom Creek, which is upstream from the camp). Today, Phantom Ranch includes a ranger station, dorms, canteen, cabins, and other structures. Hikers can get water, information, and first aid at Phantom Ranch. Phantom Ranch is 13.8 miles from the North Kaibab Trail trailhead.

Bright Angel Campground is 0.3 mile beyond Phantom Ranch, and the Colorado River is another 0.1 mile beyond Bright Angel Campground. Ribbon Falls is 6.3 miles away, Cottonwood Camp 7 miles away, Roaring Springs 8.8 miles away, and North Kaibab Trail trailhead 13.8 miles away.

The 1.7-mile-long Colorado River Trail connects South Kaibab Trail with Bright Angel Trail. This trail, which was built in 1933–1936 by the CCC, completed today's trans-canyon trail system and features towering walls made of Vishnu Schist. CCC workers used 40,000 pounds of explosives to blast out the River Trail; the last blast happened on January 20, 1936. (Before the River Trail was completed, hikers reached Phantom Ranch via Burro Spring along the Tonto Trail.) The first mules along the trail were led by prospector Shorty Yarborough in 1936 as they carried hay for the Fred Harvey Company to Phantom Ranch.

During construction of the River Trail, the NPS and CCC agreed that the trail would be named for the first person to die during the building of the trail. Because no one was killed, the trail was named the Colorado River Trail, more commonly known as the River Trail.

Hikers going from Phantom Ranch on the North Kaibab Trail cross the Colorado River at Silver Bridge, which is downstream from the Kaibab Suspension Bridge. Silver Bridge was built in the 1960s to carry a water pipeline from Roaring Springs (on the North Rim) to the South Rim. This bridge, which is just below the mouth of Bright Angel Creek, is used only by hikers; mules cross upstream at the Kaibab Suspension Bridge. From the River Rest House at Pipe Creek (elevation 2,526 feet), it is 1.3 miles to the Silver Suspension Bridge (elevation 2,491 feet) and 2 miles to South Kaibab Trail. Camping is not allowed along the Colorado River Trail.

The North Kaibab Trail, which drops 5,660 feet from the trailhead to the river, passes through every major life-zone found between Mexico and Canada. Like the South Kaibab Trail, the North Kaibab Trail is part of the Arizona Trail System, which crosses Arizona from Mexico to Utah. North Kaibab Trail is maintained and patrolled by NPS and has water available (in some instances, only seasonally) at Supai Tunnel, the Roaring Springs residence, Cottonwood Campground, Phantom Ranch, and Bright Angel Campground. A rim-to-rim hike along Bright Angel Trail and North Kaibab Trail spans 23.4 miles.

In 1981, Secretary of the Interior James Watt made the North Kaibab Trail, South Kaibab Trail, Bright Angel Trail, and Colorado River Trail part of the National Trails System to recognize their roles in tourism at Grand Canyon.

See also: North Rim; Phantom Ranch

Further Reading

Thybony, Scott. 2005. *Grand Canyon: North Kaibab Trail Guide*. Grand Canyon, AZ: Grand Canyon Association.

NORTH RIM

Although the arrival of the railroad at the South Rim in 1901 dramatically increased tourism there, the North Rim remained far from the railroad, and entrepreneurs could not convince the Utah Parks Company (a subsidiary of Union Pacific Railroad) to build a spur to the North Rim. (At the North Rim, Utah Parks Company never made significant profits nor paid any franchise fees.) By the 1920s, the Grand Circle bus tours included Zion, Bryce, and the North Rim, but tourist services at the North Rim remained rustic. Moreover, the North Rim's higher altitude produces more snow, reducing the tourist season to only about six months. Today, Grand Canyon's North Rim remains a unique destination that differs dramatically from the South Rim.

Grand Canyon's North Rim is in the northwest corner of Arizona, about 80 miles from Kanab, Utah; 44 miles from Jacob Lake, Arizona; 90 miles from Lees Ferry, Arizona; 208 miles from Flagstaff, Arizona; and 124 miles from Page, Arizona. The North Rim, which is at the end of Arizona Highway 67 South (also known as the North Rim Parkway), does not have an airport or railroad depot (the closest rail lines ever established were in Utah); the only way to get there is via car, bus or the North Kaibab Trail. GCNP's entrance station at the North Rim, which is 13 miles from the rim, has an elevation of 8,827 feet, which is higher than any other viewpoint in GCNP. The North Rim's entrance station is similar to how it looked when it was built in 1928. The North Rim, which was not included in GCNP's budget until 1924, was originally called Buckskin Plateau because of the area's many mule deer.

The North Rim is a southern extension of the Kaibab Plateau and features beautiful green meadows, coniferous forests, and expansive growths of white-barked aspen. Unlike the South Rim, the North Rim does not offer a shuttle bus to viewpoints, but several of its most scenic views (e.g., Bright Angel Point, Cape Royal, Roosevelt Point, and Point Imperial) are near the road.

Greenland Lake

This "lake" on the right side of the road on the way to Cape Royal is actually a sinkhole formed by water from a natural spring that is percolating through and dissolving Kaibab Limestone. The name "Greenland Lake" probably commemorates the area's juniper and ponderosa pine forests. However, some historians speculate that the name may have been bestowed on the area because it reminded many of the 19th-century Mormon settlers of their Scandinavian origins. The short hike to this lush area also features aspen trees, sedges, and rushes.

The North Rim receives only about 5 percent of GCNP's visitors. This is not because the North Rim's scenery, hiking, accommodations, or adventures are inferior to those at the South Rim, but instead because it takes more effort to reach the North Rim than the South Rim. Indeed, the remote North Rim—which is near the Utah border—is farther from interstate highways than the South Rim, and the closest major airport is 266 miles away in Las Vegas. Furthermore, most services (e.g., food, lodging) operate only from May 15 to October 15. Day-use visitors are allowed at the North Rim until around December 1, unless snow closes State Route 67 from Jacob Lake. All operations close at the North Rim on December 1.

There are rewards for visiting the North Rim, an area that one travel magazine described as "the Connoisseur's Grand Canyon." Whereas visitors to the congested South Rim often fight massive crowds and scorch in the heat, the smaller crowds at the North Rim enjoy more solitude. Moreover, because it is more than 1,000 feet higher than the South Rim, the North Rim is cooler than the South Rim. Many visitors are surprised to learn that, although the North Rim is only about 10 miles (as the condor flies) from the South Rim, driving from the North to the South Rim covers more than 200 miles.

The arrival of the railroad at the South Rim in 1901 stimulated a flood of tourism and economic development. In contrast, the North Rim—which was far from any railroad line—retained a rustic, primitive atmosphere. The first large-scale accommodation for tourists at the North Rim was the Wylie Way Lodge (more commonly referred to as Wylie Camp), which opened near today's Bright Angel Point in 1917. The lodge, which was actually a campground of 10 tents (plus a dining tent), was managed by Elizabeth Wylie McKee, who was the first woman to own and operate a tourist concession at Grand Canyon. While tourists at the South Rim stayed in the luxurious El Tovar Hotel and enjoyed a variety of activities, those at the North rim slept in outdoor, canvas tents.

NPS director Stephen Mather had big plans for the North Rim and encouraged the Union Pacific Railroad to improve access to the North Rim and to build a modern lodge there for visitors. By 1927, when more than 14,000 visitors came to the North Rim, Elizabeth McKee was forced to sell her camp to the Utah Parks Company for $25,000. The last of her tent platforms was removed by GCNP workers in 1946.

In 1928, the Union Pacific Railroad opened Grand Canyon Lodge, which was designed by architect Gilbert Stanley Underwood. Like pioneering architect Mary Colter at the South Rim, Underwood came to Grand Canyon with relatively little experience (he would later become famous for constructing buildings with reinforced concrete). Underwood also built lodges for the Utah Parks Company at Yosemite, Zion, and Bryce National Parks. When Underwood's original plans for the lodge at the North Rim were rejected by Mather because they were too imposing, Underwood—probably influenced by Colter—produced plans for a smaller lodge. His 225-by-250-foot Grand Canyon Lodge, which was commissioned by Union Pacific Railroad for $550,000 and built by a crew of 125 workers paid 50¢–85¢ per hour, was accompanied by more than 100 cabins separate from the main lodge building. The cabins were made of ponderosa pine logs, and the lodge was

Grand Canyon Lodge is the center of activity on the North Rim. The original lodge, which was opened in 1928 by Union Pacific Railroad, burned down in 1932, and the slightly modified lodge that today's visitors enjoy opened at the same site five years later. In this photo of the original lodge from 1929, workers serenade departing guests as part of the "Sing-Away" ceremony. (National Park Service)

made of pine and Kaibab limestone designed to look like natural rock outcroppings so that it fit into its wooded landscape. Union Pacific Railroad hoped that Grand Canyon Lodge would outdo the Northern Pacific Railroad's Old Faithful Inn at Yellowstone. Grand Canyon Lodge was advertised in *National Geographic* as the "New Way to See Grand Canyon" and the "finest [lodge] of its type in any national park."

On September 1, 1932, at the height of the Great Depression, Grand Canyon Lodge and four nearby cabins burned down. The charred remains stood undisturbed for four years. On June 1, 1937, a slightly modified lodge (e.g., without the three-story observation tower and second floor) was rebuilt on the original foundation; today's visitors can still see damage from the fire near the lodge's chimney. The complex also includes large south-facing terraces, more than 100 of the original cabins (some on the edge of the rim), and a deli, saloon/coffee shop (formerly an ice cream parlor), gift shop, post office, restaurant, and a rustic motel. Nearby is a bookstore and gift shop operated by Grand Canyon Association. The North Rim also includes an 83-space campsite, and its 230 employees live about a mile north of the lodge. Today, Grand Canyon Lodge is the most intact rustic lodge in NPS from the railroad era. It also includes the only complete cabin complex in NPS. The North Rim's concessionaire is Forever Resorts, a Scottsdale, Arizona–based company that also manages properties at Glacier, Zion, and other national parks.

Thanks to prompting by NPS officials Stephen Mather and Horace Albright, the completion of lodges at Bryce Canyon, Zion, and the North Rim produced the so-called Grand Circle Tour, in which buses brought tourists to each destination for a short visit. However, Union Pacific Railroad ended its bus tours to the North Rim in 1971, after which the railroad's facilities there were donated to GCNP. Today's lodge—which has changed relatively little since 1937—continues to appear that it "has sprung full-grown out of the canyon wall itself."

Unlike the Historic District at the South Rim, which includes iconic structures such as Kolb Studio, Hopi House, and El Tovar Hotel, no buildings have housed independent businesses at the North Rim of GCNP. Except for a general store (formerly the North Rim Inn), gas station, and laundry/shower facility a mile from the lodge, no other business-related buildings are at the North Rim. Virtually everything at the North Rim is focused at Grand Canyon Lodge, which remains an excellent example of rustic, railroad-sponsored architecture from a bygone era.

See also: North Kaibab Trail; Owens, James "Jim"; Wylie Way Camps

Further Reading

Aitchison, Stewart. 2008. *Grand Canyon's North Rim and Beyond: A Guide to the North Rim and the Arizona Strip*. Grand Canyon, AZ: Grand Canyon Association.

Swanson, Frederick H. 2007. *David Rust: A Life in the Canyons*. Salt Lake City: University of Utah Press.

O

O'NEILL, WILLIAM OWEN "BUCKEY"

William Owen "Buckey" O'Neill (1860–1898) was an Arizona sheriff, politician, judge, school superintendent, and newspaper editor who was instrumental in bringing the railroad to the South Rim. In 1897, when the Grand Canyon Railways formed, O'Neill was president and on the board of directors and was instrumental in securing funds that would later bring the railroad to the South Rim. Construction of the rail line to Grand Canyon began two years after O'Neill's death, but the company went bankrupt when tracks ended a few miles short of the South Rim. Years later, the Santa Fe Railway bought the line, renamed it the Grand Canyon Railway, finished the track, and brought the railroad to the South Rim in 1901. The arrival of this railroad, which was largely due to O'Neill's earlier efforts, permanently changed the South Rim.

Buckey O'Neill, who was instrumental in bringing the railroad to Grand Canyon, built this two-room, rim-side cabin in the 1890s at the South Rim. The Buckey O'Neill Cabin, which is now Suite #6101 of Bright Angel Lodge, is the oldest continuously standing structure on the South Rim. (Randy Moore)

The restless O'Neill, whose short-lived careers also included prospecting in Grand Canyon in 1890, was named for his ability to "buck the odds" in a card game called Faro. Unfortunately, his good luck with cards did not extend to his military service. After O'Neill left for Cuba to fight as Captain of Troop A (part of the 1st United States Volunteer Cavalry) in Teddy Roosevelt's Rough Riders, he bragged that "the Spanish bullet is not made that will get me" (Arlington Cemetery, 2017). On July 1, 1898, O'Neill was killed by a sniper the day before Roosevelt's famous charge up Kettle (i.e., San Juan) Hill. (This battle made Roosevelt famous and helped him become president.) Roosevelt lamented O'Neill's death as "the most serious loss that I and the regiment could have suffered" (Arlington Cemetery, 2017). O'Neill was buried by his men on the slope of San Juan Hill but was later exhumed and reinterred in Section 1 of Arlington National Cemetery. There, he rests under the epitaph, "Who Would Not Die for a New Star in the Flag."

In the mid-1890s, O'Neill built a two-room cabin at the South Rim. This wood frame building on a low stone foundation is the oldest continuously standing structure on the South Rim. Today, O'Neill's cabin, which is not open to the public, is Suite 6101 of Bright Angel Lodge.

Exactly nine years after his death, a bronze, equestrian sculpture by sculptor Solon Hannibal Borglum ("America's First Cowboy Sculptor") honoring O'Neill and other Rough Riders was unveiled in the courthouse plaza in Prescott, Arizona, atop a 28-ton pedestal. A larger and more impressive monument to O'Neill is O'Neill Butte, which towers above the early miles of the South Kaibab Trail.

See also: Grand Canyon Village; Railroad; Roosevelt Point

Further Reading

Arlington National Cemetery. 2017. William Owen "Buckey" O'Neill. Retrieved January 4, 2017, at http://www.arlingtoncemetery.net/wooneill.htm

OWENS, JAMES "UNCLE JIM"

James T. "Uncle Jim" Owens (c.1848–1936) was born in Texas probably in the late 1840s. After running away from home at age 11, Owens worked on ranches and cattle drives before becoming manager of the Goodnight Ranch (near Goodnight, Texas) in the 1880s. In 1897, Owens met Charles Jesse "Buffalo" Jones, who was visiting the ranch while looking to buy buffalo to pull a sled for an exhibition. While working at Goodnight Ranch, Owens met Pat Garrett, Billy the Kid (Henry McCarty), and Buffalo Bill Cody, who visited Grand Canyon in 1892 while representing British investors considering building hunting lodges on the North Rim.

In 1902, when President Theodore Roosevelt appointed Owens to succeed Jones as game warden at Yellowstone, Owens went to Yellowstone to manage the buffalo there. In 1905, Jones resigned his job at Yellowstone and went to Grand Canyon, where he was joined by Owens the following year. Jones and Owens then imported buffalo from Yellowstone and Goodnight Ranch that they used in a failed commercial attempt to breed "cattalo." Owens became a guide and hunter who killed more than 500 lions on the North Rim that were killing cattle. (One such two-week hunt for lions in 1913 included Teddy Roosevelt and Roosevelt's sons Quentin and Archie, and cousin Nicholas.) This led to an explosion of the deer population, with thousands starving to death in 1924–1927. Owens and his colleagues killed even more wolves, which were gone from the North Rim by the mid-1930s. Wolves were not seen again at the North Rim for 75 years.

Although Owens earned only about $100 per month at Yellowstone, he sometimes earned more than $350 per day at the North Rim collecting bounties paid for predators. (Advertisements in magazines such as *Ladies Home Journal* described Grand Canyon as a place where "bold hunters win bounties for mountain lions.") Owens's hunts with disillusioned dentist Zane Grey inspired Grey to write *Riders of the Purple Sage*, *Tales of Lonely Trails*, and *Roping Lions in the Grand Canyon*. Grey included Jones in *Last of the Plainsmen*. One lion that Owens trapped alive with Grey was displayed for a short time at El Tovar Hotel on the South Rim.

At the North Rim, Owens spent his winters at Fredonia. However, during other times of the year, Owens lived in cabins at the head of Bright Angel Canyon and Harvey Meadow. (Beginning around 1930, NPS began using Owens's cabin at Harvey Meadow as an office.) Outside his homes, Owens displayed signs advertising "Jim Owens Camp, Guiding and Hunting Parties. Specialty—Cougars Caught to Order. Rates Reasonable." Owens planned a tourist camp at Cape Royal to lodge 20 visitors who would be entertained by a nearby population of 40–50 buffalo, but his application was denied. The friendly burro Brighty made famous by Marguerite Henry's *Brighty of the Grand Canyon* (1953) often visited Owens at the North Rim. Owens is mentioned 274 times and appears in 21 sketches in Henry's book.

Owens, a North Rim frontiersman, cowboy, and rancher, retired to House Rock Valley to raise buffalo and preserve wildlife. In 1921, NPS director Stephen Mather urged Owens to give the buffalo to GCNP, where they would be known as the James T. Owens Memorial Buffalo Herd. Instead, Owens sold them in 1926 to Arizona for $10,000. Owens died at age 87 in Afton, New Mexico, on May 11, 1936, and was buried in Masonic Cemetery in Las Cruces, New Mexico, only about 90 feet from the grave of Pat Garrett. Owens's grave was unmarked until November 16, 2013, when admirers (wildlife manager Al LeCount of Albany, Ohio, and veterinarian Harley Shaw of Hillsboro, New Mexico, and a former resident of the Kaibab Plateau) placed a stone on Owens's grave.

In 1932, Owens was honored in GCNP when the National Board on Geographic Names changed the name of Nachi Point (above Roaring Springs) to Uncle Jim Point.

See also: Brighty, the North Rim Burro; Cattalo, Buffalo, and Pearl Grey; Roosevelt Point

Further Reading

Brown, Dick. 2017. Uncle Jim Owens—Grand old man of the North Rim. *Grand Canyon Historical Society*, 27 (2), 3–11.

P

PARACHUTING INTO GRAND CANYON

In the early morning hours of June 21, 1944, a four-engine U.S. Army B-24 Liberator 107 bomber flying on a training mission from Tonopah, Nevada, to Tucson at an altitude of 23,000 feet developed engine trouble while flying over Grand Canyon. Three of the crew—flight officer Maurice "Mo" Cruickshank, bombardier Lt. Charles "Goldie" Goldblum, and flight engineer Corporal Royce "Roy" Embanks—bailed out of the plane's bomb bay over the Tonto Platform south of Point Sublime on the north side of the Colorado River. This area, which is near River Mile 99, is about 18 miles northwest of Grand Canyon Village. After the men bailed out, the plane dropped another 2,000 feet into Grand Canyon before the pilot was able to restart the plane's engines and land the plane safely at Kingman Army Air Field.

Cruickshank broke his foot during the landing, and Goldblum's parachute caught in a tree, hanging him over a cliff overnight. The following morning, the three men reunited, and, after sizing up their predicament, Embanks and Goldblum went to the Colorado River for water, some of which they brought back to Cruickshank in their boots. They then spread their parachutes on the ground to signal their whereabouts to search planes.

Three days later, another B-24 bomber spotted the group and dropped water, food, two-way radios, blankets, shoes, and whiskey to the men, along with a note telling them that they were in Grand Canyon and instructing them to stay at their location. Their rescue was problematic; because of the unpredictable air currents in Grand Canyon, no one knew if it was safe to fly helicopters into the canyon. After a rescue from the South Rim failed because of a flood-swollen river, Emery Kolb offered to try to retrieve them in a boat, but NPS rejected his offer.

When the men were located, Alan MacRae—a well-known hiker in Grand Canyon—was on a backpacking trip with his new wife at the North Rim. When he learned about the stranded crew, MacRae and NPS ranger R. E. "Ed" Lawes hiked down the canyon to the men. The group then retraced their steps to the North Rim, where they were met by newspaper and radio reporters. After 10 days in Grand Canyon, the stranded men were saved. Their story appeared in the July 19, 1944, issue of *Time* magazine.

After the much publicized incident, all of the men returned to their jobs. Two of the airmen died in the South Pacific, and the other died in a car crash in the United States. Lawes died of food poisoning less than a year after the rescue.

In September 1980, stuntman Dar Robinson—who had doubled for Steve McQueen in the movie *Papillion* (1973)—asked for permission to parachute into Grand Canyon. When his request was denied, he approached the Hualapai Indian Reservation, who agreed to let him do his stunt. He raced his Bradley GT sports car to the edge of Grand Canyon and, as it went over the edge, Dar jumped from the car and parachuted into Grand Canyon. The stunt was later shown on the television show *That's Incredible!* Similar stunts have included Bob Burnquist skateboarding down a ramp and into the canyon (2006) and Yves Rossy flying and gliding across a Hualapai canyon near Guano Point in a jet suit (2011).

Not all such stunts turned out so well. For example, in 1993, two men base-jumped into the Lower Colorado River gorge. When their parachutes tangled, one man was killed when he hit the canyon's rocks.

See also: Airplanes and Helicopters

Further Reading

Taylor, Brenda. 2014. Days Past: WWII plane crew bails out in darkness into Grand Canyon. *Daily Courier* [Prescott, Arizona], December 7. Retrieved July 2, 2017, at https://www.dcourier.com/news/2014/dec/07/days-past-wwii-plane-crew-bails-out-in-darkness-i/

In 1929, Pauline "Polly" Mead became the first female ranger-naturalist at Grand Canyon National Park (GCNP). Her uniform was a custom-tailored version of men's uniforms, complemented with a Fred Harvey courier hat chosen by GCNP superintendent Miner Tillotson. Polly's future husband, Preston Patraw, became GCNP superintendent in 1954. This photo from 1930 shows Polly standing by the Yavapai Observation Station. (Grand Canyon National Park Museum Collection)

PATRAW, PAULINE "POLLY"

Pauline "Polly" Mead Patraw (1904–2001) was the first female naturalist-ranger at GCNP and only the second naturalist-ranger in the entire NPS. Polly first saw Grand Canyon at the North Rim in 1927 as a student at the University of Chicago. She was interested in botany, and she soon returned to GCNP to do a master's thesis on why the tree line ends so abruptly in meadows adjacent to the canyon. Her research entailed climbing trees, setting up weather stations, and sleeping in the canyon, protected only by her gun.

Mead applied for a job with the U.S. Forest Service but was rejected because the Forest Service did not hire women as rangers. When Polly then applied for a job with NPS, she was hired at GCNP and was sworn in on August 1, 1929, by her future husband, Preston Patraw. There was no official uniform for female rangers, so GCNP's superintendent (Miner Tillotson) decided Patraw should wear a riding habit topped with a wide-brimmed hat like those worn by the courier girls for Fred Harvey tours.

After two years at GCNP, Patraw left the area, but she returned to Grand Canyon in 1954 when her husband became GCNP's superintendent. Patraw died in 2001 in Santa Fe, New Mexico.

See also: Tillotson, Miner

Further Reading

Leavengood, Betty. 2004. *Grand Canyon Women: Lives Shaped by Landscape.* Grand Canyon, AZ: Grand Canyon Association.

PATTIE, JAMES

In 1826—that is, nearly 300 years after the Spanish reached Grand Canyon—American trapper and frontiersman James Ohio Pattie (c.1804–c.1851) became one of the first known non–Native Americans to visit Grand Canyon. Pattie was part of a French American trapping party (that included Ewing Young and "Old Bill" Williams, the namesake of Williams, Arizona) that hiked for 13 days up the Colorado River, which they called Red River. Pattie described Grand Canyon as a place "where the mountains shut in so close upon its shores" that they had "to climb a mountain, and travel along the acclivity, the river still insight, and at an immense depth below" (Hughes, 1972). Pattie, who published his stories in his *Personal Narrative of James O. Pattie* (1833), estimated Grand Canyon to be less than a mile wide and 300 miles long.

Pattie, who kept notes during his trip, was disappointed that the canyon blocked him from getting to beavers in the Colorado River. (These valuable pelts, which were called "hairy bank notes," were an important part of the local economy.) To Pattie, Grand Canyon was little more than "horrid mountains, which so cage [the river] up, as to deprive all human beings of the ability to descend to its banks, and make use of its waters." When Pattie returned to Santa Fe, all of his furs were seized because Pattie had not obtained a license. Pattie ended his life tired and broke, sad that "the freshness, the visions, the hopes of my youthful days are all vanished, and can never return" (Aker, 2011).

In 1827, a year after Pattie's visit, George C. Yount led a trapping party to the lower Colorado River and the following year descended to the bottom of Grand Canyon. These may have been the first non-Natives to descend Grand Canyon.

See also: Ives, Joseph Christmas

Further Reading

Aker, Andrea. 2011. James O. Pattie: Arizona's first storyteller. *In Old Arizona*, December 11. Retrieved July 2, 2017, at http://arizonaoddities.com/2011/12/james-ohio-pattie-arizonas-first-storyteller

Hughes, J. Donald. 1972. *The Story of Man at Grand Canyon*. N.p.: Grand Canyon Natural History Association Bulletin #14. Retrieved on July 1, 2017, at http://npshistory.com/handbooks/cooperating_associations/grca/14/bulletin.pdf

Pattie, James O. 2001 (1833). *The Personal Narrative of James O. Pattie*. Fremont, CA: Narrative Press.

PHANTOM RANCH

Phantom Ranch (elevation 2,460 feet) is a cluster of buildings, trails, ruins, and historic sites at the bottom of Grand Canyon just north of the Colorado River's confluence with Bright Angel Creek (near River Mile 87.5). This relatively flat area, which was used intermittently by Native Americans from around 1050 to 1140, also includes ruins of a kiva and pit houses by the river. John Wesley Powell camped at the river here on August 15, 1869, during his pioneering expedition through Grand Canyon. He named the creek Silver Creek but later—when he was writing about his voyage—changed the name to Bright Angel Creek.

The first commercial use of this area for tourism occurred in 1903, when Utah businessman Edwin Dilworth "Uncle Dee" Woolley and four friends—Thomas Chamberlain, a Kanab businessman; Jim Emmett, a northern Arizona cattleman who had operated Lees Ferry; E. S. Clark, a Flagstaff attorney; and Tim Hoyt, another Kanab businessman—each fronted $10,000 to form Grand Canyon Transportation Company. Woolley set up his operations at the North Rim at Woolley's Station, which was located near today's Harvey Meadow. Woolley hired

Weary hikers are happy to see this sign welcoming them to Phantom Ranch, a 14-acre resort designed by Mary Elizabeth Jane Colter at the bottom of Grand Canyon. Rooms at Phantom Ranch, which provide the only lodging below the rim in Grand Canyon National Park, are the most sought-after accommodations in the park. (Randy Moore)

newspaperman and politician David Rust (1874–1963), his son-in-law, to build a trail into Bright Angel Canyon from the North Rim to the river. Parts of Rust's trail were later made part of North Kaibab Trail. Rust gave up his $125-per-month salary as a school principal in Fredonia to work for Woolley for $50 per month and one-twelfth of the company stock.

After Rust applied for and received a U.S. Forest Service Special Use Permit in 1907, he established a six-tent camp for 12 "sturdy and adventurous" tourists (plus a cook's tent) on 2 acres of land 0.3 mile upstream along Bright Angel Creek at the bottom of the canyon. There, he raised sheep and planted a garden and several cottonwood trees. Rust and Nathan Galloway were mining partners in Glen Canyon, and Galloway asked Rust to help him get his boat displayed at El Tovar. (Rust tried, but the display never happened.) Although Rust never ran the Colorado River through Grand Canyon (he turned down offers from Nathan Galloway, Bert Loper, and the Kolb brothers), he did some boating around his Spartan camp, which was the precursor of Phantom Ranch.

By late 1907, Rust—who spent 33 years as a North Rim guide—had added an aerial tramway across the river linking the camp to the south side of the river. The first people to ride the tram were Rose Evans and Lida Bilveal, who were guests at the South Rim's El Tovar Hotel; these were the first paying customers to be hauled across the Colorado River in Grand Canyon. NPS nullified Rust's concession permit when GCNP was created in 1919, and in December, 1920, his tram was replaced by a 420-foot-long swinging wood-and-rope suspension bridge 56 feet above the Colorado River. This was the first suspension bridge in GCNP. In 1928, the suspension bridge was replaced by the Kaibab Suspension Bridge; this bridge was made of 67 tons of materials and cost $39,474 to build. Ever since the Kaibab Suspension Bridge was painted black in 1933–1934 by CCC workers, it has been called the Black Bridge. The wider Bright Angel (Silver) Suspension Bridge was built in 1970 to carry hikers and the trans-canyon water pipeline across the Colorado River. In 1966, the original pipeline was destroyed by a flash flood.

Bright Angel Suspension Bridge

Bright Angel Suspension Bridge, also called Silver Bridge, is just downstream from the Kaibab Suspension (Black) Bridge. Bright Angel Suspension Bridge was built in the late 1960s to support the trans-canyon water pipeline and connect Phantom Ranch and Bright Angel Trail. Bright Angel Suspension Bridge, which is at River Mile 88, is for foot traffic only; mules cross the Colorado River upstream on the Kaibab Suspension (Black) Bridge.

The trans-canyon water pipeline, carried across the Colorado River by the Bright Angel Suspension Bridge, supplies water from Roaring Springs to GCNP's more than 5 million visitors each year. The pipeline, which was installed in 1965, is also GCNP's largest source of deferred maintenance. The 16-mile-long aluminum pipeline springs 5–30 leaks per year that sometimes lose more than 35 percent of the water carried in the pipe. Its repair and replacement, which are top priorities for GCNP, are projected to cost $160 million.

At Phantom Ranch, Rust grew alfalfa (for livestock), peaches, and plums and raised chickens and rabbits as food for guests. The camp, which was located near today's mule corals, was originally known as Rust Camp, but after former president Teddy Roosevelt visited the camp in July 1913, it became known as Roosevelt Camp. Little is left of Rust Camp except for the towering cottonwood trees that Rust planted and that continue to provide visitors with refreshing shade. As Rust noted later in his life, these cottonwoods "are my green memorial, the only thing left of my work" (Swanson, 2008).

In 1922, the Fred Harvey Company was awarded the concession for the ramshackle camp. Harvey hired Mary Colter, who immediately began to redesign the camp as a place for "food, lodging, and comfort against an austere backdrop." The 53-year-old Colter and her sister Harriet rode mules to Phantom Ranch and signed the guest registry when the camp—then consisting of a lodge with a kitchen, a dining area, and four cabins made of rounded river rocks—opened on November 9, 1922; this was the first lodging facility to open at Grand Canyon after the canyon's designation as a national park. (Today's cabins 8 and 11 are two of the original cabins.) Harvey wanted to call the camp Roosevelt Chalet, but Colter insisted on Phantom Ranch. David Rust, whose camp began tourism at the now famous Phantom Ranch, never returned to see Colter's ranch.

In the late 1920s, Colter added a recreation hall (which is now a bunkhouse for employees), a larger dining room, and eight more cabins to the ranch; those cabins, which have less stone than the original buildings, rented for $6 per night (for comparison, rooms at the El Tovar Hotel then cost $7–$10). A three-day mule trip from the South Rim, which also included Roaring Springs and Ribbon Falls, cost $33. In 1946, Colter designed a laundry room at the lodge. By 1977, four dorms that could accommodate 92 hikers were added to Phantom Ranch.

In the 1930s, the 200 members of CCC Company 818, who camped near today's Bright Angel Camp, built the Trans-Canyon Telephone Line, Clear Creek Trail (from Phantom Ranch to the Tonto Platform), River Trail, a mule coral (built in 1935), and a 35-by-70-foot pond-shaped swimming pool that was fed by Bright Angel Creek. The pool was damaged by flooding in December 1964, and new public hygiene laws in 1972 closed the pool. (The pool was located at the large depression near cabins 5 and 6.) Although initially an exclusive hideaway for the rich and famous (some of whom stayed for weeks), Phantom Ranch later became very popular. After more than 800 visitors appeared at the ranch on Easter weekend in 1964, NPS began requiring reservations for overnight stays at Phantom Ranch. In 1981, a new sewage treatment plant was built near the river that could serve up to 225 people. At the river, a USGS gaging station is on river right. The Fred Harvey Company lost money on Phantom Ranch through the 1960s.

Today, Phantom Ranch—which was never a working ranch—covers 14 acres and includes cabins, dormitories, a shower building, a dining hall (serving meals only to people who have made reservations), a mule corral, a medical facility, a noncommercial heliport (for NPS use), a beach frequented by boaters resupplying their crews with water, and the 32-site Bright Angel Campground, which is about 0.5 mile from Phantom Ranch. When Mountain Bell installed a microwave transmitter at Phantom Ranch in 1982, NPS removed the telephone lines (but not the telephone poles) at the Phantom Ranch campground. Along the trail from the river to Phantom Ranch is a monument to the CCC's construction of the telephone line; a similar plaque is in Grand Canyon Village on the South Rim.

At the river near Phantom Ranch is a River Ranger Station (built in 1934); there is also an NPS ranger station (built in 1966) between the river and the Phantom Ranch canteen. Unlike in Supai, Arizona, which is also in Grand Canyon, there is no official mail service at Phantom Ranch. However, visitors can buy postcards at Phantom Ranch and stamp them "Mailed by Mule at Phantom Ranch—The Bottom of the Grand Canyon." These cards are carried by concessionaire's mules to the South Rim, where they are then mailed. On their way back to Phantom Ranch, mules bring food (about 2.5 tons per week) and supplies.

Temperatures at Phantom Ranch are significantly warmer than at either rim. For example, the average yearly temperature at the ranch is 68°F; the average high/low temperature in July at Phantom Ranch is 106°F/78°F, while that in January is 56°F /36°F. There is less than 1 inch of snow per year at Phantom Ranch.

While at Phantom Ranch, visitors can enjoy the camaraderie of the cantina, visit the NPS rangers' station, and—if hiking to and from Phantom Ranch isn't enough—hike along several beautiful trails:

The most leisurely of these trails is the 1.7-mile River Loop, which crosses both the Black (Kaibab) and Silver (Bright Angel) Suspension Bridges. You'll enjoy excellent views of Bright Angel Creek, Inner Gorge, and the Colorado River as you pass ancestral Puebloan ruins near the Kaibab Suspension Bridge and, just past Boat Beach, the grave of trail worker Rees Griffiths.

More strenuous is the 3.0-mile (round-trip) hike to Phantom Overlook. This hike features the CCC's stone bench (made of slabs of Vishnu Schist) and a panoramic view of the Colorado River and South Kaibab Trail. There is no water or shade on this trail.

The strenuous 12-mile hike to Ribbon Falls will take hikers up the North Kaibab Trail through The Box, a narrow 40-foot-wide, 2.5-mile-long section of Bright Angel Canyon just above Phantom Ranch. Whereas Bright Angel Creek is about 2 feet deep at Phantom Ranch, it is usually more than 20 feet deep in The Box. Hikers also see where Phantom Creek (the namesake of Phantom Ranch) empties into Bright Angel Creek less than a mile upstream from Phantom Ranch. After crossing several bridges, hikers reach Ribbon Falls, which is in a grotto on the west side of Bright Angel Creek.

Phantom Ranch is the only lodging below the rim in GCNP. Visitors get to Phantom Ranch only by foot, mule, or boat; there is no commercial helicopter service to Phantom Ranch. Phantom Ranch, which uses an online lottery-based system for obtaining accommodations, is GCNP's most sought-after destination.

See also: Civilian Conservation Corps; Colter, Mary Elizabeth Jane; Tramways; Woolley, Edwin Dilworth "Uncle Dee"

Further Reading

Anderson, Michael F. 2000. *Polishing the Jewel: An Administrative History of Grand Canyon National Park*. Grand Canyon, AZ: Grand Canyon Association.

Audretsch, Robert W. 2012. *Images of America: Grand Canyon's Phantom Ranch*. Charleston, SC: Arcadia Publishers.

Swanson, Frederick H. 2008. *David Rust: A Life in the Canyons*. Logan: Utah State University Press.

Thybony, Scott. 2001. *Phantom Ranch*. Grand Canyon, AZ: Grand Canyon Association.

PIMA POINT

Pima Point (elevation 6,798 feet), which is one of the best places to see the Colorado River from the South Rim, is named for the Pima Indians (now Tohono O'odham) of south-central Arizona. From Pima Point, visitors can see one of the best examples in the world of what John Wesley Powell called a "geological section." Such spectacular views of Earth's history prompted Powell to note, "All about me are interesting geological records. The book is open, and I can read as I run" (Pyne, 2001, in Dutton, 1882).

Abyss

"Abyss" is a name applicable to countless places in Grand Canyon, but the site officially named Abyss is on the South Rim between Mohave Point and Pima Point, approximately halfway between Hermits Rest and Grand Canyon Village. At this 6,884-foot-high overlook, the Great Mohave Wall drops more than 2,600 feet to the greenish Bright Angel Shale that covers the Tonto Platform. East of Abyss is Isis Temple, which rises 7,012 feet above sea level and is capped by white Coconino Sandstone. At Abyss, you can hear Boucher and Granite Rapids below.

Pima Point overlooks Granite Rapids, Hermit Rapid, and more than 40 miles of Grand Canyon. To the east is Bright Angel Canyon, to the west is Powell Plateau, Eremita Mesa, and Hermit Creek, and to the northwest is Mount Trumball, the highest peak in a cluster of dormant volcanoes. From Pima Point, visitors can see parts of Hermit Trail along Cope Butte, across Hermit Creek, and down Travertine Canyon. Also to the east are Monument Creek and the Alligator. Pima Point is the South Rim's closest place to the Colorado River. Grand Canyon Lodge on the North Rim is 11 miles away. On quiet days, visitors at Pima Point can hear Granite Rapid 5,000 feet below.

In 1926, a 6,000-foot-long tramway began taking visitors on a 20-minute ride from Pima Point to the Santa Fe Railway's Hermit Camp, which consisted of 11 cabins (plus showers, restrooms, stables, a dining room and a blacksmith's shop) built in 1912 by the Santa Fe Railway. The tram, which was carried on 7/8-inch steel cable without any intermediate supporting towers, was the longest single-span tram in the world at the time. By 1930, however, the camp and tramway were abandoned, and by 1936 they were gone.

See also: Hermit Trail; Tramways

Further Reading

Pyne, Stephen J. Foreword. In Dutton, Clarence E. 1882. *Tertiary History of the Grand Canyon*. Washington, DC: U.S. Government Printing Office. (Reprinted 2001 by the University of Arizona Press, with an introduction by Wallace Stegner and a new foreword by Stephen J. Pyne.)

PIONEER CEMETERY AND OTHER GRAVESITES

The South Rim's Pioneer Cemetery is the only cemetery in a national park. The cemetery probably originated with the burial of Grand Canyon pioneer John Hance, who was buried on January 6, 1919, in a relatively open area on the South Rim, about 1.7 miles west of today's Grand Canyon Visitor Center. Hance's grave was soon joined by others, and in 1922 the site was fenced in by NPS and became Grand Canyon Pioneer Cemetery. Several sources claim that Hance's 112-inch-long grave

Pioneer Cemetery on the South Rim includes the graves of many Grand Canyon pioneers, including John Hance, William and Ada Bass, John Verkamp, Ralph Cameron, and Emery and Ellsworth Kolb. (Randy Moore)

is the longest grave in the cemetery in honor of the tall tales Hance told visitors, but several graves in the cemetery are longer.

In 1928, GCNP partnered with John Ivens American Legion Post 42 (chartered on January 19, 1923) to care for the cemetery. Private John W. Ivens (1893–1918), a former employee of the Fred Harvey Company and the Kolb brothers, was killed by German mustard gas in 1918; his was the first reported casualty from northern Arizona in World War I. Ivens was originally buried in France but, on April 7, 1921, was reinterred at Arlington National Cemetery. In 2017, when Pioneer Cemetery closed to new plots, more than 390 people were interred there. Graves and memorials in the cemetery include a variety of Grand Canyon's pioneers, famous residents, and others:

Park rangers—Hubert Reginald "Bert" Lauzon, a stage driver, trail guide (for William Bass), toll collector at Bright Angel Trail (for Coconino County), and ranger who helped the Kolb brothers film part of their river voyage through Grand Canyon, is buried with both of his wives (one of those wives, Ida, was a daughter of William and Ada Bass, for whom Lauzon worked for $60 per month); Fred Johnson, who drowned in Horn Creek Rapid in 1929 and whose body was never recovered; Kenneth Patrick, an equipment operator and trail guide who was the first NPS ranger killed in line of duty; Dan Davis, the first ranger at the scene of the 1956 air crash over Chuar and Temple Buttes; Glen Sturdevant, the first park naturalist at GCNP, who drowned and was replaced by Edwin McKee (Sturdevant's wife Eileen was buried beside him 56 years later); David Orville Karraker, the only person to serve as superintendent of both the Albright Training Center at GCNP and Mather Training Center at Harper's Ferry.

Scientists—David White, a USGS geologist whose headstone is adorned by fossilized ferns; Edwin Dinwiddie "Eddie" McKee, the second ranger-naturalist at GCNP and a founder in 1932 of the Grand Canyon Natural History Association, which is today's Grand Canyon Association. McKee, who first came to Grand Canyon in 1927 as a summer intern with David White (and C. Hart Merriam), is honored by McKee Amphitheater between the Rim Trail and GCNP headquarters. McKee's small gravestone in Pioneer Cemetery is made of Tapeats Sandstone from Pumpkin Spring.

Residents of Grand Canyon Village—John Bradley, who operated the Fred Harvey mule barn; Velma Franklin, a musician who often performed at Bright Angel Lodge; Harriet Sinyella, who worked at El Tovar Hotel and helped record the Havasupai alphabet; Samuel Turner, who was an agent for the Santa Fe Railway.

Artists—Gunnar Widforss, a famous artist whose paintings are displayed in Bright Angel Lodge and El Tovar Hotel. Widforss died near El Tovar Hotel while preparing to leave Grand Canyon.

Pioneers—Ellsworth and Emery Kolb (Emery was a member of Post 42); John George Verkamp and his family and descendants; William Wallace and Ada Diefendorf Bass, the first white couple to raise a family (four children) at Grand

Canyon; Bass was cremated and, as per his wishes, his ashes were spread over Holy Grail Temple (a rock spire along Shinumo Creek), which was subsequently called Bass Tomb. Ada, who died in 1951 (18 years after her husband), is entombed at Pioneer Cemetery.

Military veterans—A memorial commemorates veterans of World War II.

Prospectors—Pete Berry, who came to Grand Canyon around 1887 to settle the estate of his brother John, is buried near John Hance. For many years, Berry was in business at Grandview Point with Niles and Ralph Cameron. Berry is buried beside his second wife, Martha.

Trail guides and workers—George Cravey; Elmer Watahomigie, a Havasupai Indian and "A Great Havasupai Cowboy" who helped build the Kaibab Suspension Bridge in 1928. On his way from guiding tourists to Supai in December 1967, Elmer's truck broke down in a snowstorm; his body was found seven years later under a ridge and was interred in Pioneer Cemetery.

Auto mechanics and drivers—Charles Oscar Sinclair Lovenskiold; "S. M. R.," who drove trucks for the railroad; Emmett Myron "Curly" Ennis, a friend of architect Mary Jane Colter, worked at Grand Canyon as a wrangler, mechanic, garage foreman, and as head of transportation for the Fred Harvey Company. (In 1927, Curly shared the reward for capturing bank robber Matt Kimer at Grand Canyon.) In 1940, Curly's son, Robert E. Ennis, was the youngest professional trail guide ever hired at Grand Canyon. In 1928, Curly—the vice commander of American Legion Post 42—turned over the ownership of Pioneer Cemetery to GCNP Superintendent Miner Tillotson.

Hotel workers—German immigrant Rosemarie Acosta, who also managed a hotel in Tusayan.

Harvey Girls—Ida Lemay Cummings worked as a Harvey Girl and lived at Grand Canyon for more than 50 years; Edith Blanche Kolb Lehnert, the daughter of Emery and Blanche Kolb, and the first person to cross the Kaibab Suspension Bridge; Edna Lemmons "Was a Harvey Girl"; Bette Rich was "Just an Old Harvey Girl at Rest in the Place She Loved."

Politicians—Ralph Cameron, a powerful politician and businessman whose epitaph claims "Arizona Can Never Forget Him," fought making GCNP a national park for several years.

Park superintendents—Miner Tillotson, who was also the chief engineer for South Kaibab Trail and Black Bridge; Carol Anne "Carla" Martin, who became only the fourth woman to be a superintendent and first woman in the NPS Western Region at NPS (at Tuzigoot National Monument, which is southwest of Flagstaff, Arizona). Martin's epitaph reads "Catch the wind."

Victims of the airplane crash—On the morning of June 30, 1956, a midair collision occurred between TWA Flight 2 (with 70 people aboard) and United Airlines Flight 718 (with 58 people aboard) over Chuar and Temple Buttes. There were no survivors of this crash, which at the time was worst disaster in commercial aviation history. The remains of 29 people aboard the United flight were identified and returned to their families; unidentified remains were interred in four coffins buried beneath the memorial near the entrance to Pioneer Cemetery. The names of two relatives of the 29 unidentified people were added to the memorial, bringing the total number of names to 31. (There is another monument for passengers aboard the TWA flight at Citizens Cemetery in Flagstaff.) Both planes were traveling east from Los Angeles, California at about 300 mph. The last message probably came from Robert W. Harms, the United DC-7's first officer: "Salt Lake . . . United seven-eighteen . . . we're going in . . ."

Telephone operator—Eagle Mae Smith Jones, who came to Grand Canyon in 1950 and was working a 16-hour shift on the day of the 1956 air crash. Jones later worked for the Fred Harvey Company and Verkamp's Curios before retiring to Williams.

CCC workers—Julius A. Gaustad and Roy Lemons, who helped install the trans-canyon telephone line; John McLaughlin, on whom writer Dorr Yeager—the first naturalist-ranger in Rocky Mountain National Park—modeled his character Bob Flame.

Trail workers—Donna Fritts, a member of the first female trail crew in GCNP; William Lockridge; Edwin Cummings, who helped Eddie McKee choose the rocks from Grand Canyon's various geologic sediments that were used to build the fireplace in today's History Room at Bright Angel Lodge, came to Grand Canyon in 1919 and worked as a trail guide for the Fred Harvey Company for 30 years. He also helped burn Hermit Camp in 1936 and made most of the wood-and-rawhide furniture at Phantom Ranch, Hermits Rest, and Desert View Watchtower.

Teachers—Katherine Wilcox, who taught at Grand Canyon School.

Blacksmiths at GCNP—Rudolph Muller, who helped organize the 55 mules used to carry supplies used to build Phantom Ranch; John W. Harris; James Hopkins.

Nurses—Lillian Tomesina Catti was a nurse who, after hearing about Rees Griffiths's injury while working on the South Kaibab Trail, hiked through the night on February 6, 1922, from the South Rim to Phantom Ranch to help him (he died just before she arrived). In 1934, Catti married Ray Tankersley, who helped find Glen and Bessie Hyde's boat in 1929.

Surveyors and engineers—Millard Wilcox, who helped engineer the Grand Canyon pipeline; Maurice Castagne, who managed Orphan Mine.

Prospectors—William Henry Ashurst, who died around January 17, 1901, in a rockslide while assessing his Arkansas Traveler Mining Claim near River Mile 86. Ashurst, who had staked the claim a year earlier, was wrapped in a blanket and buried under a pile of rocks in mid-February by his friend John Hance. Ashurst's remains were moved by Niles Cameron (Ralph Cameron's older brother) to the head of Bright Angel Trail in 1908 and then by CCC workers to Pioneer Cemetery (at the request of his son and U.S. Senator Henry Fountain Ashurst) on May 7, 1937. Henry Fountain Ashurst, who—with Congressman Carl Hayden—helped pass legislation to create GCNP in 1919, wanted his father's grave near that of his friend, John Hance. The relocation cost $38.50, including $13.50 for the headstone's brass plaque and $25 for the permits and coffin.

An especially intriguing grave is that of housekeeper Elsie M. Cox Reed, who was found dead of a gunshot on July 4, 1924, in the bedroom of South Rim hotelier Martin Buggeln. Her death was ruled a suicide. She was buried in Pioneer Cemetery with little fanfare, and her grave remained unmarked for more than 90 years (a tombstone was finally placed there on August 14, 2015). Many people believed that Elsie was murdered by her husband Frank for her alleged affair with Martin Buggeln.

Shrine of the Ages

In 1952, fund-raising began for construction of an interfaith chapel at Grand Canyon. The idea for an interfaith chapel had originated around 1917, and its architect (Harold E. Wagoner) had incorporated the architectural style of a kiva. When donations failed to meet the amount required to build the shrine (originally planned just west of the Hermits Rest shuttle bus stop), authorities decided to build the shrine at its current location adjacent to Pioneer Cemetery. Shrine of the Ages was completed in 1970, and NPS took ownership of the shrine five years later. Today, Shrine of the Ages hosts a variety of activities, ranging from funerals and religious services to educational talks presented by NPS rangers.

In 2017, Grand Canyon Pioneer Cemetery—which is listed on the National Register of Historic Places—was closed to new plots because it is full. However, it remains an often overlooked story of Grand Canyon's history that is open to the public.

Other Graves in GCNP

Not all graves in Grand Canyon are in Pioneer Cemetery. For example, Drift Fence Cemetery at Supai was once part of GCNP but is now part of the Havasupai Reservation and off-limits to visitors. (The Grand Canyon National Park Enlargement Act in 1975 expanded the reservation and gave the land to the Havasupai.) Also, several river runners—for example, Willie Taylor, David Quigley, and Peter Hansbrough—who drowned in the Colorado River are buried along the river. Famed NPS ranger John Riffey is buried with his wife at Toroweap, where he worked for almost 40 years. Several death certificates list Grand Canyon as burial places, but the sites of these burials are unknown. Countless people have had their ashes spread in Grand Canyon.

One of the most unusual gravesites in GCNP is that of Charles (1859–1921) and Olga (1866–1920) Brant. Charles began managing El Tovar Hotel in 1904. Both Charles and Olga wanted their graves to overlook the hotel. To find their graves, start at the Hermits Rest Transfer and hike west on the Rim Trail toward Trailview Overlook. At the 3,310 Million Years Old marker, turn left and walk 40 feet up the hill. Then go left again for 65 feet toward the low wall surrounding the graves. The Brant's dog, Razzle Dazzle, outlived his owners, who left him $5,000. NPS

As per their wishes, Charles and Olga Brant were buried at this site overlooking El Tovar Hotel, where they worked. Their gravesite, which is just off of Rim Trail, includes the grave and tombstone of their faithful dog, Razzle Dazzle (rear left). (Randy Moore)

director Stephen Mather issued a special permit for Razzle Dazzle to remain in the park. (To protect wildlife, dogs and other pets in GCNP can be taken only on named trails and developed areas above the rim and must always be leashed.) In 1928, seven years after Charles's death, Razzle Dazzle died and was buried beneath his own headstone, which describes him as the "Faithful and Beloved Airedale Pet" of his owners.

To be interred in the cemetery, one must have lived at Grand Canyon for at least three years and have made significant contributions to the development of public knowledge about, understanding of, or appreciation of Grand Canyon. In the years before it was closed to new plots in 2017, Pioneer Cemetery averaged about four burials per year.

And the Rest . . .

Although Pioneer Cemetery and other graves in GCNP provide an interesting history of the area, they do not include everyone whose final resting place is in Grand Canyon. Many people have disappeared while in GCNP, and in some instances their remains have not been found. Although the most famous of these people were Glen and Bessie Hyde, there have been many others. For example, in 2015, a group of river runners hiking near Emerald Canyon (River Mile 104) found the wreckage of a homebuilt RV-6 experimental airplane (with human remains). In 2011, the plane's pilot, Joseph Radford, had told friends that he planned to kill himself, and he turned off his plane's radio shortly after takeoff. When he did not return to Grand Canyon Airport, NPS helicopters flew 2,000 air miles scouring 600 square miles but did not find Radford or his plane.

Grand Canyon has also been a favorite place for people to commit suicide. Indeed, in the year that the movie *Thelma and Louise* (1991) was released on video, there were three independent suicides involving people driving off the rim at Grand Canyon. Since then, about one-fourth of all suicides in GCNP have involved drives off the South Rim.

The fates of many other people in Grand Canyon remain unknown. For example, acerbic American writer and suicide advocate Ambrose Gwinnett Bierce (1842–1914?) visited Grand Canyon and claimed that when he died, "nobody will find my bones." (Bierce had also written a collection of short stories called *Mysterious Disappearances*.) According to Bierce's friend and biographer, Walter Neale, Bierce killed himself in Grand Canyon.

See also: Hance, John; Kolb, Ellsworth and Emery; Widforss, Gunnar

Further Reading

Anderson, Parker. 2015. *Images of America: Grand Canyon Pioneer Cemetery*. Charleston, SC: Arcadia Press.

Nuttall, Kern. 2016. *In a Better Place: Cemeteries and Gravesites of Grand Canyon*. Flagstaff, AZ: Vishnu Temple Press.

PLACE-NAMES IN GRAND CANYON

Sites in Grand Canyon have been named for a variety of reasons. Some sites have been named to document stories and create points of reference, whereas other names personalize landscapes, invoke ideas and images, describe views, convey a sense of history, or honor events and people who have gone before us. Regardless of their origin, these names are part of our understanding of Grand Canyon, for they underlie how Grand Canyon is discussed, experienced, understood, appreciated, and described.

Pioneers at Grand Canyon, like those elsewhere, organized their views of the landscapes by identifying prominent formations with names. At Grand Canyon, several of these formations were larger than the mountains east of the Mississippi. Americans named few places for Hopi and Zuni Indians, despite the fact that these Native Americans had lived in Grand Canyon and that they considered Grand Canyon a sacred place of creation. An exception to this was John Wesley Powell, the one-armed Civil War veteran who was the first person to intentionally float through Grand Canyon. Powell named several sites with Indian words and named several rock layers for their colors.

Later, cartographers François Matthes and Richard Evans invoked mythological names (especially of Arthurian legends) for places that impressed them, and, in the 1920s, surveyor Claude Birdseye named sites according to their distance from Lees Ferry (e.g., Birdseye's 25 Mile Rapid is 25 miles from Lees Ferry). Native Americans, hikers, river runners, mapmakers, prospectors, tour operators, and National Park Service officials have named hundreds of other places in Grand Canyon.

The names of some places have changed. For example, Three Castles of the early 1900s was renamed Sinking Ship in 1930. Similarly, Lees Ferry was previously called Pahreah Crossing, Bass Camp was called Camper's Paradise, South Bass Trail was called Mystic Springs Trail, Pearce (often misspelled Pierce) Ferry was called Colorado Crossing, Hopi Point was called Rowes Point, Bright Angel Creek was called Silver Creek, and Bright Angel Trail was called Cameron's Trail. Even the Colorado River has had several names, such as El Rio del Tison (Firebrand River). Forgotten names such as Sentinel Point, Eternity Point, and Sunset Point no longer appear on maps of Grand Canyon. Several people who have named iconic places in Grand Canyon, such as Frank Bond (who named Blacktail Canyon and Mount Hayden, among others), have no sites named for them.

Today's names of sites in Grand Canyon came mostly from Anglos having enough political power to bestow names and make maps. These names are as diverse as the grand views (e.g., Grandview Point) they often describe. Secular-minded geologist Clarence Dutton, who saw similarities between the buttes in Grand Canyon and the temples of India and China, felt that only the names of gods could do justice to Grand Canyon; that's why he used names such as Shiva Temple, Vulcans Throne, Zoroaster Temple, and Buddha Temple to refer to several of Grand Canyon's most impressive formations. Nearby is the Roman Section of Grand Canyon, which includes Jupiter, Apollo, and Venus Temples. Other places are named for historic people and groups (e.g., Mohave Point, Shoshone Point, Yuma Point, Cochise Butte, Apache Point, some of whom never lived at Grand Canyon), legends and deities (e.g., Isis Temple, Krishna Shrine), Camelot (e.g., King Arthur Castle, Lancelot Point), presidents (President Harding Rapid, Roosevelt Point), scientists (e.g., Darwin Plateau, Lyell Butte, Newton Butte), early settlers (Mount Wodo, Yumtheska Point, Peshlaki Point, Matkatamiba Canyon, Towago Point), explorers and local personalities (e.g., Ives Point, Powell Plateau, Kolb Arch), and versions of other names (e.g., Stephen Aisle, Chuar). Other names need no explanation; for example, Breezy Point is usually breezy, Crystal Creek has clear water, Redwall Canyon is a canyon with red walls, Fossil Canyon has many fossils, Milk Creek has milky, lime-rich water, and Tilted Mesa is, well, a tilted mesa. Finally, the names of some sites are due to happenstance. For example, House Rock Valley at River Mile 17 was named after someone seeking shelter wrote in charcoal atop a rock overhang "Rock House Hotel." As river runner Frederick Dellenbaugh noted in 1904, "From this came House Rock Valley, and the name was soon a fixture, and went on our maps. And this easily are names established in new country" (Brian, 2015).

The U.S. Board on Geographic Names, which was created by President Benjamin Harrison in 1890, provides standardized geographic names and resolves questions about geographic questions. As per the Board, place-names in the United States do not use the possessive apostrophe on federal maps or signs. There are only five exceptions to this rule (e.g., Carlos Elmer's Joshua View, Arizona; Martha's Vineyard, Massachusetts; Ike's Point, New Jersey; Clark's Mountain, Oregon;

John E's Pond, Rhode Island), none of which are in Grand Canyon. If you see an unnamed site in Grand Canyon, you can suggest a name for the site by contacting the Board.

See also: Birdseye, Claude; Evolution

Further Reading

Brian, Nancy. 2015. *River to Rim: A Guide to Place Names along the Colorado River in Grand Canyon from Lake Powell to Lake Mead*. Second edition. Flagstaff, AZ: Earthquest Press.

McNamee, Gregory. 1997. *Grand Canyon Place Names*. Boulder, CO: Johnson Books.

POINT IMPERIAL

Point Imperial (elevation 8,803 feet), which terminates an 11-mile drive on a paved road from Grand Canyon Lodge, is the highest and northernmost viewpoint on either rim in GCNP that is easily accessed by car. From the parking lot, a short walk on a flat, paved trail takes hikers to an overlook from which they see black and red Precambrian rocks not visible at most other North Rim viewpoints, as well as the vivid orange Mount Hayden. Mount Hayden honors Charles Trumbull Hayden (1825–1900), an Arizona pioneer who operated a mill and ferry at what is now Tempe, Arizona. Hayden's son Carl was Arizona's first U.S. senator.

The Colorado River is not visible from Point Imperial, but visitors can see Echo and Vermilion Cliffs, Marble Canyon, Kwangut Butte, Nankoweap Mesa, and the pastel-hued Painted Desert, which is part of the Navajo Reservation. No rocks of the Grand Canyon Metamorphic Suite are visible from Point Imperial (the Colorado River has not yet cut that deeply here), but visitors can see the Grand Canyon Supergroup. Desert View is 18 air miles and the Colorado River 7 air miles from Point Imperial. The Colorado River is more than a mile below Point Imperial.

Point Imperial Trail provides an easy, 2-mile hike along GCNP's northern boundary that features large areas burned by the Outlet Fire in May 2000. By the time almost 1,000 firefighters contained the blaze, more than 14,000 acres had been burned within GCNP and the adjacent Kaibab National Forest. Beyond these areas are connections to Kaibab National Forest roads and the Nankoweap Trail, a wilderness route that crosses the lower slopes of Boundary Ridge on its 14-mile, 6,050-foot drop to the river. (This trail, whose construction in 1882 was supervised by John Wesley Powell and Charles Walcott, enabled pack animals to take equipment and supplies to the remote Marble Gorge.)

To the east, the trail from Point Imperial meets Saddle Mountain Trail, from which hikers can access Upper Seiber Point. There, mounted inconspicuously on a flat rock just below the rim, is a plaque honoring Lois Webster. The plaque, which is not approved by NPS, is "In Memory of Lois Webster, Jinn of all the canyons, Once here in body, Now here in spirit! Placed by one who loved her."

Point Imperial is near River Mile 50 of the Colorado River. Although Point Imperial is the highest point along the North Rim, it is not the highest point in GCNP. For example, the North Rim entrance station has an elevation of 8,827 feet.

See also: North Rim

Further Reading

Aitchison, Stewart. 2008. *Grand Canyon's North Rim and Beyond: A Guide to the North Rim and the Arizona Strip*. Grand Canyon, AZ: Grand Canyon Association.

POINT SUBLIME

Point Sublime (elevation 7,458 feet) is a remote, rocky peninsula at the end of a 17.8-mile road from near Grand Canyon Lodge through the Kaibab Plateau forest. This overview offers a view of the narrowest section of Grand Canyon, where the South Rim is less than 5 miles away. From a sheer cliff, hikers can also see Boucher Rapids, but there are no maintained trails into the canyon or to the river from Point Sublime. Desert View on the eastern end of the South Rim is 26 air miles from Point Sublime. The unpaved, challenging road to Point Sublime was completed in 1924 to help fight forest fires. Point Sublime is the westernmost of the North Rim viewpoints accessible by vehicles.

Most visitors to Point Sublime consider it to be one of the premier overlooks at GCNP. To the east are Shiva Temple and Cape Royal, to the west are Mount Trumbull and Powell Plateau, and to the south are the Tower of Babel and a view of the Colorado River.

Clarence Dutton used Point Sublime to summarize Grand Canyon in Chapter 8 of his *Tertiary History of the Grand Cañon District, with Atlas* (1882): "The earth suddenly sinks at our feet to illimitable depths. In an instant, in the twinkling of an eye, the awful scene is before us." In 1870s, American explorer, artist, and geologist William Henry Holmes (1846–1933)—who illustrated Dutton's *Tertiary History*—was inspired by Point Sublime to produce *Panorama from Point Sublime*, a masterpiece of American scientific illustration. Thomas Moran's (1837–1926) famous painting titled *The Chasm of the Colorado* was also inspired by the view from Point Sublime. This painting hangs in the Smithsonian Institution Museum of American Art in Washington, DC.

See also: Dutton, Clarence; North Rim

Further Reading

Dutton, Clarence E. 1882. *Tertiary History of the Grand Cañon District, with Atlas*. Washington, DC: U.S. Government Printing Office. (Reprinted 2001 by the University of Arizona Press, with an introduction by Wallace Stegner and a new foreword by Stephen J. Pyne.)

POSTAGE STAMPS

The first postage stamps to feature national parks were issued on July 24, 1934. These stamps included a 2¢ stamp featuring Grand Canyon's Bright Angel Canyon and Deva Temple, Brahma Temple, and Zoroaster Temple viewed from the North Rim. These NPS-themed stamps were selected by President Franklin D. Roosevelt, an avid stamp collector, who—at the urging of Secretary of the Interior Harold Ickes—declared 1934 National Parks Year.

Grand Canyon was next featured on a 5¢ stamp celebrating the 50th anniversary of NPS in 1966. This stamp, which was part of the NPS's Mission 66 rebranding, was meant to announce NPS's new logo, but a public outcry ensued and NPS retained its long-standing arrowhead logo. In 1990, a 25¢ stamp featured a Grand Canyon landscape, and in 1998 a 32¢ stamp labeled "Grand Canyon National Park 1919" was used for the *Trailblazing: 100 Years of Our National Parks* program sponsored by the Smithsonian National Postal Museum.

In 2000, the U.S. Postal Service embarrassed itself by issuing a 60¢ stamp featuring Lipan Point labeled "Grand Canyon, Colorado." (In 2000, as it is today, Grand Canyon is in Arizona.) After noting the mistake, the Postal Service spent $500,000 to destroy all 100,000,000 of the stamps, after which it reissued a new stamp correctly labeled "Grand Canyon, Arizona." However, when a GCNP ranger pointed out that the image on the reprinted stamp was mirror-reversed, an again embarrassed Postal Service spokesperson said, "We're sticking with it. No matter how you look at it, the Grand Canyon makes a beautiful postage stamp" (Associated Press, 2000). In 2010, a 28¢ stamp featured the correct orientation of the image.

The most recent postage stamp honoring Grand Canyon was issued in 2016 as the eighth in a series of 16 stamps featuring NPS; that so-called Forever stamp featured a painting by Thomas Moran (commissioned in 1912 by the Atchison, Topeka and Santa Fe Railway) titled *The Grand Canyon of Arizona, from Hermit Road*. This painting capped 20 years of Moran painting images of Grand Canyon for the railroad.

Grand Canyon has been featured on more postage stamps than any other national park.

See also: Thematic Essays section: John Wesley Powell; Mission 66; Moran, Thomas

Further Reading

Associated Press. 2000. Licked again; Postal Service sticks with latest flaw in Grand Canyon stamp. Retrieved July 1, 2017, at http://lmtribune.com/nation/world/licked-again-postal-service-sticks-with-latest-flaw-in-grand/article_46eb6e86-98bd-5cc2-aaab-6cf610312622.html

POWELL POINT

A 10-minute walk from Maricopa Point leads to Powell Point (elevation 6,778 feet), which is named in honor of Grand Canyon explorer John Wesley Powell. In August 1869, the course of the Colorado River through Arizona was unknown; on maps, this area was blank. The self-taught Powell and a small crew made the first intentional trip through Grand Canyon. During that trip, Powell and his crew passed below Powell Point on August 17, 1869.

Powell Point was originally called Sentinel Point. In 1909, Congress appropriated $5,000 to memorialize Powell at Powell Memorial with an altar-like platform declaring Powell to be "the first explorer of the Grand Canyon." The original plan for the memorial was a Roman chair facing the canyon, but this plan was scrapped in favor of an altar-like, truncated stone pyramid. On May 20, 1918, just nine months before the creation of GCNP, Secretary of the Interior Franklin K. Lane dedicated Powell Memorial. Atop the granite memorial is a bronze plaque featuring a likeness of Powell and this inscription:

> Erected by the Congress of the United States to Major John Wesley Powell first explorer of the Grand Canyon who descended the river with his party in row boats traversing the gorge beneath this point August 17th 1869 and again September 1st 1872.

Powell Monument is the largest and most prominent monument in GCNP.

Powell Point offers good views of Isis Temple, the Grand Canyon Supergroup, and, in the east, Brahma Temple, which rises 775 feet above Powell Point. Ironically, visitors to Powell Point cannot see the Colorado River; it is obscured by the inner gorge. Powell Point is a favorite place for watching sunsets.

See also: Thematic Essays section: John Wesley Powell

Further Reading

Anderson, Michael. 2001. *Along the Rim: A Guide to Grand Canyon's South Rim from Hermit's Rest to Desert View.* Grand Canyon, AZ: Grand Canyon Association.

R

RAILROAD

The Atchison and Topeka Railroad was chartered in 1859 to connect Kansas and California. It later became the Atchison, Topeka and Santa Fe Railroad, which reached Flagstaff in 1882. In 1887, the 15th Territorial Legislature of Arizona promised six years of tax exemptions to any railroad that would build a line to Grand Canyon. This later prompted Grand Canyon pioneer and jack-of-all-trades William O. "Buckey" O'Neill—a newspaperman, lawyer, miner, author, and mayor of Prescott, Arizona—to raise $1,000 for a survey for a prospective railroad from Williams to Grand Canyon. Three years later, after Congress authorized a right-of-way through the Grand Canyon Forest Reserve, the Santa Fe and Grand Canyon Railroad Company was organized to build a railroad and telegraph line from Williams to the Grand Canyon. Although he did not live to see the railroad at Grand Canyon, O'Neill was the person most responsible for bringing the railroad to Grand Canyon.

The Santa Fe and Grand Canyon Railroad Company was incorporated by Lombard, Goode, and Company in 1897, with Lowry W. Goode as president. When the company later went bankrupt when profits from shipping copper and other minerals from the canyon never materialized, the Santa Fe Railway bought the company for $150,000 and renamed it the Atchison, Topeka and Santa Fe Railway. Congress granted a 200 foot-wide right-of-way through Grand Canyon Forest Reserve to a 20-acre depot near what became Grand Canyon Village. The railroad, which took over and finished building track to the South Rim, was operated as a new company directed by Edward Payson Ripley (1845–1920) but was controlled by the Atchison, Topeka & Santa Fe Railway Company. Today, Ripley's name adorns the steam locomotive at Disneyland—Locomotive #2—that runs through the park's "Grand Canyon" exhibit.

On September 17, 1901, engineer Harry Schlee guided Locomotive #282 into the newly renovated Santa Fe Railway Station at the South Rim, thereby opening a new era in tourism in Grand Canyon. All 22 residents of Grand Canyon Village lined the track to welcome the train to the South Rim. The first train carried 30 passengers and mail. The $3.95 ticket (more than $95 today) and 3-hour trip replaced the jarring, uncomfortable stagecoach ride, which took all day, dropped tourists at Hance Ranch or Bass Camp, and cost $20.

The Santa Fe line, which naturalist John Muir praised for its limited impact on the environment, billed itself as a luxury train. Muir had concerns about the railroad reaching Grand Canyon but excused it because "in the presence of such stupendous scenery [the railroads] are nothing. The locomotives and trains are mere beetles and caterpillars, and the noise they make is as little disturbance as the hooting of an owl in the lonely woods" (Lago, 2015). This was also the first train service to Grand Canyon; it transformed the South Rim into the bustling Grand Canyon Village. Three months after the first train arrived, the first automobile arrived at Grand Canyon.

The arrival of this steam engine at the South Rim—and the first railroad car at Grand Canyon—signaled the end of Grand Canyon's frontier days. The railroad advertised itself as "The First Train to Carry Passengers to the Grand Canyon of Arizona," and one of its popular brochures asked, "Won't you be one of the 25,000 visitors at the Grand Canyon of Arizona this summer? It is the world's scenic wonder—nothing like it." For decades, there was no bank on the South Rim, so the railroad carried about $30,000 in cash to and from the rim each day. Although today's Grand Canyon Railway stages mock robberies on the way to Grand Canyon, the original train was never robbed.

Stephen Mather, NPS's first director, knew that for his agency to have a political constituency, he had to get people to the parks. And if he wanted to increase tourism at national parks, he needed the railroads as partners. These partnerships proved to be successful, as did the Grand Canyon Railway's hiring of several prominent artists to interpret Grand Canyon. These artists' paintings, which were featured in the railroad's advertisements, lured visitors to the canyon. The railroad also advertised Grand Canyon with large models of the canyon that it displayed at worlds' fairs (e.g., the 1901 Pan American Exposition in Buffalo). A 306-foot-long, 34-foot-tall diorama of Grand Canyon produced in 1915 remains on display at Disneyland. The advertising campaign was successful; within a few years, the railroad was the primary way for visitors to get to Grand Canyon. By 1910, when the railroad advertised Grand Canyon as the "chief attraction of a trip to California," a round-trip ticket from Williams to Grand Canyon cost $7.50. In 1904, the railroad introduced the Fred Harvey Company to Grand Canyon.

The arrival of the railroad at Grand Canyon rearranged local economies at the Canyon's South Rim. During the next 50 years, the railroad built many of the village's shops, overlooks, public utilities, hotels, and restaurants, which Harvey managed. In 1919, when Grand Canyon became part of GCNP, the U.S. Railroad Administration described the park as a geologic marvel and a spiritual emotion, and it urged people to visit the park "to make hearty your affection for America." In its first year of operation, the Santa Fe Railway brought more tourists to Grand Canyon than had seen it in all previous years combined.

In ensuing years, the railroad faced competition as tourists enjoyed more options for travel. Automobiles soon replaced the train as the preferred way to get to Grand Canyon. Whereas 84 percent of visitors to Grand Canyon in 1920 arrived aboard the Santa Fe Railway, in 1927—for the first time—more visitors came to Grand Canyon in cars (77,422) than aboard the train (70,382). The following year, the first regularly scheduled buses delivered 224 passengers to Grand Canyon, and in 1933—when 4,614 people came to Grand Canyon on buses, only 11,239 passengers came aboard the train. That same year, the first regularly scheduled airplane service to Grand Canyon brought 107 passengers to the canyon, and by 1966, this number increased to 9,842 passengers. In the 1940s, the Santa Fe Railway cut its passenger service to only the summer months.

In 1954, Santa Fe Railway sold its tourist facilities (e.g., Hermits Rest, Desert View Watchtower, Lookout Studio, El Tovar Hotel, Bright Angel Lodge, Phantom Ranch, Hopi House) to the Fred Harvey Company for $1.5 million and donated its service-related buildings to NPS.

By 1950, only 9 percent of the visitors to Grand Canyon arrived aboard the train. In 1967, when fewer than 5,000 visitors come to Grand Canyon on the train, the railroad ended its regularly scheduled passenger service to Grand Canyon; the following year, the train closed its station at Grand Canyon Village. Indeed, only 150 passengers rode the trail in June 1968, and the railroad had also lost its contract to carry mail to and from the South Rim. The final passenger train that arrived at Grand Canyon had one coach car (carrying three paying passengers) and one baggage car. Later, the railroad's freight shipments ended, and train passengers arriving in Williams, Arizona, went to Grand Canyon's South Rim via bus. The Santa Fe Railway retained its depot and right-of-ways until 1982, when courts determined that they had been abandoned and their ownership reverted to the federal government.

At the North Rim, at which there was never direct railroad service, the Union Pacific Railroad ended its bus-based passenger service to nearby towns in 1971. The railroad donated its Grand Canyon Lodge, cabins, and other North Rim facilities to NPS, which now leases those facilities to a concessionaire.

In 1987, Grand Canyon Railway announced plans to revive the defunct Santa Fe Railway and resume service from Williams, Arizona, to and from the South Rim. Two years later, after rehabilitating more than 60 miles of track, the first steam train since 1953 arrived in Grand Canyon. The train was cheered by 10,000 people in Williams and 1,500 at the South Rim. Two years later, and 88 years to the day after the first train arrived at Grand Canyon, the Santa Fe Railway resumed passenger train service to the South Rim. The following year, Grand Canyon Railway carried more than 100,000 passengers to Grand Canyon.

Today, Grand Canyon Railway carries more than 200,000 passengers per year to Grand Canyon. During warmer months, on so-called Steam Saturdays, the train is pulled by one of the railroad's two historic steam locomotives (#4960 and #29) instead of a diesel locomotive. The two newest railway cars are the Fred Harvey Luxury Dome (added in 2012) and the Mary Colter Luxury Dome (added in 2017), both of which have a lounge, downstairs bar, and upstairs seating offering 360° views. Within GCNP, Grand Canyon Railway maintains the railway's yards, depot, and crossings, as did the original Santa Fe.

See also: Grand Canyon Village; Mather, Stephen; O'Neill, Buckey

Further Reading

Lago, Don. 2015. *Grand Canyon: A History of a Natural World and National Park.* Reno: University of Nevada Press.

Richmond, Al. 2007. *Rails to the Rim: Milepost Guide to the Grand Canyon Railway.* Eighth edition. Williams, AZ: Grand Canyon Railway.

Richmond, Al, and Pearsall, Marc. 2007. *The Grand Canyon Railway: Sixty Years in Color.* Second edition. Flagstaff, AZ: Grand Canyon Railway.

RAPIDS

Most of the more than 160 rapids in Grand Canyon have been formed by debris flows from side canyons. Debris flows, which are 60–80 percent sediment, have a slurry consistency resembling that of wet concrete; they are usually produced in Grand Canyon by flash floods caused by heavy rains on steep slopes, and they can move car-sized boulders through canyons to the river. Rapids, which account for about 9 percent of the total distance of the Colorado River in GCNP, produce half of the 8-foot-per-mile drop in GCNP. Most rapids are less than 150 feet long, but there are exceptions; for example, Lava Falls is almost twice that length.

Although John Wesley Powell's trips through Grand Canyon made him famous, boating through Grand Canyon was slow to catch on. Indeed, during the 80 years after Powell's original trip through Grand Canyon, only 345 people would boat all or part of the Colorado River through Grand Canyon. But boating through Grand Canyon then became popular; whereas only 70 people ran the river in 1955, more than 16,400 did so in 1972. More people rafted through Grand Canyon in 1967 alone than had rafted in all of the pre–Glen Canyon Dam years combined. Today, boating through Grand Canyon thrills more than 25,000 adventure seekers per year.

They Swam the Grand Canyon

The first (and only) people to swim the entire length of Grand Canyon were insurance salesmen Bill Beer and John Daggett in 1955. Beer and Daggett wore life jackets and held two river-bags linked by a small piece of rope that they sat on. Beer and Daggett spent about 5 hours per day in the river and ran every rapid except Hance Rapid (Beer later swam that rapid). Their 26-day stunt made them celebrities, and movies of their swim were shown on television across the United States. The following year, Beer reenacted parts of the swim for *ABC's Wide World of Sports*. The publicity generated by Beer and Daggett helped produce the largest single boost in the number of people wanting to run the Colorado River through Grand Canyon.

Today's trips along the Colorado River through Grand Canyon last from a few days to two weeks. More than 70 percent of today's river runners are aboard large, motorized rigs, and most of the rest are aboard oar-powered rafts or dories. These trips are coordinated by GCNP's 16 licensed outfitters, who choreograph the boats' schedules in GCNP to ensure maximum use without too much overcrowding. Costs range from approximately $1,200 for a motorized, three-day trip through part of Grand Canyon, to $4,000 for a two-week, oar-powered adventure that spans the river's entire 277 miles in GCNP. Although Glen Canyon Dam tamed the Colorado River and profoundly changed the Colorado River's flow and Grand Canyon's natural habitat, more than 90 percent of today's river runners describe their trip down the river as a "wilderness" adventure.

Only July 15, 1889—just five days after he inscribed a memorial to drowning victim Frank Brown near River Mile 12—Peter Hansbrough (and Henry Richards) drowned just a few miles downstream. When Hansbrough's body was found several months later, his remains were buried below President Harding Rapid (near River Mile 44) beneath this carving of his initials ("PMH 1889") in a towering wall of Muav Limestone. In 1974, P.T. Reilly proposed changing the name of 25 Mile Rapid to Hansbrough-Richards Rapid, but this change has not yet been approved. (Randy Moore)

GCNP's major rapids are listed here, along with their difficulty ratings (i.e., how difficult each rapid is to navigate). In Grand Canyon, 9 of the 26 rapids rated 5–10 are in the 61 miles of Marble Gorge, 6 are between the Little Colorado and Phantom Ranch, and 11 are between Phantom Ranch and Lake Mead. River Miles are the distances (in miles) from Lees Ferry, which is at the eastern end of the Colorado River in GCNP, where most boating expeditions start.

Badger Creek Rapid at River Mile 8.0 (Rating: 4–6) is the first major rapid in GCNP. This rapid, which is most difficult when water is low, drops 15 feet in one-sixth of a mile. Mormon scout and missionary Jacob Hamblin—who was exploring the area for fellow church members hoping to settle the area—shot a badger in Badger Canyon near here. When Hamblin took the badger to the next creek downstream and boiled the carcass, that creek's alkaline water—combined with fat in the badger—formed soap, which is now the namesake of the downstream creek (i.e., Soap Creek). Jacob Lake, the "gateway city" to the North Rim, is named after Hamblin.

Soap Creek Rapid at River Mile 11.4 (Rating: 7) follows a left turn in the river, causing some boats to veer too far to the right. Adjacent Soap Creek is usually dry, but the rapid can be challenging. The first of many attempts to run this rapid was in 1911 by the Kolb Brothers from the South Rim. Soap Creek Rapid, the last rapid to be run in Grand Canyon, was successfully run for the first time 16 years later by Clyde Eddy and Parley Galloway (the son of famed river runner Nathan Galloway), who mistook it for Badger Creek Rapid. Soap Creek Rapid, which is more than twice as long as Badger Rapid, has changed significantly since the 1930s; flash floods have made it easier to navigate.

House Rock Rapid at River Mile 17.1 (Rating: 4–7), which is named for House Rock Wash, is easier to run in the afternoon when water released from Glen Canyon Dam is higher than normal. In the rapid, the river turns sharply to the right around a debris fan, and waves that can flip rafts are near the end of the rapid. River runners usually run this rapid along its right side because there is a large hole on the left. Famed river runner Emery Kolb considered Soap Creek Rapid as difficult to navigate as Hance, Horn Creek, Granite, Hermit, and Lava Falls Rapids. House Rock Rapid is the most respected rapid in Marble Canyon.

24.5 Mile Rapid at River Mile 24.7 (Rating: 5–6) is where, on July 8, 1949, legendary boatman Albert "Bert" Loper had a heart attack and died while rowing his boat *Grand Canyon* through Grand Canyon. Loper was 79 years old. A memorial to Loper is downstream near the remains of his boat at River Mile 41.6.

25 Mile (Hansbrough–Richards) Rapid at River Mile 25.1 (Rating: 6) is a left-turning rapid named for Peter Hansbrough and Henry Richards, who were with the Brown–Stanton expedition scouting land for installing a railroad through Grand Canyon. On July 15, 1889, after doing some surveying, Hansbrough (a boatman) and Richards (a cook's helper)—who were not wearing life jackets—drowned just beyond this rapid. The railroad was never built. Interestingly, in 1976, river historian Otis "Dock" Marston protested the naming of Hansbrough–Richards Rapid, claiming that the deaths of these people were due to stupidity and that continuing to name rapids for such people would make the course of the Colorado River a memorial to inept boaters.

President Harding Rapid at River Mile 44.0 (Rating: 4) was originally called Boulder Rapid for a boulder in the center of the rapid, but it was renamed in 1923 by USGS surveyor Claude Birdseye, upon hearing by radio at Soap Creek of the death of President Warren G. Harding. Most river runners avoid the rocks on the right side of this rapid by going left of the boulder that splits the river. A favorite campsite is on river left.

Hance Rapid at River Mile 77.1 (Rating: 7–8) is named for Grand Canyon pioneer "Captain" John Hance, who first saw Grand Canyon in 1883. To successfully run this long, wide rapid, which drops 30 feet, boats make several maneuvers to avoid boulders pushed into the rapid from Red Canyon, which enters on river left. Near this rapid, river runners might see backpackers at Hance Camp at the end of New Hance Trail. Hance Rapid, which has the largest overall drop of any rapid in Grand Canyon, is 4,700 feet below the rim.

Grand Canyon's basement rocks come into view for the first time at river level just past Hance Rapid.

Sockdolager Rapid at River Mile 79.1 (Rating: 5–7) is a long rapid named for an outdated term meaning "knockout blow." In 1869, Powell's crew described this rapid as "a perfect hell of waves." After Powell's boat was swamped by the rapid, he wrote "we drift down another hundred yards through breakers—how, we scarcely know" (Ghiglieri, 2003). Sockdolager Rapid is the first major rapid in the Upper Granite Gorge.

Horn Creek Rapid at River Mile 90.8 (Rating: 7–9) is a rapid that, at low water, has large waves and can be one of the Inner Gorge's most difficult rapids. Two rocks in the middle and left produce horn-like waves, but the rapid does not get its name from these two waves; instead, this rapid was named by Henry Gannett for American scout, miner, and Apache interpreter Tom Horn (1860–1903), who was executed in Cheyenne, Wyoming, for allegedly murdering 14-year-old Willie Nickell. River runners usually go right to left through this rapid in order to miss the rocky shore on the right. This rapid is one of the shortest and steepest drops along the river in GCNP.

Granite Rapid at River Mile 93.9 (Rating: 7–8) is a long, difficult rapid formed by debris from Monument Creek, which enters on the south side of the river. The rapid is bordered on the right by a cliff and waves, and on the left by a large beach popular with river runners for camping. River runners are often photographed in this rapid by people at the Abyss overlook on the South Rim.

Hermit Rapid at River Mile 95.5 (Rating: 7–8), which boaters run down the middle, has the biggest waves to this point on the river. The roller coaster–like standing waves peak at 15 feet in the fifth wave. Although Hermit Rapid is not technically difficult, it can flip small boats. Hermit Trail ends near the head of this rapid. Invasive species such as tumbleweed (*Salsola*), camel thorn (*Alhagi*), and tamarisk (*Tamarix*, the most common tree along the river)—all of which were absent or relatively rare before Glen Canyon Dam—are especially abundant along the river here.

Crystal Rapid at River Mile 98.2 (Rating: 7–10) is an intimidating and respected rapid, named for the clear water that enters on river right from Crystal Creek. This rapid formed in 1966 when a flash flood of Crystal Canyon washed debris into the river. In 1983, during an emergency release of water at Glen Canyon Dam, Crystal Rapid had a three-story-high hydraulic wall and was flipping big, motorized boats, prompting GCNP—for the first time ever—to close part of the Colorado River to river runners. Crystal Rapid is one of the few places in GCNP where two side canyons converge; Crystal Creek enters from the north, and Slate Creek enters from the south. The debris fans from these canyons create challenges for boaters, and help explain why Crystal Rapid is the most dangerous rapid in Grand Canyon. Indeed, more than 15 percent of major on-river injuries have happened here, and six people—all of whom were wearing life jackets—have died here. Recalling stories of the many rafts that have flipped at Crystal Rapid, superstitious guides will not flip pancakes or fried eggs at breakfast on the morning they are to run Crystal Rapid.

Deubendorff Rapid at River Mile 132.3 (Rating: 5–8) was named (and misspelled) for Seymour Sylvester "Dubie" Dubendorff, a Vernal, Utah, newspaper handyman and boatman whose boat flipped here in 1909 during the first pleasure run through Grand Canyon with Nathan Galloway and Julius Stone. (Galloway and Stone are memorialized at nearby Galloway Canyon and Stone Creek.) After he recovered from his accident, Dubendorff—who Stone claimed "doesn't seem to know how to get scared"—defiantly claimed, "I'd like to try that again. I know I can run it!" Deubendorff Rapid is notorious for damaging the props of boat motors.

Upset Rapid at River Mile 150.2 (Rating: 7–8) is a large rapid named in 1923 when USGS boatman Emery Kolb flipped his boat here. (Kolb, who was not injured, was rescued.) In 1967, on the day after his 44th birthday, river guide Jesse Edwin "Shorty" Burton drowned here when his boat capsized and his life jacket was caught underwater. Burton, who was famous for the pies and cobblers that he served to his passengers, is honored by a "pie pan memorial plaque" just past this rapid on river right. River runners usually run Upset Rapid along its left side.

Lava Falls (Vulcan) Rapid at River Mile 179.7 (Rating: 8–10) is a relatively short, intimidating rapid that is one of the most difficult and dangerous rapids in Grand Canyon. Lava Falls Rapid, which is named for the nearby lava flows that created the dark basalt formations of the area, was formed by debris from Prospect Canyon and is one of the last major rapids in GCNP. Running Lava involves a 15-foot drop over 100 yards. Most of the drop occurs in two short increments, hence the name "Lava Falls." Lava Falls was regularly portaged until the 1950s, and pioneering Grand Canyon river runner Robert Stanton considered it to be Grand Canyon's most difficult rapid. In 2017, 89-year-old Dick Griffith became the oldest person to ever row this rapid (Griffith was four days older than the previous record holder, Martin Litton).

232 Mile (Honeymoon) Rapid at River Mile 232.5 (Rating: 4–7) is where Grand Canyon historians speculate that honeymooning Glen and Bessie Hyde were thrown from their boat and drowned in 1928 during a boat trip through Grand Canyon. Days earlier, Glen and Bessie had hiked to the South Rim to visit Emery Kolb, whose offer of life jackets was refused. The Hydes' boat and belongings were found below River Mile 237, but no trace of the Hydes was ever found. Had the Hydes succeeded, Bessie would have been the first known woman through Grand Canyon.

See also: Thematic Essays section: How the Colorado River Carved Grand Canyon

Further Reading

Belknap, Buzz, et al. 2017. *Grand Canyon River Guide*. Evergreen, CO: Westwater Books.

Crumbo, Kim. 1994. *A River Runner's Guide to the History of Grand Canyon*. Boulder, CO: Johnson Books.

Ghiglieri, Michael, ed. 2003. *First through Grand Canyon: The Secret Journals and Letters of the 1869 Crew Who Explored the Green and Colorado Rivers*. Flagstaff, AZ: Puma Press.

REILLY, PLEZ "P. T."

Plez Talmadge Reilly (1911–1996) was born in Dallas, Texas, but soon thereafter moved to California, where he was educated and excelled at sports. In 1936, Reilly began working for the U.S. General Land Office, and he married his high school friend Mary Elizabeth (Susie) MacLean the next year. In 1940, Reilly began a successful 29-year career with Lockheed Aircraft Corporation.

Reilly became interested in the Colorado River and Grand Canyon after reading an article in *Desert Magazine*. After beginning his Grand Canyon boating as a boatman for Norman Nevills in 1949, Reilly later began offering private trips in Grand Canyon. Reilly, who ran Grand Canyon 14 times, refused to offer commercial trips, instead claiming that he did the trips for his own enjoyment. In 1955, he took Martin Litton through Grand Canyon and then hired Litton as a boatman the following year. Two years later, Reilly was on the river during its flow of over 125,000 cubic feet per second (cfs); he was also on the river when the gates of Glen Canyon Dam closed, reducing the flow to 1,000 cfs. Reilly, who organized one of the first trips to publicize the impact of Glen Canyon Dam on Grand Canyon, also began designing and building modified cataract boats having a narrow, pointed stern and greater buoyancy than those used by Nevills. He named his first boat *Susie R*, after his wife.

In 1962, Litton convinced Reilly to try McKenzie River Drift Boats, which had flared sides. These, in turn, were replaced by Rogue River drift boats, which are now the most common type of dory in Grand Canyon. Thanks to help from Reilly, the Sierra Club published *Time and the River Flowing: Grand Canyon* (1964), which was distributed to every U.S. senator and representative to help preserve Grand Canyon. During a river trip in 1964, Reilly discovered the grave and inscription of Peter Hansbrough, who died with Henry Richards on the river on July 15, 1889.

Reilly continued leading noncommercial trips through Grand Canyon until 1964 and then returned to run his last trips through Grand Canyon in the early 1980s. One of these trips, using motorized boats in 1982, was his first and only trip through Grand Canyon in anything other than an oar-powered boat. By the time he left the river, Reilly was concerned about the growing popularity of river running in Grand Canyon. Reilly described the changes along the river as "monumental" and warned that the Colorado River's ecosystems were too fragile to withstand the impact of 15,000 river runners per year. In 2017, more than 25,000 people boated through Grand Canyon.

On April 16, 1956, Reilly and Martin Litton flew over Grand Canyon, during which Reilly took some aerial photos of Sinyala Fault. When Reilly later examined the photos with a magnifying glass, he discovered a 400-by-40-foot arch in Redwall Limestone about 1.5 miles up a tributary of 140 Mile Canyon. After hiking to the arch on June 29, 1956, Reilly named the arch Keyhole Natural Bridge. Reilly's discovery of this seldom visited bridge, which was one of the last major land bridges to be identified in or around Grand Canyon, was announced by the *Los Angeles Times* as "Big Natural Span Found in Colorado River Area."

Reilly later sold *Susie Too* (the second boat he named after his wife) to Litton, who renamed it *Music Temple* and made it part of his Grand Canyon Dories commercial river running company. Today, that boat is in the historic boat collection of GCNP. In 1972, P. T. and Susie moved to Arizona. P. T. died of congestive heart failure in 1996, and Susie died in 2001. Reilly's authoritative *Lee's Ferry: From Mormon Crossing to National Park* was published in 1999 (Reilly had paid for several new headstones in the cemetery there). Susie and P. T. are commemorated by the Elizabeth M. and P. T. Reilly Internship at the Cline Library at Northern Arizona University.

See also: Lees Ferry; Litton, Martin; Sierra Club

Further Reading

Marston, Otis Reed "Dock." 2014. *From Powell to Power: A Recounting of the First One Hundred River Runners through the Grand Canyon.* (Edited by Tom Martin.) Flagstaff, AZ: Vishnu Temple Press.

RELIGIOUS BOOKS AND SCRIPTURE PLAQUES

Grand Canyon has a long association with religious beliefs and groups. For example, in the early 1960s, the Grand Canyon Natural History Association (now the Grand Canyon Association, a nonprofit organization supervised by NPS) began selling Rev. George Wahlin booklet titled *The Bible Creation Story and Grand Canyon* (1962) in its bookstores. Within a year, the 16-page, 35¢ booklet was outselling everything except Edwin McKee's *Ancient Landscapes of the Grand Canyon Region* (1931). Wahlin's book claimed that "the grandest thing about the Canyon is that it confirms the creation story of the Bible. These five chapters, or eras, fit perfectly into the periods of creation presented in the beginning of the Book of Genesis." In the early 1970s, the Grand Canyon Natural History Association bought the publication rights to the booklet for $1,000. By 1974, the Grand Canyon Natural History Association had stopped selling it.

In 1970, a Phoenix convent known as Canaan of the Desert (a branch of the Evangelical Sisterhood of Mary) donated three bronze plaques featuring Bible verses to GCNP. The plaques, which were meant to minister "to millions of visitors," were displayed at Hermits Rest (Psalms 68:4), Lookout Studio (Psalms 104:24), and Desert View (Psalms 68:4) without incident for decades. In 2003, however, the American Civil Liberties Union questioned the display of the plaques in GCNP. To avoid controversy, GCNP Superintendent Joe Alston had the plaques taken down. This, however, produced its own controversy. After the *Arizona Republic* headlined the story with "Bible Verses Out at Canyon," there was a public outcry, and even a protest by some members of Congress. Nine days later, after it was noted that religious symbols are displayed in several national parks (e.g., Yosemite, Sitka National Historical Park), NPS Deputy Director Donald Murphy ordered the plaques restored at their original sites in GCNP. As of 2018, the plaques remained on display at those sites in GCNP.

See also: Thematic Essays section: Grand Canyon and Modern Christian Fundamentalism; Dinosaurs; Mckee, Edwin "Eddie"

Further Reading

Wahlin, George W. 1960. *The Bible Creation Story and Grand Canyon*. Fifth edition. Riverside, CA: Privately printed.

RIM TRAIL

More than 85 percent of people who visit GCNP do not hike into Grand Canyon, preferring instead to walk, ride, or drive along the North and South Rims while admiring the views. The most popular trail in GCNP is the South Rim's Rim Trail, which stretches 14 miles from the South Kaibab Trailhead to Hermits Rest. The trail is ideal for people seeking quiet views of the canyon. Along the well-defined trail are stops and viewpoints including (from east to west) Pipe Creek Vista, Mather Point, Yavapai Point, Park Headquarters Junction, Trailview Overlook, Maricopa Point, Powell Point, Hopi Point, Mohave Point, Abyss, Monument Creek Vista, and Pima Point. Rim Trail is well-defined, relatively flat (the average grade is 4 percent), and mostly paved (between Monument Creek Vista and Powell Point, the trail is a 3-foot-wide dirt trail). Elevations along the trail vary by approximately 200 feet. Only a few sections of the trail (e.g., between Pipe Creek Vista and Bright Angel Lodge) have grades that exceed accessibility standards.

Rim Trail, which can be accessed from any viewpoint in Grand Canyon Village or along Hermit Road, provides a great introduction to Grand Canyon. The trail is open year-round and includes several shuttle bus stops, enabling visitors to pick the areas they want to explore. Leashed pets are allowed on the trail but not on shuttle buses.

Most visitors at the South Rim access the Rim Trail at the Canyon View Information Plaza, which adjoins the South Rim's largest parking lots. This plaza includes a large bookstore and gift shop, restrooms, food and drinks, a theater, a coffee shop, shuttle bus service, and informative displays about Grand Canyon's history, trails, wildlife, and weather. On busy days, more than 15,000 people visit Canyon View Information Plaza. In front of the prominent Visitors Center is a plaque noting Grand Canyon's designation as a World Heritage Site. West of the Grand Canyon Visitor Center are a general store, campground, bank, post office, and Yavapai Lodge.

See also: Grand Canyon Village; Hopi Point; Maricopa Point; Mather Point; Powell Point

Further Reading

Anderson, Michael. 2001. *Along the Rim: A Guide to Grand Canyon's South Rim from Hermit's Rest to Desert View*. Grand Canyon, AZ: Grand Canyon Association.

ROADS

When GCNP opened in 1919, there were four roads to the park: one to the North Rim from near Jacob Lake, Arizona, and three to the South Rim from Flagstaff, Williams, and Ash Fork, Arizona. None of these roads was paved. Indeed, when Grand Canyon became a national park in 1919, the only paved road at Grand Canyon was Hermit Rim Road, which stretched from Grand Canyon Village to

Hermits Rest. This road, which had been built in 1911–1912 by the Santa Fe Railway, had speed limits of 20 miles per hour on straight stretches of the road where no car was visible for 200 yards, 8 miles per hour when vehicles were passing animals on the road, and 12 miles per hour along all other parts of the road. Hermit Rim Road was repaved in 1920 and then rebuilt in the 1930s as increased amounts of traffic began to degrade the road. A paved road to Grandview Point and Desert View (called East Rim Drive) was not completed until 1926, and the entrance road to the South Rim from Williams was not paved until 1931. The completion of East Rim Drive began the practice of including roadside picnic sites.

In 1935, NPS and the Bureau of Public Roads began building the 35-mile so-called mountain highway from Cameron (elevation 4,370 feet) to Desert View (elevation 7,438 feet). This road opened the eastern end of GCNP to cars and gave tourists a second, scenic road to and from the South Rim. The road, which cost $1,060,000, was completely paved by 1937. At this point, more than 200,000 people per year were visiting GCNP. The road to the North Rim opened in 1913 but was not paved until 1926. The North Rim road to Cape Royal and Point Imperial, which opened in 1926, was paved in 1932.

Today, roads to the North and South Rims are paved, safe, and busy. Traffic at the South Rim was reduced somewhat by the NPS seasonal shuttle bus to and from Tusayan; this bus annually carries more than 125,000 people (the equivalent of more than 20,000 vehicles) into GCNP. However, the South Rim often remains congested and, at peak times during the summer, parking places are difficult to find.

See also: Gateways to Grand Canyon

Further Reading

Lago, Don. 2015. *Grand Canyon: A History of a Natural World and National Park*. Reno: University of Nevada Press.

ROOSEVELT POINT

Roosevelt Point (elevation 8,470 feet), which is near River Mile 56, honors President Theodore Roosevelt (1858–1919), who on November 28, 1906, created the Grand Canyon Game Preserve and, in the process, partially protected what later became Grand Canyon. This proclamation protected the area's deer and other "game animals" but did not apply to predators such as coyotes, bobcats, and mountain lions. On January 11, 1908, Roosevelt's Presidential Proclamation 794 established Grand Canyon National Monument, the precursor of GCNP. The monument was tangled in a lawsuit until 1920, when the U.S. Supreme Court ruled that Roosevelt had acted properly because Grand Canyon was of "scientific interest" specified by the Antiquities Act of 1906. Roosevelt's proclamation ended new claims (i.e., "private entry") of Grand Canyon's land but did not eliminate existing claims of private ownership. A plaque at Roosevelt Point (and a smaller one at GCNP headquarters) commemorates Roosevelt's now famous declaration on May 6, 1903 (during his first of three visits to Grand Canyon):

> In the Grand Canyon, Arizona has a natural wonder which, so far as I know, is in kind absolutely unparalleled throughout the rest of the world. I want to ask you to do one thing in connection with it in your own interest and in the interest of the country—to keep this great wonder of nature as it now is . . . Leave it as it is. You cannot improve on it. The ages have been at work on it, and man can only mar it. What you can do is to keep it for your children, your children's children, and for all who come after you, as the one great sight which every American . . . should see. (Roosevelt, 1903)

Roosevelt's speech became famous and continues to be quoted often by conservationists. What is seldom reported, however, is that Roosevelt gave his moving speech while on the porch of the Bright Angel Hotel, which was affiliated with the Santa Fe Railway; his speech was aimed at the railroad, which was planning to build a variety of buildings and accompanying facilities on the rim, including the luxurious El Tovar Hotel. Roosevelt felt that building such commercialism at the rim of Grand Canyon was shameful, and he gave his famous speech while incorrectly believing that the railroad had cancelled their plans. Indeed, here are the seldom mentioned words that preceded Roosevelt's "leave it as it is" comments:

> I was delighted to learn of the wisdom of the Santa Fe Railroad in deciding not to build their hotel on the brink of the cañon. I hope you will not have a building of any kind, not a summer cottage, a hotel or anything else, to mar the wonderful grandeur, sublimity, the great loneliness and beauty of the cañon.

Roosevelt wanted every American to see Grand Canyon but did not want the area to have any trace of Americans having done so.

Roosevelt, who became the 26th president of the United States on September 14, 1901 (i.e., after President William McKinley was assassinated), established conservation as national policy. Although Roosevelt was the first president born and raised in a big city (in Roosevelt's case, New York City), he embraced conservation. (Although many writers in the early 1900s described Grand Canyon, virtually none of them emphasized its conservation.) For example, in 1910 he noted that "of all the questions which can come before this nation . . . there is none which compares in importance with the great central task of leaving this land even a better land for our descendants than it is for us" (National Park Service, 2017). Roosevelt, who claimed "there can be no greater issue than that of conservation in this country" (National Park Service, 2017), was the first president to speak extensively about conservation. His later refusal during a hunting trip to shoot a bear gave rise to Teddy Bears.

Roosevelt Point is 14.8 miles from Grand Canyon Lodge. From Roosevelt Point, visitors enjoy a pretty but somewhat restricted view of the canyon but can hike a scenic 0.2-mile loop through a secluded woodland for more perspectives of the area. The most prominent peak visible from Roosevelt Point is Tritle Peak, an 8,388-foot-high summit capped by Kaibab Limestone. Roosevelt Point also offers excellent views of Kwagunt Creek, which flows parallel to Nankoweap Creek.

Roosevelt, who first visited Grand Canyon on May 6, 1903, died on January 6, 1919 (New Mexico Statehood Day), just 50 days before the creation of GCNP. Roosevelt did much to protect and popularize Grand Canyon and other national parks and monuments. Indeed, Grand Canyon was one of the 152 national forests, 51 federal bird reserves (now national wildlife refuges), four national game preserves, and 18 national monuments (with a combined area exceeding 220 million acres) protected by Roosevelt. Roosevelt, with Congressional approval, also created five national parks. This was a remarkable political accomplishment, especially since conservation was strongly opposed by U.S. House of Representatives Speaker Joe Cannon, who famously promised that he would "put people above scenery" and that he'd spend "not one cent for scenery" (Bolles, 1951). Indeed, between 1901 and 1911, Congress rejected more than 40 bills that would have established national forests in the east. Although the Raker Act of 1913 established the precedent that economic concerns could supersede conservation (even in national parks), that same year witnessed Secretary of the Interior Franklin Lane's proclamation that scenery, like forests, land, and minerals, is a natural resource that warrants legal protection and economic development.

Roosevelt also extended to the U.S. Board on Geographic Names the authority to assign, change, and standardize geographic names in the United States. Nevertheless, no trail or viewpoint officially honored Roosevelt until the dedication of Roosevelt Point in 1996.

Roosevelt died of a heart attack on January 6, 1919—five weeks before Grand Canyon became a national park—in Cove Neck, New York. He is buried in Youngs Memorial Cemetery in Oyster Bay, New York. He is commemorated by plaques at Roosevelt Point and at Jacob Lake's Kaibab Plateau Visitor Center, the latter of which was erected in 1958 by the Arizona Game Protective Association and the U.S. Forest Service. Roosevelt is also honored throughout the United States and elsewhere on stamps, plaques, and countless other monuments (e.g., on Mount Rushmore, with George Washington, Abraham Lincoln, and Thomas Jefferson).

See also: Antiquities Act of 1906; O'Neill, Buckey

Further Reading

Bolles, Blair. 1951 (1974). *Tyrant from Illinois: Uncle Joe Cannon's Experiment with Personal Power.* Westport, CT: Greenwood Press.

National Park Service. 2017. Theodore Roosevelt quotes. Retrieved June 1, 2017, at https://www.nps.gov/thro/learn/historyculture/theodore-roosevelt-quotes.htm

Roosevelt, Theodore. 1903. At Grand Canyon, Arizona, May 6, 1903. Retrieved June 29, 2017, at http://www.theodore-roosevelt.com/images/trenvpics/trgrandcanyonspeech.pdf

ROWE, SANFORD

Sanford H. Rowe (1853–1929) was a buffalo hunter and one of the first European Americans to arrive at the area that would become Grand Canyon Village. Rowe, who operated a livery stable in Williams, Arizona, transported tourists and others to Grand Canyon. In 1892, Rowe bought William Bass's livery business in Williams and began coordinating stagecoaches and freight wagons to the South Rim.

In 1890, while being given a tour of the South Rim by a Havasupai Indian named Big Jim, Rowe discovered a spring above Shaski Wash (today's Bright Angel Wash) 3 miles south of the rim. Because he had already filed a homestead claim in Williams, Rowe could not patent the land, so instead he filed three mining claims—the Lucky Strike, Little Mamie, and Highland Mary claims—around the spring. Rowe mined small amounts of ore from the area but later (with partner Ed Hamilton) opened an automobile center called Rowe Well Camp near his claims that included $1-per-day cabins, a bowling alley, a bar, a dance hall, and other facilities. Rowe, the first tourism operator at the south-central rim, was the first person to use the Bright Angel Trail as a tourist path to the Tonto Platform. He also built a trail for tourists to Hopi Point (which was then called Rowes Point). Rowe died in 1929 at Grand Canyon. The first GCNP ranger station was near Rowe Well.

By the 1950s, Rowe's buildings were crumbling and considered an eyesore, and in 1956 they were purchased by GCNP for $42,500. By 1961, all of Rowe's buildings were gone, except for a barbecue pit built by Kenneth B. "Red" Wright. Today, visitors pass Rowe's site (now a picnic area) when they travel along Rowe Well Road from West Rim Drive to Forest Road 328.

See also: Grand Canyon Village

Further Reading

Lago, Don. 2014. *Canyon of Dreams: Stories from Grand Canyon History.* Salt Lake City: University of Utah Press.

RUSSELL, CHARLES, AND THE *ROSS WHEELER*

Charles Silver Russell (1877–1925) was a prospector in Prescott, Arizona, who boated through Grand Canyon in 1908 with Edwin Monett while looking for potential mines. His trip failed to uncover any promising sites. Later, however, Russell was surprised at the publicity received by the film made by Ellsworth and Emery Kolb during their 1911–1912 trip through Grand Canyon. Russell considered the film poor and amateurish and believed that he could make a better one. In 1914, he decided to try.

To help him get down the Colorado River, Russell hired his old partner Bert Loper. For the trip, Loper had a boat made of galvanized steel that he named *Ross Wheeler* in honor of his murdered friend and steamboat pilot, Roswell "Ross" Wheeler. Loper's boat was modeled after the wooden boats used during the Galloway–Stone expedition through Grand Canyon in 1909. On June 19, 1914, Russell and Loper left Green River, Utah, carrying provisions and $5,000 worth of cameras and related gear for the movie. When one of their boats sank in Cataract Canyon, they stored their gear and hiked out of the canyon. When a disagreement caused Loper to quit the project, Russell took *Ross Wheeler* and hired two new crewmembers—Bill Reeder, a resident of Green River, and August Tadje, an amateur filmmaker—to restart the job. Neither of the new crewmen had any experience in Grand Canyon.

Russell recovered the boat that he had stored but again sank it and hiked out of the canyon. When Reeder quit the expedition, Russell went to Salt Lake City

Ross Wheeler **is the only surviving boat from Charlie Russell's ill-fated attempts to make a movie about river running in Grand Canyon. Today, Russell's steel boat resides near the base of South Bass Trail at River Mile 108.5.** ***Ross Wheeler*****, which was built by Bert Loper, is 66 miles downstream from the remains of** ***Grand Canyon*****, the boat in which Loper died in 1949. Russell never made his movie. (Randy Moore)**

and ordered a new boat that he named *Titanic II*. When that boat was lost in Upper Granite Gorge in Grand Canyon, Russell hiked out of the canyon along Bright Angel Trail, ordered another boat, and hauled that boat on a week-long trip down Bright Angel Trail to the river. After more lost boats and hikes into and out of Grand Canyon, Russell finally gave up; he abandoned his only surviving boat—*Ross Wheeler*—at the base of South Bass Trail and hiked out of the canyon for the last time. His movie was never made, and Russell later ended up in an asylum.

A miner named John Waltenberg later found *Ross Wheeler* and pulled it up to its present site along the river. (*Ross Wheeler* is 66 miles downstream from the remains of Bert Loper's boat.) Although thieves have stolen the boat's oars and parts, it remains at that site today, where it is a popular stop for history-loving river runners.

See also: Bass Trails and Bass Camp; Loper, Bert

Further Reading

Grand Canyon River Heritage Museum. 2017. Ross Wheeler. Retrieved June 1, 2017, at http://www.gcrivermuseum.org/river-heritage/the-boats/ross-wheeler/#none

S

SHOSHONE POINT

This seldom-visited viewpoint, which is 2 miles southeast of Yaki Point, is not marked by a sign but can be accessed from a parking lot on the north side of Desert View Drive between mileposts 244 and 245 (about 1.3 miles east of the Yaki Point turnoff and 6.3 miles west of the Grandview Point turnoff). The quiet viewpoint, which terminates a relatively flat, 1-mile unpaved road through ponderosa pine woodland, offers 180° views of the canyon. To the west are Yaki Point, Cremation Creek, and Phantom Creek from the North Rim. In the foreground are Lyell and Newton Buttes, and in the distance are Zoroaster Temple and Thor Temple. To the east are Grapevine Creek, Hance Rapid, and Vishnu Temple, one of Grand Canyon's most striking peaks. The Colorado River is visible near Tanner Canyon in the northeast, where it curves north toward Marble Canyon.

Cremation Creek

Cremation Creek, at Mile 85.8 below Lyell Butte, is named for an ancient ash pit that Grand Canyon habitué Emery Kolb believed contained human remains spread there by Havasupai Indians. The first pilots to fly airplanes into Grand Canyon entered the canyon here in 1927 as they searched for Bessie and Glen Hyde, who disappeared in Grand Canyon while on their honeymoon. Cremation Point is now known as Shoshone Point.

Shoshone Point has picnic tables, toilets, and fire grills. It is especially popular from May through October for weddings, birthdays, and other events, but the use of these facilities requires advance permission from the NPS. There is no water or electricity at Shoshone Point.

See also: Grandview Point

Further Reading

Anderson, Michael. 2001. *Along the Rim: A Guide to Grand Canyon's South Rim from Hermit's Rest to Desert View.* Grand Canyon, AZ: Grand Canyon Association.

SIERRA CLUB

Sierra Club was founded in 1892 by 182 charter members, who elected Scottish American naturalist and writer John Muir (1838–1914) the club's first president.

The club began as a small hiking fraternity but later—in its first work related to conservation—fought to defeat a proposed shrinkage of Yosemite National Park. Muir, the so-called Father of National Parks who helped increase the public's awareness of Grand Canyon, voiced a common reaction to Grand Canyon:

> It is a hard job to sketch it even in scrawniest outline . . . I cannot tell the hundredth part of the wonders of its feature." America's leading nature writer John Burroughs, who often accompanied Muir on his travels and came to Grand Canyon with Muir in 1909, had similar feelings: "Of the many descriptions of it, none seems adequate. To rave over it, or to pour into it a torrent of superlatives, is of little avail . . . No matter how far you have wandered hitherto, or how many famous gorges and valleys you have seen, this one, the Grand Canyon of the Colorado, will seem as novel to you, as unearthly in the color and grandeur and quantity of its architecture, as if you had found it after death, on some other star. (Muir, 2017)

By the early 1960s, the government was intent on damming the Colorado River. Indeed, by 1962 it had identified several specific sites for the dams. The dams even appeared on many maps produced at the time. In 1966, the Sierra Club—inspired and prodded by Grand Canyon river runner Martin Litton—sprang into action by buying the first of several full-page ads in *The New York Times*, *The Washington Post*, and other prominent newspapers protesting the government's plans to dam Grand Canyon ("This time it's the Grand Canyon they want to flood. *The Grand Canyon.*"). The ads pled for the public to act, claiming "Now Only You Can Save Grand Canyon from Being Flooded . . . For Profit." When the Bureau of Reclamation claimed that the dams and their resulting reservoirs were good because they would make the canyon more accessible to the public, the Sierra Club responded with one of its most effective ads, asking, "Should we also flood the Sistine Chapel so tourists can get nearer the ceiling?" (Brower, 1970a).

Politicians such as Morris Udall (a great-grandson of Lees Ferry pioneer John D. Lee) attacked and vilified the Sierra Club, claiming that its ads were "dishonest" and "distorted." David Brower, who headed the Sierra Club and who considered Litton to be his "conscience," responded by asking, "If we can't save the Grand Canyon, what the hell can we save?" (Brower, 1970a, b). The day after these ads appeared, the Internal Revenue Service (IRS) revoked Sierra Club's tax-exempt status, ostensibly because of the club's disruptive political activities.

Sierra Club was the only environmental group willing to challenge the governmental erosion of constitutional principles (e.g., free speech and the right to petition for grievances). Stewart Udall (another great-grandson of Lee), who served as Secretary of the Interior from 1961 to 1969, instructed the directors of GCNP and NPS not to object to plans for the dams. However, after rafting through Grand Canyon with Jack Currey, Udall yielded to political pragmatism, reversed his earlier position, and began opposing the building of dams in Grand Canyon. The Sierra

Club's letter writing campaign—the last major John Muir–style preservation campaign—was successful. The dams planned for Grand Canyon—that is, Marble Canyon Dam and Bridge Canyon Dam—were never built.

Meanwhile, the public was outraged by the government's attempt (through the IRS) to silence the Sierra Club's efforts to protect Grand Canyon. By 1968, the Sierra Club was viewed as the "savior" of Grand Canyon, and its membership began to swell rapidly. Membership in Sierra Club ($14 per year in 1966) increased by 12,000 to 33,000 in 1966, to 45,000 in 1967, to 58,000 in 1968, and to 135,000 in 1971. These increases far exceeded those of any other conservation organization.

In 2004, 48 years after his first trip through Grand Canyon as a boatman, the Sierra Club's 87-year-old Martin Litton became the oldest boatman to run Grand Canyon and row all of the rapids himself. Litton, who was one of the last pioneers of the modern environmental movement, was one of the few people who understood the damage that the Marble Canyon Dam would permanently inflict on Grand Canyon. As he noted later, "We certainly aren't sorry that we kept the dams out of the Grand Canyon, and if we lied to do it, fine."

See also: Litton, Martin; Water at Grand Canyon

Further Reading

Brower, David. 1970a. Grand Canyon battle ads. Retrieved July 2, 2017, at http://content.sierraclub.org/brower/grand-canyon-battle-ads

Brower, David. 1970b. *Grand Canyon and the Living Colorado.* New York: Sierra Club and Ballantine Books.

Muir, John. 2017. *John Muir: Selected Writings.* (Introduction by Terry Tempest Williams.) New York: Everyman's Library.

Pearson, Byron E. 2001. *Still the Wild River Runs: Congress, the Sierra Club, and the Fight to Save Grand Canyon.* Tucson: University of Arizona Press.

SIPAPU

Sipapu (also *Sipapuni*), the Hopi word for "place of emergence," is a small spring atop an 80-foot-high travertine dome 4.4 miles upstream from where the Little Colorado River joins the Colorado River in Grand Canyon (River Mile 61.5). The site is scared to Hopis, who believe that it is where they entered the world (the fourth world) from the third world. When Hopis die, their spirits return to Grand Canyon through the *Sipapu* to emerge in the afterlife. According to legend, the Hopi Tribe is relatively small because a pregnant woman became lodged in the *Sipapu*. In kivas of Hopi settlements, a small hole in the floor represents a symbolic *Sipapu*.

See also: Desert View; Little Colorado River

Further Reading

Courlander, Harold. 1987. *The Fourth World of the Hopis: The Epic Story of the Hopi Indians as Preserved in Their Legends and Traditions.* Albuquerque: University of New Mexico Press.

SOUTH KAIBAB TRAIL

In the early days of Grand Canyon, trails into the canyon were controlled by mining claims. The most popular of these, the Bright Angel Trail, was a toll trail; people had to pay $1 to hike on the trail into Grand Canyon. To avoid these tolls, in 1912 Santa Fe Railway built a trail into the canyon for its customers; this became Hermit Trail. When Ralph Cameron demanded $150,000 for Bright Angel Trail, NPS officials began studying how they could build their own toll-free trail into the canyon for visitors at Grand Canyon. Much of their communication between NPS and GCNP was sent in code because Ralph Cameron's friend (hotel promoter Lannes L. Ferrall) was postmaster at Grand Canyon.

The projected cost of South Kaibab Trail was only $40,000 (it eventually would cost $73,000), so construction began on a new NPS-owned trail that is today's South Kaibab Trail. South Kaibab Trail remains the only major trail built by NPS as an alternative to a privately owned trail (in this case, Bright Angel Trail).

South Kaibab Trail was originally known as Yaki (Yaqui) Trail because it originates at Yaki Point, which is about 5 miles east of Grand Canyon Village. (The trail's current name honors Kaibab National Forest.) Except for parts of the trail below Tipoff, no trail existed here before the NPS built South Kaibab Trail. Whereas Bright Angel Trail's trailhead is 9.5 miles from the river, South Kaibab's is only about 7 miles to the river. Along South Kaibab Trail you're likely to encounter wood rats, rock squirrels, and century plant (i.e., Utah agave).

Unlike other major trails in GCNP, which were developed from trails made by animals and Native Americans, South Kaibab Trail was extensively engineered, often by dynamiting the trail out of sheer cliffs. South Kaibab Trail has a uniform width of about 4 feet, has a grade less than 17 percent, and follows ridges and canyon walls rather than a fault or side canyons. As a result, all but less than a mile of the trail is usually in the sun, which minimizes the accumulation of snow and ice on the trail during winter. At the Colorado River, South Kaibab Trail connects with the North Kaibab Trail, thereby providing hikers and mule trains with the most direct access to Phantom Ranch on the opposite side of the river.

Construction of South Kaibab Trail began in December 1934. CCC crews, working simultaneously from the rim and river, finished the work when they met near the middle of the trail the following June. Construction was supervised by Miner Tillotson, who in 1927 became superintendent of GCNP. The trail goes through eight geologic formations before ending in Vishnu Schist in the Inner Gorge.

South Kaibab Trail's trailhead (elevation 7,200 feet) is on the west side of Yaki Point. There, hikers find a mule barn, water, bus stop, parking lot, and a residence for NPS rangers, which were built between 1925 and 1929. You cannot access the trailhead by car; to get there, you must either hike or take a GCNP shuttle bus.

The trail begins with a steep descent along switchbacks called The Chimney. At the bottom of the first set of these switchbacks, the trail turns north through the Toroweap Formation. There is then a steep descent through 330 feet of Coconino Sandstone to Ooh Aah Point, which provides a panoramic view of Grand Canyon. At this point, you are about 600 feet below the rim. To the west, hikers see how Pipe Creek has cut through the gorge.

Unlike Bright Angel Trail, which takes hikers through Bright Angel Canyon, South Kaibab Trail offers spectacular ridgeline views of Grand Canyon. In this photo, a mule train taking supplies to Phantom Ranch descends switchbacks near the top of South Kaibab Trail. In the background is O'Neill Butte. South Kaibab Trail, which was built by the National Park Service (NPS) to bypass Ralph Cameron's Bright Angel Toll Road, is one of the premier trails in the NPS. (Randy Moore)

Hikers then descend through the Hermit Formation and enjoy a variety of vistas before reaching a relatively flat area atop the Supai called Cedar Ridge. Interestingly, there are no true cedars at Cedar Ridge; most of the trees here are pinyon pine. At Cedar Ridge, there are bathrooms and a hitching rail for mules. To the north is O'Neill Butte, and to the east is Vishnu Temple. At Cedar Ridge, which is a popular turnaround point for day hikers, you have hiked 1.5 miles and are 1,140 feet below the rim.

Hikers then go east through the Supai Group and then down along the east side of O'Neill Butte. This formation is named for Buckey O'Neill, a pioneer at Grand Canyon's South Rim. South Kaibab Trail then levels off as it crosses Mormon Flat, an area named for the crew of Mormons who built this section of the trail. To the east, hikers see Cremation Canyon, where Havasupai cremated their dead.

Hikers then reach Skeleton Point, a site atop Redwall Limestone 2040 feet below the rim. West of here is the first view of the Colorado River along South Kaibab Trail. You can also see Bright Angel Trail entering the Inner Gorge and, across the river, the Phantom Ranch Ranger Station and campground, Cheops Pyramid, Isis Temple, and Shiva Temple in the distance. To the east is the North Rim's Wotans Throne and Cape Royal.

Switchbacks called the Red and Whites then descend 500 feet through Redwall Limestone, followed by Temple Butte, Muav Limestone, and Bright Angel Shale formations before the trail levels off at the Tonto Platform at Tipoff. This is the intersection of South Kaibab Trail with the Tonto Trail, the longest trail in Grand Canyon. Hikers can take a short trail from here to Panorama Point, a viewpoint 1200 feet above the Colorado River that provides views of Zoroaster Temple, the two suspension bridges crossing the river, and Bright Angel Creek. Tipoff is a 4.6-mile hike from Indian Garden. There is an emergency phone at Tipoff.

South Kaibab Trail then enters the 1,600-foot-deep Granite Gorge. Along the way, hikers pass through Tapeats Sandstone atop the Grand Canyon Supergroup. Near the river, the trail intersects the Colorado River Trail; from here, hikers can go west to Bright Angel Trail and the silver Bright Angel Suspension Bridge. A last set of switchbacks on South Kaibab Trail leads to a short tunnel and the black Kaibab Suspension Bridge, whose construction by NPS in 1928 depended on 42 Havasupai workers who carried 550-foot-long cables on their shoulders from the South Rim to the River. After crossing the river, hikers are on the North Kaibab Trail, which leads to Bright Angel Campground (less than a half mile away), Phantom Ranch, and the North Rim. On busy days, South Kaibab Trail often hosts more than 500 hikers.

Mule trains use South Kaibab Trail to take supplies to and from Phantom Ranch. Near the bridge are a cableway and gage house that mark the former site of the USGS gaging station (NPS manually monitored the Colorado River until an automated system was installed in 1977). Hikers also find a trail that passes Boat Beach

The 440-foot-long Kaibab Suspension Bridge, which carries hikers and supplies-laden mules across the Colorado River at River Mile 87.5, is about 0.5 miles from the junction of the South Kaibab Trail and the River Trail. After Civilian Conservation Corps workers painted the bridge black in 1933–1934, the Kaibab Suspension Bridge was called Black Bridge. About 0.75 miles downstream is Bright Angel (Silver) Suspension Bridge. (National Park Service)

> **Kaibab Suspension Bridge**
>
> Kaibab Suspension Bridge, also called Black Bridge, is a 440-foot-long suspension bridge near River Mile 87.5 that replaced Rust's Cable (and a subsequent replacement completed in 1921) as a main route linking the North and South Kaibab Trails. Although mules carried most of the 122 tons of materials needed to build the bridge into Grand Canyon, the 550-foot-long suspension cables—each weighing 2,030 pounds—were carried more than 9 miles (and down 4,000 feet) along the South Kaibab Trail on the shoulders of 42 Havasupai workers. Kaibab Suspension Bridge is 65 feet above the river and about 5 feet wide.
>
> Kaibab Suspension Bridge was completed in 1928, the same year as the Yavapai Observation Station. The first person to cross the bridge was Edith Kolb, the daughter of South Rim pioneers Emery and Blanche Kolb. An archaeological site near the Kaibab Suspension Bridge includes pit houses and a ceremonial kiva from about 1050 CE.

and the ruins of a 900-year-old pueblo, which consisted of five connected rooms and a separate kiva (i.e., round ceremonial room). Fewer than 20 people lived here.

Also nearby is the grave of Rees Bladen Griffiths (1873–1922), a NPS trail foreman who was killed in a rockslide in 1922 after dynamiting a nearby area while building the Kaibab Trail. Immediately after the accident, Chief Ranger Charles J. Smith and nurse Lillian Catts made a nighttime descent from the South Rim down Bright Angel Trail to help Griffiths. Smith and Catts arrived just before midnight, but Griffiths had died a few minutes earlier. Griffiths, who had worked earlier as a guide for David Rust, was wrapped in sheets from El Tovar Hotel, placed in a makeshift coffin, and buried in an alcove on the north side of the river a few hundred feet downstream from the Kaibab Suspension Bridge "in the Grand Canyon he loved so well."

South Kaibab Trail is one of the best trails in NPS. Between its trailhead and the river, South Kaibab Trail drops 4,740 feet. The trail is maintained and patrolled by GCNP, but there is no water along the trail. As a result, most rim-to-river-and-back hikers descend to the river along the South Kaibab Trail, and ascend the longer Bright Angel Trail, along which water is available.

See also: North Kaibab Trail; Phantom Ranch

Further Reading

Thybony, Scott. 2004. *Grand Canyon: South Kaibab Trail Guide*. Grand Canyon, AZ: Grand Canyon Association.

SQUIRRELS

Among the most interesting and entertaining animals at Grand Canyon are the Abert's and Kaibab squirrels. Abert's squirrels (*Sciurus aberti aberti*) live throughout most of the Colorado Plateau (and especially in New Mexico, southwestern Colorado, and the Grand Canyon). These vocal, engaging, tassel-eared squirrels are gray with a white underbelly and are closely associated with and largely confined to cool and dry forests of ponderosa pine. Unlike most squirrels, Abert's squirrels do not hibernate or store food, instead continuing to forage through the winter.

Kaibab squirrels (*Sciurus aberti kaibabensis*) live on at the North Rim of Grand Canyon in a 20-by-40-mile rectangle around Jacob Lake. These squirrels have a dark gray belly and a white tail, and live only in the ponderosa pine forests of the Kaibab Plateau. Both these squirrels' ancestors were once in the same population, but the formation of Grand Canyon created a barrier that isolated parts of this population. These isolated populations then evolved in different conditions to produce today's two distinctive types of squirrels. This is a classic example of allopatric speciation, in which biological populations of the same species become geologically isolated from each other and, as a result, do not interbreed.

In 1965, largely because of the Kaibab squirrel, NPS and the U.S. Forest Service designated 200,000 acres of the Kaibab Plateau at the North Rim the Kaibab Squirrel Area National Natural Landmark because of its "exceptional value as an illustration of the nation's natural heritage and [its] contribution to a better understanding of the environment." You can see Kaibab squirrels and a plaque designating this landmark at the U.S. Forest Service Kaibab Plateau Visitor Center just a few yards south of Jacob Lake Inn.

See also: North Rim

Further Reading

Friends of the Kaibab Squirrel. 2017. Retrieved June 3, 2017, at http://kaibabsquirrel.org

Whitney, Stephen R. 2010. *A Field Guide to the Grand Canyon*. Second edition. Seattle, WA: Mountaineers Books.

STAGECOACHES

Before the arrival of the railroad at Grand Canyon in 1901, stagecoaches were a common means of travel to and along the canyon. In 1885, William Bass began advertising the first stage line (two trips per week) to Grand Canyon from Ashfork, Arizona, and Sanford Rowe later offered service from Williams. However, most of the early stagecoach service to Grand Canyon originated in Flagstaff. There, the Atlantic and Pacific Railroad offered three stagecoach rides per week (each taking 4–6 hours and costing $20) to the Bright Angel Trailhead and John Hance's ranch near Grandview Point. By 1895, this service was taken over by James Thurber, who (with Lyman Tolfree) bought Hance's hotel and trail. After 1901, when the railroad provided a faster, cheaper, and more comfortable way to get from Williams to the South Rim, stagecoach service to and from other cities to Grand Canyon soon disappeared.

However, stagecoaches remained popular along the South Rim. For example, in 1911, Santa Fe Railway built and opened Hermit Trail, Hermit Camp, and Hermit Rim Road along the canyon's edge. The railroad used stagecoaches to carry tourists along "a most unique scenic roadway . . . to the head of Hermit Trail, a distance of about 9 miles." The road had been designed by architect George E. Kessler and the U.S. Forest Service for horses and open-top coaches. Cars, which frightened the stagecoach's horses, were not allowed until 1919, when Hermit Camp became a premier destination at Grand Canyon. The Fred Harvey Company offered Harvey Car Tours to Hermits Rest for $3 and to Desert View for $8.

The large number of cars on Hermit Road after 1919 quickly damaged the road. This prompted GCNP to repair and widen the road in the 1920s, and in 1934–1935 to rebuild the 12-foot-wide road (then called West Rim Drive). New parts of the road joined the refurbished old road near Hopi Point. Parts of the original road that were bypassed are now part of the rim trail between Hopi Point and Hermits Rest. The so-called West Rim Road was the last major road built at GCNP.

At the North Rim, the Union Pacific Railroad offered stagecoach tours that also included Bryce and Zion National Parks, and the Parry Brothers operated stagecoach rides to Cape Royal. A model of one of the stagecoaches common at Grand Canyon sits atop the fireplace in El Tovar Hotel.

Stagecoaches remained common along the South Rim until the 1920s. In 1973, after more than 1.9 million visitors came to GCNP, park officials began a shuttle bus service along Hermit Road. This was one of the first uses of a transit service in NPS.

See also: Gateways to Grand Canyon; Railroad; Roads

Further Reading

Lago, Don. 2014. *Canyon of Dreams: Stories from Grand Canyon History.* Salt Lake City: University of Utah Press.

STANTON, ROBERT

Robert Brewster Stanton (1846–1922) was a mining consultant and civil engineer who, from 1872 to 1884, helped plan and build railroads in the American South (through the Cumberland Mountains from Cincinnati to Chattanooga) and West (through the Rocky Mountains). In 1889, he was hired by Frank Brown as a private consultant to Brown's Denver, Colorado Canyon and Pacific Railroad Company to determine the feasibility of building a railroad near water level along the Colorado River through Grand Canyon. Stanton considered that expedition, which was only the second complete trip through Grand Canyon, as an ordinary railway survey but dangerous because of its route.

Brown was impatient; he gave Stanton only 10 days to organize and launch the expedition. He also offered Stanton $10,000 if he could complete his report about the feasibility of the railroad within 30 days. For his part, Stanton questioned Brown's appreciation of the risk associated with boating through Grand Canyon. Stanton urged Brown to use sturdy boats and life jackets, but Brown provided cheap boats and no life jackets. This especially concerned Stanton, who like John Wesley Powell, was disabled (Stanton had a nearly nonfunctional left arm) and could not swim. The underprepared expedition was postponed after three crewmembers (including Brown) drowned.

Six months later, Stanton returned to Grand Canyon to finish his expedition. He returned with life jackets and better boats, whose construction in Illinois was supervised by Harry McDonald. These three boats, each 22 feet long, were made of oak and had several airtight compartments. On this trip, with McDonald as head boatman, Stanton was thrown out of his boat at Separation Canyon and was saved by his life jacket. Stanton and six others completed the expedition. Like Powell, Stanton was amazed and inspired by what he was seeing, noting that he stood "spellbound in wonder and admiration, as firmly as I was fixed in the first few miles in surprise and astonishment" (Stanton, 1932).

Stanton knew that he would need photographs of Grand Canyon to convince people to invest in the proposed railroad, so he took photographer Franklin Asa Nims on the expedition to systematically document the proposed route of the railroad. However, when Nims broke his jaw, right leg, and both feet in a 20-foot-fall on New Year's Day, 1890, Stanton hiked more than 20 miles to get help and then became the expedition's photographer. Nims had taken fewer than 300 photographs, and Stanton took hundreds more, despite the fact that he knew relatively little about photography. Stanton's photographs provide almost the only existing data for how the canyon looked then as compared to now. Stanton named several sites and formations in Grand Canyon, including the following:

Elves Chasm is a nearly tropical chasm filled with ferns and orchids at Mile 116.5 named for its colorful travertine formations. The chasm is a short, steep hike from the river, where hikers can see the falls (water from Royal Arch Creek) flowing from Tapeats Sandstone. Early settlers fancied that the area was mined by elves.

Hansbrough Point at Mile 44 honors Peter Hansbrough, who drowned at 25 Mile Canyon with Henry Richards in the ill-fated Brown Expedition of 1889. Hansbrough's remains were found the next year and buried below President Harding Rapid (and opposite what is now called Hansbrough Point, formerly Small Point) under a nearby headstone of Muav Limestone in which his friends carved Hansbrough's initials, "PMH 1889."

Upstream from Hansbrough's grave is a ring of stones denoting the grave of 15-year-old David Quigley, a Boy Scout from Monticello, Utah, who drowned 60 miles upstream in Glen Canyon and whose body was found in driftwood near here on July 10, 1951. Quigley, whose body was found by Laphene "Don" Harris, was not wearing a life jacket.

Stanton left Grand Canyon believing that a railroad could be built through the canyon: "The line as proposed is neither impossible nor impracticable, as compared to some other transcontinental railroads, could be built for reasonable cost" (Stanton, 1987). According to Stanton's plan, which required 20 miles of tunnels, the railroad was to run on electricity generated by dams on the river, its bridges were to be built from wood harvested from the rim, and its cost was to be $51 million. Powell opposed the railroad, but the optimistic Stanton bought stock in the company and served on the board of directors. Stanton's optimism was unfounded, however, and the financial panic of 1890 ended prospects for raising the money needed for the project. Stanton's book titled *Availability of the Cañons of the Colorado River of the West for Railroad Purposes* (1892) outlined many of his ideas, but the railroad was never built; discoveries of oil in California and elsewhere eliminated the need for a railroad to take coal to California for heating and other purposes.

Despite his contributions, Stanton was hesitant to claim discoveries at Grand Canyon; as he noted:

> I went down the Colo. in the ordinary discharge of my everyday duties as a railroad engineer for the purpose of locating a RR line along its banks and through its cañons. We were not exploring the Grand Cañon, nor seeking adventure in it, nor looking for anything except that [railroad] line. Therefore I think it out of place to speak even of the somewhat startling adventures that we did meet with which resulted in the death, by drowning, of three of my companions, and the maiming of a fourth. (Stanton, 1987)

Stanton's expedition through Grand Canyon, like John Wesley Powell's, did little to promote whitewater boating in Grand Canyon. In future decades, however, boating would become a popular adventure for tens of thousands of tourists per year.

On his trip downstream, Stanton and McDonald (like Powell) had noted placer gold in some of the Colorado River's sandbars. In 1897, Stanton created Hoskininni Mining Company, a gold mining operation in Glen Canyon that operated from 1897 to 1902. This venture was financed by industrialist Julius Stone, the first person to go down the Colorado River through Grand Canyon for sport, and was named after a Navajo Chief who claimed there was gold along the Colorado River. The company built a giant 180-ton, 82-bucket dredge powered by five gas engines in Glen Canyon. However, the gold particles were too small to extract, and Hoskininni Mining Company failed. The $100,000 investment provided by Stone and others produced revenues of $66.95. The company's assets sold for $29.86.

McDonald staked the Morning Star Mine and other claims in 1891 along Chuar Creek near River Mile 65.5, and Stanton eventually worked mines in the United States, Canada, Mexico, and the West Indies. Stanton's report in 1890 that Grand Canyon contained many minerals triggered a rush of miners to the area.

After his trips through Grand Canyon, Stanton became keenly interested in the history of the region. In 1907, he began writing what became a 1,039-page manuscript originally titled *The River and the Canyon: The Colorado River of the West, and the Exploration, Navigation, and Survey of Its Canyons, from the Standpoint of an Engineer* about the history of the Colorado River. The manuscript describes the history of the Colorado River (until 1871), Stanton's expedition on the river, Stanton's interview with James White (Stanton doubted White's story about being the first person to raft through the canyon in 1867), and the expedition through Grand Canyon by Emery and Ellsworth Kolb in 1911–1912. Stanton finished the manuscript in 1920, but portions of it were not published until 1932 (10 years after Stanton's death) as *Colorado River Controversies*. Stanton, like South Rim pioneer William Bass, distrusted John Wesley Powell and criticized Powell by dedicating his book "To All Truthful Colorado River Voyagers." In that book, for which Julius Stone wrote the foreword, Stanton criticized John Wesley Powell's exaggerated claims about his two expeditions through Grand Canyon. Stanton also felt that the plaque at Powell Memorial was a disservice to the three men who left Powell's first expedition two days before it ended. Stanton's friend Julius Stone later helped post a plaque at Separation Canyon that honored the men and cited Stanton's book.

More of Stanton's manuscript was published in 1965 as *Down the Colorado* (edited by Dwight L. Smith), and his diary of the 1889 and 1889–1890 expeditions (edited by Smith and historian C. Gregory Crampton) in 1987 as *The Colorado River Survey*. A manuscript by Franklin A. Nims, the photographer with Stanton who was injured in Grand Canyon and left the expedition, was also edited by Smith and published in 1967 as *The Photographer and the River, 1889–90*. Much of the material in Stanton's original manuscript has now been published.

Stanton is honored by the South Rim's Stanton Point, a spire of rock on the South Rim just northwest of Mile 127 Creek, and by Stantons Cave, a cave in Redwall Limestone on river right at River Mile 31.8 where Stanton cached his boats and supplies after ending his ill-fated expedition in 1889. In 1934, river runner Bus Hatch and his companions found several 4,000-year-old, 5-inch-long split-twig figurines in this cave, which were made from willow twigs, split lengthwise, and then wrapped and twisted to resemble animals hunted by residents. (Small spears pierced some of the figurines.) The figurines were assumed to be children's toys. Subsequent excavations at the cave unearthed bones of ancient horses, Harrington goats, humpback chubs, California condors, extinct vultures, and other birds (and more than 100 more figurines).

Stanton died of pneumonia on February 23, 1922, in New Canaan, Connecticut, and is buried in Fairfield Cemetery there.

See also: Thematic Essays section: The Colorado River's Historic Sites; Thematic Essays section: John Wesley Powell; Brown, Frank

Further Reading

Nims, Franklin A. 1967. *The Photographer and the River, 1889–90: The Colorado Cañon Diary of Franklin A. Nims with the Brown–Stanton Railroad Survey Expedition*. Edited by Dwight L. Smith. Santa Fe, NM: Stagecoach.

Stanton, Robert Brewster. 1932. *Colorado River Controversies*. Edited by James M. Chalfant, with a foreword by Julius F. Stone. New York: Dodd, Mead.

Stanton, Robert Brewster. 1987. *Colorado River Survey: Robert B. Stanton and the Denver, Colorado Canyon & Pacific Railroad*. Edited by Dwight L. Smith and C. Gregory Crampton. Salt Lake City, UT: Howe Brothers.

Stanton, Robert Brewster. 1890. *Through the Grand Canyon of the Colorado*. New York: C. Scribner's Sons.

STONE, JULIUS

Julius Frederick Stone (1855–1947) was raised as one of 11 children in Lenawee County, Michigan. Although he had only six years of formal education, he enjoyed a successful career as an executive with railroads, mining companies, and other businesses.

In 1909, Stone hired boatman Nathaniel Galloway to guide him and others (C. C. Sharp, Seymour Dubendorff, and Raymond Cogswell) down the Colorado River through Grand Canyon. Whereas previous expeditions through Grand Canyon had been conducted for exploration, trapping, and scouting a route for a potential railroad, Stone's expedition was simply for sport. This expedition made Stone the first river runner to run the Grand Canyon just for fun. On that trip, Stone rowed his own boat but was extremely impressed with Galloway's boating skills. The Stone–Galloway boat (the so-called Stone Boat), which was made of pine and elm, is the oldest extant boat to have run the Colorado River through Grand Canyon. The Stone boat was donated in 1950 (with the permission of Stone's daughter) to GCNP and is now part of the park's collection of historic boats. The Stone–Galloway expedition, which included photographer Raymond Cogswell and 27-year-old boatman Seymour Dubendorff, was the seventh complete transit of Grand Canyon.

In early 1932, after promising not to write a book about Grand Canyon, Stone published *Canyon Country, The Romance of a Drop of Water and a Grain of Sand.* Stone's book, the jacket of which advertised Robert Stanton's upcoming *Colorado River Controversies* (1932), lamented "the undeserved stigma of cowardice" that Stone believed had been attributed to the three members of John Wesley Powell's 1869 expedition through Grand Canyon. Stone's book also included an edited version of Stone's diary, numerous photographs, and a chapter about erosion written by Willis Lee of USGS. Stone dismissed James White's claim of having traversed the Grand Canyon as being supported only by ignorant, irrational people. In 1938, Buzz Holmstrom ran Grand Canyon in the same boat he used the previous year, but he renamed it *Julius F* to honor benefactor Julius Stone.

On August 28, 1939, on the 70th anniversary of the splitting of John Wesley Powell's expedition, Stone—accompanied by his son George, river runner Willis Johnson, Frank Swain, and Russell Frazier—helped mount a plaque replacing the one that was wired in place by members of the Dusty Dozen expedition in 1934 at Separation Rapid. That plaque recognized the names of William Dunn, Oramel Howland, and Seneca Howland, who completed all but the last two days of Powell's 1869 expedition through Grand Canyon and whose names are not listed on the plaque displayed at the Powell Memorial. (Frazier took home the original plaque from Separation Canyon, which he had helped mount five years earlier.) Today, Separation Rapid is covered by Lake Mead. In 1943 and 1944, Stone's plaque was raised 40 feet to its current location. The holes used to mount the earlier plaque remain visible nearer the river.

The 92-year-old Stone died on July 25, 1947. He is honored by Stone Creek, which enters the Colorado River from river right near River Mile 132.4.

See also: Thematic Essays section: John Wesley Powell; Galloway, Nathan; Stanton, Robert

Further Reading

Stone, Julius F. 1932. *Canyon Country, The Romance of a Drop of Water and a Grain of Sand.* New York: G. P. Putnam's Sons.

STRONG, HARRIET

The famed USGS Birdseye Expedition in 1923 identified more than 20 potential sites for dams in Grand Canyon. These dams would have created stair-step "cash register" reservoirs that could have generated hydroelectric power used to pay the costs of the dams and future reclamation.

The first stair-step dams in Grand Canyon were proposed to Congress in 1918 by 73-year-old inventor Harriet Williams Russell Strong (1844–1926). Strong opposed the commercial development of Grand Canyon, including the construction of power-generating and flood-control dams by for-profit companies. Instead, Strong urged the House Committee on Water and Power to set aside Grand Canyon for government projects, including dams to provide water for irrigation.

Strong's proposals, which were documented in her "Dam and Reservoir Construction" (U.S. Patent #374,378) and elsewhere, were never implemented. Strong continued to promote women's education. She also founded the Business League of America and was the first woman delegate to the Chamber of Commerce of the United States.

Strong died in an automobile accident on September 6, 1926, in Los Angeles, California. She is buried in the Great Mausoleum's Sanctuary of the Beatitudes (Crypt 3035) at Forest Lawn Memorial Park in Glendale, California.

See also: Tramways

Further Reading

National Women's Hall of Fame. 2017. Harriet Williams Russell Strong. Retrieved May 3, 2017, at https://www.womenofthehall.org/inductee/harriet-williams-russell-strong/

STURDEVANT, GLEN

Glen Ernest Sturdevant (1895–1929), a geology graduate of the University of Arizona and a veteran who served in World War I, joined GCNP in 1925 as the park's first naturalist. While at GCNP, Sturdevant organized nature walks and campfire lectures for visitors, and—with the help of his wife Eileen—began publishing the monthly bulletin *Grand Canyon Nature Notes*. Sturdevant also began what would become GCNP's interpretive program.

On the morning of February 20, 1929, Sturdevant, GCNP Ranger Fred Johnson, and Chief Ranger James P. Brooks were returning from a 10-day trip in Grand Canyon in which they had collected a variety of biological and geological specimens for study. Just above Horn Creek Rapid, their flimsy boat became caught in an eddy, and Johnson was thrown into the river. Brooks jumped in to rescue Johnson but was carried downstream. During the rescue attempt, Sturdevant was also thrown from the boat.

Brooks washed ashore downstream on the north side of the river but could not find either Johnson or Sturdevant. After Brooks climbed out of the canyon and told authorities about the accident, GCNP organized four search parties (one of which Brooks was a member of) to search for Johnson and Sturdevant. Sturdevant's body was recovered 2 miles below Horn Creek. GCNP authorities and others searched for Johnson's body for several days and even dynamited Horn Creek Rapid (a practice no longer allowed in GCNP), but Johnson's body was never found.

On February 26, 1929—the 10th anniversary of the creation of GCNP—GCNP Superintendent Miner Tillotson spoke at a military funeral for Sturdevant at the Community Building. A similar service was held for Johnson at Powell Memorial Point. Sturdevant (a charter member of John Ivens American Legion Post 42) was buried at Pioneer Cemetery on the South Rim. A cenotaph honoring Johnson (1898–1929) stands beside Sturdevant's grave. Sturdevant and Johnson are honored in Grand Canyon by Sturdevant Point and Johnson Point (southward projections of Buddha Temple) near River Mile 87.5. (*Note:* The Johnson Point near Lees Ferry is named for Warren Marshall Johnson, who—with Joseph L. Foutz—operated Lees Ferry from 1876 to 1896.)

Glen Sturdevant, shown here with his son in 1926, was Grand Canyon National Park's (GCNP) first naturalist. In 1929, while returning from a 10-day research trip, Sturdevant and fellow ranger Fred Johnson drowned near Horn Creek (River Mile 90.5). Sturdevant, who is buried in the South Rim's Pioneer Cemetery, began what became GCNP's interpretive program. (Grand Canyon National Park Museum Collection)

See also: Pioneer Cemetery and Other Gravesites

Further Reading

Anderson, Michael F. 2000. *Polishing the Jewel: An Administrative History of Grand Canyon National Park.* Grand Canyon, AZ: Grand Canyon Association.

SUPAI AND THE WATERFALLS OF THE HAVASUPAI

Grand Canyon includes many spectacular waterfalls, but seeing them often requires significant effort and time. The easiest falls to see are those at Roaring Springs, which is 4.7 miles down the North Kaibab Trail. These 400-foot-high waterfalls at the head of Bright Angel Creek is the source of all water used at GCNP, including that at the North and South Rims, Phantom Ranch, and Bright Angel Campground. As its name suggests, Roaring Springs is often loud.

The waterfalls most accessible from the South Rim are not in GCNP but instead are near the village of Supai on the Havasupai Reservation in southwestern Grand Canyon. Supai is the only permanent community within Grand Canyon, and the Havasu Trail through Hualapai Canyon is the only public, maintained trail to Supai. Compared to most other trails in Grand Canyon (e.g., Tanner Trail, Lava Falls Trail), the 8-mile Havasu Trail to Supai is relatively easy. Except for its descent to the wide canyon floor, the trail follows a creek bed and stream.

Havasu Trail begins at Hualapai Hilltop (elevation 5,200 feet), which is at the end of Indian Road 18, 193 miles (by road) from Grand Canyon Village and 68 miles north of Peach Springs, Arizona (on old Highway 66, north of Interstate I-40). From Hualapai Hilltop, hikers traverse steep switchbacks through Coconino Sandstone and then across a long talus slope covering Hermit Shale. Where there is enough soil, hikers see blackbrush (*Coleogyne ramosissima*). Far down the canyon, Mount Sinyella comes into view. There are no services at Hualapai Hilltop.

From Hualapai Hilltop, hikers go 6.5 miles down the Havasu Trail through Havasu Canyon (a tributary canyon on the south side of the Colorado River) to Havasu Creek. The water at Supai is laden with calcium carbonate, magnesium carbonate, calcium sulfate, and other minerals that give the water of Havasu Creek a beautiful bluish-green color (Havasupai means "people of the blue-green waters"). These colors, when contrasted with the maroon and travertine landscapes, provide spectacular views. At its junction with Havasu Creek, the dry desert habitat is transformed into a green oasis dominated by cottonwood, seep willow (*Baccharis*

Havasu Falls, which is one of Grand Canyon's premier destinations, is a short hike from Supai, Arizona. The beautiful, 90-foot-high waterfall is one of Grand Canyon's most photographed sites. (Terry Eiler/National Archives)

salicifolia), and arrowweed (*Pluchea sericea*). After turning left, visitors hike 1.5 miles along the creek to a beautiful oasis called Supai. In this village, the trail widens to a sandy lane shaded by cottonwoods. This "main street" is wide enough for two packhorses to pass. At Supai, which is approximately 3,000 feet below the rim, the elevation is 3,195 feet.

The Havasupai are the only indigenous people of Grand Canyon who continue to live in the canyon; their current reservation abuts the western side of GCNP. Supai, which the Department of Agriculture has referred to as the "most remote community" in the contiguous United States, has a population of approximately 500 enrolled tribal members (when Garcés met the Havasupai in 1776 on his way to see the Hopis, there were 34 families). Supai is guarded by two towering pillars of stone called *Wii gl'iiva* (also *Wig-Li-I-Wa*); these spiritual landmarks are the guardian spirits of the Havasupai (they are also called the Prince and Princess, and the God and Goddess). According to legend, the settlement here will disappear when these pillars fall.

Supai, which is an oasis bordered by walls of Havasu Canyon, has a school, stables, clinic, rodeo grounds, post office, more than 120 houses, church, lodge, café, and a sandy main street. Supai is also the administrative and social center of the Havasupai tribe. The area has often been damaged by flashfloods; for example, floods in 1910 and 1990 increased the flow of the creek from 65 cubic feet per second (cfs) to more than 22,000 cfs and, in the process, uprooted trees, rerouted waterfalls, and damaged buildings.

Supai, which covers 1.7 square miles, is open year-round, but visitors can get there only by foot, horseback, or helicopter. Supai, whose economy is based on tourism, was named in 1906 for its post office in the mistaken belief that *supai* was a Havasupai word. The post office at Supai (Zip code 86435), which opened in 1901, is the last post office in the United States that transports mail by horses or mules. (Mail to and from Phantom Ranch is transported by concessionaires, not the U.S. Postal Service.) There are no cars in Supai.

Most of the more than 20,000 visitors who visit Supai each year come to see the nearby waterfalls, none of which can be accessed by automobile. The closest of these falls are Upper Navajo Falls (50 feet high, also called New Navajo Falls), which is 1.5 miles away, and Lower Navajo Falls (40 feet high). (The Navajo Falls are named for Chief Navajo, a Havasupai leader who died in 1898.) Less than a mile beyond Navajo Falls is Havasu Falls (also called Havasupai Falls and Bridal Veil Falls), at which a 90-foot-high waterfall pours into a luminous blue pool; this is one of GCNP's best backpacking destinations. A trail built by CCC workers descends the west side of the falls, and, below the falls, travertine dams have formed several stepped pools. This waterfall, which hikers hear before they can see it, is the most photographed waterfall in Grand Canyon.

Hiking another 2 miles leads visitors to Mooney Falls (196 feet high), the highest of the waterfalls in Havasu Canyon. Mooney Falls was named for James (also referred to as Daniel) Mooney, a sailor-turned-prospector who died here in 1880 while trying to descend 330 feet to the bottom of Havasu Canyon. (Mooney had earlier found lead and silver near the base of Havasu Falls.) Near Mooney Falls, which is taller than Niagara Falls, vanadium was mined until 1910, when a flash-flood destroyed the mining operation. Getting to the azure water of Mooney Falls includes some narrow ledges and two tunnels with steep steps. The trail here was upgraded in 2009 when a chain railing and stanchions replaced the chain that had been there since 1883. Mooney Falls was the most sacred waterfall of the Havasupai, who called it ha *Gv'oma* ("End Waterfall"). Although no camping is allowed below Mooney Falls, the Havasupai Campground was named the best campground destination in 2017 by *Travel and Leisure* magazine.

Four often challenging miles beyond Mooney Falls is a series of cascades collectively called Beaver Falls, named for a beaver that once lived there. The Hualapai Reservation ends just beyond Beaver Falls; from here on, Havasu Creek flows into GCNP. At Beaver Falls, the Hualapai Reservation ends, and hikers are back in GCNP.

Past Beaver Falls, the creek cuts into the Temple Butte Formation. Below the Temple Butte is the Tonto Group, but only the Muav limestone is exposed. The blue color of the water diminishes past Beaver Falls, and the scenery becomes dominated by bare rocks and desert vegetation.

Just before reaching the Colorado River, Havasu Canyon narrows before emptying Havasu Creek into the river just above Havasu Rapid near River Mile 157. During summer months, the mouth of Havasu Creek is often clogged with rafts. A hike from Hualapai Hilltop to the Colorado River spans 17 miles.

See also: Gateways to Grand Canyon

Further Reading

Hirst, Stephen. 2007. *I Am the Grand Canyon: The Story of the Havasupai People*. Third edition. Flagstaff, AZ: Grand Canyon Association.

Ostroff, Hannah S. 2016. In the Grand Canyon, the U.S. Postal System still delivers mail by mule. *Smithsonian Insider*, August 26. Retrieved May 16, 2017, at http://insider.si.edu/2016/08/grand-canyon-u-s-postal-service-still-delivers-mail-mule

Smithson, Carma Lee, and Euler. 1994. *Havasupai Legends: Religion and Mythology of the Havasupai Indians of the Grand Canyon*. Salt Lake City, UT: University of Utah Press.

Witt, Greg. 2010. *Exploring Havasupai: A Guide to the Heart of the Grand Canyon*. Birmingham, AL: Menasha Ridge Press.

T

TANNER TRAIL

The rugged Tanner Trail, which begins just east of Lipan Point, is a steep, 9-mile, unmaintained trail that drops 4,750 feet along its way to the Colorado River. This makes Tanner Trail the longest trail to the river from GCNP's South Rim. The trail, which originally included parts of Anasazi and Hopi Indian routes into Grand Canyon, is named for prospector Seth Benjamin Tanner (1828–1918), a Mormon frontiersman who accompanied Brigham Young to "the promised land" near Great Salt Lake in 1847. After working as a scout, Tanner settled near Grand Canyon in 1876. The following year, Tanner filed a claim for Tanner Ledge, one of the first claims made inside Grand Canyon. Tanner improved an Indian trail (then called the Bunker Trail, for Tanner's colleague Fred Bunker) and used it to access his copper and silver mines on river left near River Mile 65.5 at Palisades Creek. Just down-canyon from here is where Horsethief Trail (consisting of the Tanner and Nankoweap Trails) crosses the Colorado River. During Prohibition, this trail was used to transport booze from Arizona to Utah; it's also where GCNP naturalist Glen Sturdevant found a distillery at an abandoned campsite in 1928. Nearby are some of the oldest sedimentary rocks in Grand Canyon. These rocks, collectively known as Grand Canyon Supergroup, are colorful rocks 0.8–1.2 billion years old and identified by their 20-degree tilt.

The part of Tanner Trail from Lipan Point to the top of the Redwall Limestone was built by Tanner and Franklin H. French, who married Emma Lee (the widow of John Lee, the original operator of Lees Ferry). The trail was later known as the Tanner–French Trail and the French Trail but today is referred to as Tanner Trail. Tanner's trail was the southern part of the Horsethief Trail. Construction of the Nankoweap Trail, which is the northern part of the Horsethief Trail, was overseen in 1882 by John Wesley Powell and Charles Walcott. (The unmaintained Nankoweap Trail, which drops 6,040 feet in the 14 miles from Saddle Mountain Trailhead to the Colorado River, is one of the most challenging of the trails into Grand Canyon.) The Tanner Trail is probably where García López de Cárdenas became the first known non-Native to encounter Grand Canyon.

Tanner Trail descends quickly with steep switchbacks through Kaibab Limestone, Toroweap Formation, and Coconino Sandstone. Some of these switchbacks, especially in the Coconino Sandstone, are among the steepest in GCNP. Soon thereafter, hikers reach Escalante and Cardenas Buttes, and then the break in Redwall Limestone, Muav Limestone, and Bright Angel Shale that lead into the lower canyon. There, the trail levels out and heads north along a ridge of Tapeats Sandstone, after which it descends toward Tanner Rapid, Tanner Delta, and the

Colorado River. The final parts of Tanner Trail are picturesque and lead through Dox Sandstone to an area where unsubstantiated legends claimed that John Lee buried several pots of gold. Downriver you'll find Hance Rapid, Red Canyon Trail, an overlook of Unkar Rapids, and the start of Tonto Trail. Upriver along the Beamer Trail is the confluence of the Colorado and Little Colorado Rivers.

Tanner was the first European American (aside from García Lopez de Cárdenas and Francisco Tomás Garcés) to explore Grand Canyon's southeastern rim. In 1880, Tanner organized the Little Colorado Mining District just east of Grand Canyon. He and other prospectors found minerals there, but high costs prohibited the shipping of the ores elsewhere.

Tanner later went blind and died in Taylor, Arizona, in 1918. He is honored by Tanner Wash at River Mile 14.5, a site that was earlier called Roundy Wash (for Mormon Bishop Lorenzo W. Roundy, who drowned while trying to cross the river at Lees Ferry in May 1876). Tanner Trail was abandoned in the 1930s and today is rarely maintained. The strenuous trail, which offers great views but no water, is the primary access by foot into eastern Grand Canyon.

See also: Lipan Point

Further Reading

Abbott, Lon, and Cook, Teri. 2004. *Hiking the Grand Canyon's Geology*. Seattle WA: Mountaineers Books.

TILLOTSON, MINER

Miner Raymond "Tillie" Tillotson (1886–1955) was a civil engineer and geology major at Purdue University who, after graduation, began working for the U.S. Forest Service. He came to Grand Canyon in 1920 as an engineer and, in April, 1927, became GCNP superintendent. Two years later, he (with Frank J. Taylor) published *Grand Canyon Country* (1929). He served as GCNP's superintendent until December 1938, after which he continued his NPS career in Richmond, Virginia, and Santa Fe, New Mexico. Tillotson was the first long-term, important superintendent of GCNP.

During Tillotson's time as superintendent, GCNP built or completed (often with the help of the CCC) several major sanitation structures, roads, trails, and bridges, including the North and South Kaibab Trails, the Trans-Canyon Telephone Line, new GCNP headquarters, a relocation of part of West Rim Drive, additional housing at Grand Canyon Village, GCNP hospital, and the Colorado River Trail, Upper Ribbon Falls Trail, and Clear Creek Trail. Like Grand Canyon pioneer William Bass, Tillotson believed that Grand Canyon had been formed by an earthquake that left a gash in the earth that was later filled by the Colorado River.

Tillotson changed how GCNP manages its wildlife. For example, in 1925, GCNP moved a small population of pronghorn (*Antilocapra americana*) to the Tonto Platform. (These antelope are native to the Coconino Plateau but had never been reported in GCNP.) Thanks to feeding by GCNP, the population soon grew from

Miner "Tillie" Tillotson served as Grand Canyon National Park (GCNP) superintendent from 1927–1938, during which time he supervised construction of the North and South Kaibab Trails, village housing, GCNP hospital, sanitation facilities, Trans-Canyon Telephone Line, and other major improvements in the park. In 1931, Tillotson also ended the long-established practice of killing predators in GCNP. This photo from 1930 shows Tillotson relaxing on the patio wall of the North Rim's Grand Canyon Lodge. (Grand Canyon National Park Museum Collection)

9 to 28 animals. When Tillotson ended the feeding of all deer in GCNP in 1934, however, the population disappeared, thereby returning the Tonto Platform to its previous, more pristine condition. Similarly, Tillotson announced in 1928 that GCNP would no longer try to eliminate predators (as had wardens such as Uncle Jim Owens) but instead hunt them "as needed" as part of a "partial extermination" to minimize threats to deer and other game animals. In 1931, Tillotson ended the practice of killing predators in GCNP. However, he continued to stock streams (e.g., Bright Angel Creek, Tapeats Creek) in GCNP with invasive species such as trout to enhance fishing in GCNP; as he announced in 1931, all "favorable trout streams of the Park are now well-stocked." The stocking of these streams ended in 1964 with the building of Glen Canyon Dam.

Tillotson is buried in lot 1 of section 1 beside his wife Winifred at the South Rim's Pioneer Cemetery. Tillotson's tombstone features a quote from a 1902 poem by Rudyard Kipling titled "The Palace," which contemplates our small and temporary place in life.

See also: Civilian Conservation Corps; Pioneer Cemetery and Other Gravesites; Trans-Canyon Telephone Line

Further Reading

Anderson, Michael F. 2000. *Polishing the Jewel: An Administrative History of Grand Canyon National Park*. Grand Canyon, AZ: Grand Canyon Association.

TIYO

Tiyo was the son of a Hopi tribal chief who wondered about the fate of water of the Colorado River. With the help of his village, Tiyo set out on a river voyage in a hollowed-out trunk of a cottonwood tree. Tiyo was guided by Spider Woman, who led him to the Land of Snakes at the mouth of the Colorado River. Tiyo met The Snake People, who gave Tiyo the secrets to bringing rain for their crops that sustain life. Tiyo later married Snake Woman, and their children became the Snake Clan of Walpi. The legend of Tiyo is presented in a mural by Hopi artist Fred Kabotie at Desert View Watchtower.

The Tiyo legend is among the earliest explanations of the ceremonial Hopi Snake-Antelope Dance, which is celebrated every August (or early September) at several Hopi villages. Hopi believe that their Snake Dance honors the spirits of their ancestors who return in the form of clouds and bring rain. Snakes are the brothers of these spirits; hence the name "Snake Dance."

Tiyo is commemorated in GCNP by Tiyo Point (formerly named Juniper Point) on river right near River Mile 91.5. This overlook, which is accessible via a forested path on the North Rim, provides a spectacular view of several Grand Canyon landmarks, especially of Shiva Temple.

See also: Desert View

Further Reading

Lago, Don. 2014. *Canyon of Dreams: Stories from Grand Canyon History*. Salt Lake City: University of Utah Press.

TONTO TRAIL

The Tonto Trail, which stretches for more than 70 miles, is the longest, continuous trail in GCNP. The trail crosses the gray-green, inner-canyon Tonto Platform on the south side of the Colorado River. The name of this platform, which is a prominent erosional tableland of Bright Angel Shale 3,000 feet below the South Rim, honors Tonto Apaches who lived in north-central Arizona until they were moved to the San Carlos Reservation in the 1870s.

Tonto Platform does not terminate at either rim; instead, it stretches from Garnet Canyon (about 10 miles west of South Bass Trail) in the west to Red Canyon and New Hance Trail in the east. Tonto Trail separates the upper canyon from the Inner Gorge. Except for Grandview Trail, all of the major trails that cross Tonto Trail (e.g., South Bass, Boucher, Hermit, Bright Angel, South Kaibab, and New Hance Trails) access the Colorado River.

In 1902, Ralph Cameron filed a claim for the River View Toll Road atop the Tonto Platform from Red Canyon to Slate Creek (west of Boucher Canyon). Unlike Bright

Clear Creek Trail

Clear Creek Trail, which was built in 1934–1935 by the CCC at a cost of $50,000 as a mule trail for visitors from Phantom Ranch, is the only trail traversing the Tonto Platform on the north side of the Colorado River. The trail is 8.4 miles long but nets only 4 river miles (River Miles 84–88). Clear Creek Trail branches off of North Kaibab Trail 0.4 mile north of Phantom Ranch and ascends 1150 feet via a series of switchbacks to the Tonto Platform, which it follows for more than 8 miles east. Tonto Trail goes around the south side of Zoroaster Temple and then drops into the Clear Creek drainage.

After finishing construction of Clear Creek Trail, CCC workers stocked Clear Creek with rainbow trout. Soon thereafter, Fred Harvey advertisements promoted the "excellent trout fishing" in the area. This trout fishing ended when floods washed the trout into the Colorado River. By the 1940s, GCNP stopped stocking the creek with trout. Mule trips to the area also ended, and the trail never became very popular.

Most visitors use Clear Creek Trail to access Grand Canyon's tallest waterfall, the 800-foot-high Cheyava Falls (*cheyava* is supposedly a Hopi word for "intermittent waters") that flows only in the spring after snow melt at the North Rim. Cheyava Falls is a seasonal waterfall north of Cape Royal in Redwall Limestone.

Angel Trail, this toll road never became profitable. Cameron, his brother Niles, and Pete Berry extended the Tonto Trail from Hance Creek west to Garden Creek at Indian Garden. Although these prospectors and developers, along with Louis Boucher and Dan Hogan, traveled on Tonto Trail west toward Hermit Creek, today's Tonto Trail from Bright Angel Trail to Hermit Trail was developed primarily by the Fred Harvey Company, which led tourists on a Hermit-to-Bright-Angel Loop from 1912 to 1930.

Tonto Trail is about 1,500 feet above the Colorado River. Tonto Trail intersects eight established rim-to-river trails, including the South Kaibab Trail at Tipoff; from here, hikers can go to Yaki Point on the South Rim, continue east or west on the Tonto Trail, or go to the Colorado River Trail and Phantom Ranch. The busiest section of Tonto Trail is the 4.6 miles between Bright Angel Trail and South Kaibab Trail. This part of the Tonto Trail is easy to follow. The second busiest part of Tonto Trail is the 12 miles linking Indian Garden and Hermit Camp. The least used and most remote section of the trail is the 30 miles linking South Bass Trail and Boucher Creek.

Although the Tonto Trail ends at Garnet Canyon, other trails continue along the river to Elves Chasm near Royal Arch Point. In the east, at Red Canyon, the Escalante Route extends east and connects to the Tanner Trail and Beamer Trail across from Unkar Creek. Camping along Tonto Trail (e.g., at Indian Garden and Horn Creek) requires a permit from GCNP's Backcountry Information Center and is on a first-come/first-served basis.

Except for the intersection of Bright Angel Trail and Indian Garden, there is no drinkable water on Tonto Trail. However, hikers can treat water from Hermit Creek,

Monument Creek, and Garden Creek, which flow year-round. Do not drink water from Horn Creek, which has been contaminated by Lost Orphan Mine (now closed) below Powell Point.

See also: Bright Angel Trail, South Kaibab Trail

Further Reading

Abbott, Lon, and Cook, Teri. 2004. *Hiking the Grand Canyon's Geology*. Seattle WA: Mountaineers Books.

TOROWEAP

The iconic Toroweap (Tuweep) Overlook—a site visited often by photographers and geologists but seldom by ordinary tourists—is a remote valley and overlook near River Mile 178 on the North Rim of western Grand Canyon that towers over a sheer, breathtaking 3,000 feet cliff above the Colorado River. Toroweap is the only place along either rim in GCNP that provides such a steep, vertical view of the Colorado River. The site is only 55 miles west of North Rim headquarters (as the condor flies) but is almost 150 miles by road. The term "Toroweap" is often used interchangeably with *Tuweep*, a Paiute word for "dry or barren valley." The Toroweap area is an ancestral home of the Southern Paiutes. Frederick Dellenbaugh described Toroweap as "the edge of the world," and famed geologist Clarence Dutton wrote;

> The scenery [at Toroweap] becomes colossal . . . [W]e feel like mere insects crawling along the street of a city flanked with immense temples . . . It would be difficult to find anywhere else in the world . . . such dramatic and inspiring surroundings. (Annerino, 2017)

Toroweap was part of Grand Canyon National Monument until 1975, when the monument was merged with GCNP. The overlook, which is 70 miles upriver from the Hualapai Skywalk, is famous because it offers views not seen from developed areas of the North and South Rims. However, seeing those views extracts a price; even in ideal conditions, the dramatic, uncrowded views from Toroweap can be accessed only by negotiating extremely difficult roads. The most common route to Toroweap—called the Sunshine Route—is a turnoff (Bureau of Land Management Road 109) 7.5 miles west of Fredonia (and 6 miles east of Pipe Spring National Monument) on Arizona State Route 389. After the turnoff, you drive 61 miles on dirt roads to the rim. It takes at least 2 hours to drive these 61 miles of washboard roads, giant potholes, and sharp rocks. Six miles from the rim is a GCNP ranger station on the left; the ranger (as well as volunteers) here patrols more than 200,000 acres of the western half of GCNP.

Several trails, sparsely marked with cairns, enable visitors to see different aspects of Toroweap. The 1.6-mile (round-trip) Saddle Horse Canyon Trail along the rim at Toroweap showcases several viewpoints of a beautiful canyon. The 6-mile Tuckup

Trail, which starts 4.7 miles south of the ranger station, highlights the area's unique geology.

From the rim at Toroweap, visitors see cinder cones and lava flows and hear Lava Falls Rapid, which is just downstream. (This rapid was formed by debris washed down Prospect Canyon, not from remnants of lava flows.) Reaching the rim requires a high-clearance vehicle, patience, expert driving, and good tires. More than 20 percent of the vehicles trying to get to the overlook have at least one flat tire; those not having a spare will pay $1,200–$2,000 in towing costs. Do not underestimate the difficulty of this drive; some vehicles have gotten stuck for several days.

This upstream view of the Colorado River from Toroweap (Tuweep) Overlook (River Mile 178) features one of the most iconic landscapes in the National Park Service. Just downstream is Lava Falls Rapid, one of the last major rapids in Grand Canyon National Park. (National Park Service)

There is no water, lodging, gas, food, garbage collection, or telephone service at Toroweap Overlook. There is a primitive campground (with bathrooms) a mile north of the overlook; staying there requires permits that are available at GCNP's Backcountry Information Center (permits are not issued at Toroweap). Visitors must bring their own food, fuel, spare tires, water, tire repair kits, and portable air compressors. The Toroweap airstrip is closed.

The Lava Falls Trail from Toroweap is the shortest and steepest route to the river from the North Rim. The trail, which begins at the end of Lava Falls Road along the southern flank of Vulcans Throne, is 1.5 miles long, drops 2,540 feet, and is exceedingly difficult. The trail follows the Toroweap Fault, along which more than 150 lava flows have occurred in the past million years. Some of these lava flows, which have poured over the Supai Group and Redwall Limestone of the Inner Gorge, repeatedly dammed the river. Lava Falls Trail follows the youngest of these cascades, which happened 10,000–20,000 years ago. The trail crosses steep, often treacherous talus and is one of the scariest routes in Grand Canyon.

Among the oldest of these lava dams was Prospect Dam, which formed 680,000 years ago, eventually rose to 2,300 feet above the river, and created a lake that took more than two decades to fill and stretched more than 280 miles to Moab,

Utah. As sediment accumulated in the lake (as is happening today in Lake Powell), the water finally cascaded over the lava dam, creating a 2,000-foot waterfall. Later, the entire dam gave way, creating catastrophic floods downstream. Prospect Dam is named for Prospect Canyon, which is directly across from (i.e., is due south of) Toroweap Overlook. Debris from Prospect Canyon has created Lava Falls Rapid. This rapid, which can be seen downstream (i.e., to the west) from Toroweap Overlook, is a favorite of river runners. From Toroweap, Lava Falls Rapid looks like a riffle, but it is one of the most treacherous rapids in Grand Canyon.

Just beyond the "Metropolitan Toroweap" buildings and farm machinery is the grave of legendary GCNP ranger John Hauert Riffey (1911–1980), who—except for two years of military service—worked at Toroweap from 1942 to 1980. Riffey, who had no staff and whose nearest neighbor was 23 miles away, had a budget of just more than $200 to fund repairs, equipment, protection, and virtually everything else at Toroweap. Riffey did everything at the 200,000 acres that comprise Toroweap, including repairing the road, fighting fires, filing NPS reports, collecting trash, and helping the occasional tourist. (Visitation averaged about eight people per day.) Riffey had a legendary work ethic; when he died of a heart attack in 1980, he had more than 13 weeks of unused vacation and 90 weeks of sick leave. Riffey loved Toroweap and his work there; as he noted in 1972, "My only contribution to society is trying to keep this place just like it is" (Phelan, 1985).

Riffey's death was reported nationwide, including in an article in the July 29, 1985, issue of *Sports Illustrated* titled "Ranger John Riffey Stayed Faithful to the Land

The North Rim's Toroweap (Tuweep) Valley is one of the most remote sites in Grand Canyon National Park (GCNP). (*Tuweep* is a Paiute word for dry or barren valley.) This view includes the graves of legendary GCNP ranger John Riffey and his wife Meribeth. (Courtesy Bob Ford)

He Loved for 38 Years." John intended to be buried in a family plot in Riverside Cemetery in Denver, Colorado, by his first wife, Laura Riffey (1908–1962), but instead he was buried just west of the Toroweap Ranger Station. Riffey's burial at Toroweap required a special dispensation from NPS, and a nearby boulder was his first headstone. On its way to the graveside service at Toroweap, the airplane carrying Riffey's casket made an unauthorized flight over the Grand Canyon and Lava Falls in a final tribute to Riffey.

Riffey's second wife, Western Washington University biologist Meribeth J. Mitchell, later added a traditional headstone. When Meribeth died in 1993, she was buried beside her husband, with the NPS stipulation that no other family members could be buried there. Ironically, John Riffey was buried in his GCNP uniform, despite the fact that he never wore it at Toroweap.

See also: Dutton, Clarence; North Rim

Further Reading

Annerino, John. 2017. *Hiking Grand Canyon: A Detailed Guide to More than 100 Trails*. New York: Skyhorse Publishing.

Phelan, Richard. 1985. Ranger John Riffey stayed faithful to the land he loved for 38 years. *Sports Illustrated*, July 29. Retrieved July 4, 2017, at https://www.si.com/vault/1985/07/29/620938/ranger-john-riffey-stayed-faithful-to-the-land-he-loved-for-38-years#

TRAIL OF TIME

In October 2010, GCNP opened its Trail of Time, a 2.83-mile (4.56-kilometer) self-guided trail that features panoramas, viewing tubes, interpretive panels, and rocks to help visitors appreciate Grand Canyon's beauty, age, geology, and history (e.g., uplift of the Colorado Plateau, the Cambrian "explosion"). The rim-side trail, which was made possible by a $2-million grant from the National Science Foundation, begins just west of Yavapai Point. Along the trail, brass markers every meter designate a million years of geological history. As visitors walk west, they walk back in time as they see, touch, and examine samples of the more than 50 rocks that comprise the geologic formations in Grand Canyon. These rocks are placed near their "birthdays" in Grand Canyon's history; for example, 270 meters into the trail is a sample of Kaibab Limestone, which is 270 million years old. Five meters further along the trail (i.e., at 275 meters) is a piece of 275-million-year-old Coconino Sandstone, and at 340 meters along the trail is a piece of Redwall Limestone that is 340 million years old.

The Trail of Time showing the history of Grand Canyon stretches from Yavapai Point to just east of Verkamp's Visitor Center. A marker designating Earth's age (i.e., 4.56 billion years) is just east of Maricopa Point, and one designating the age of the universe (i.e., 13.75 billion years) is at Pima Point. The Trail of Time is the world's largest geoscience exhibit.

See also: Geology along the River; Grand Canyon Village; Hopi Point; Maricopa Point; Rim Trail

The Trail of Time is a 2.83-mile-long, self-guided trail along the South Rim that helps visitors try to comprehend Grand Canyon's vast age. The most popular parts of the trail between Yavapai Point and Verkamp's Visitor Center, feature samples of the different-aged rocks found in Grand Canyon. The Trail of Time is the world's largest geoscience exhibit. (Randy Moore)

Further Reading

National Park Service. 2010. Grand Canyon National Park to dedicate the Trail of Time. Retrieved July 4, 2017, at https://www.nps.gov/grca/learn/news/grand-canyon-national-park-to-dedicate-the-trail-of-time.htm

TRAMWAYS

There have been many tramways into Grand Canyon, as well as across the Colorado River in Grand Canyon. For example, several early prospectors built tramways across the Colorado River, and in 1907 David Rust erected a 450-foot-long cable 60 feet above the Colorado River to transport visitors and livestock across the river. This tramway, which was operated by gravity and a wench, was used until 1919. Grand Canyon pioneer William Bass had trans-river cables near River Miles 109 and 111, and in 1926 the Santa Fe Railway built a 6,000-foot-long tramway to carry visitors and supplies from the South Rim's Pima Point to Hermit Camp. In the 1950s, the U.S. Guano Corporation built a now abandoned 9,400-foot-long tram linking Guano Point (near River Mile 267) to a guano mine at Bat Cave on the opposite side of the river in Grand Canyon. River runners can see Bat Cave (now closed

The Guano Point Tramway was an 9,400-foot-long cable that stretched from Guano Point to the guano-mining operations at Bat Cave (on river right near River Mile 266) on the opposite side of the river. By the 1960s, the depleted mine had been abandoned, but the ironworks remain at Guano Point as a monument to the attempts to mine the Grand Canyon. Guano Point, which is on the Hualapai Reservation near Grand Canyon Skywalk, offers Grand Canyon West's best views of Grand Canyon. (Randy Moore)

to visitors) and remains of the tram at Grand Canyon West. Two trams still exist in GCNP at USGS gaging stations.

Although there has never been a tram from rim to rim in Grand Canyon, there were plans for one. In 1919, just weeks before GCNP became a national park, San Francisco engineer George K. Davol and 12 surveyors announced plans to build a 10-mile-long trans-canyon tram that would allow visitors to sail from rim to rim over Grand Canyon. Davol's plans included a series of cables—one linked to another—suspended on steel towers from the South Rim's Hopi Point to Dana Butte on the Tonto Platform, Tower of Set, Horus Temple, Osiris Temple, Battleship, then over Shiva Temple, and ending at the North Rim. Davol's primary backer, the Santa Fe Railway, hoped to use the tram to slow the influence of competitor Union Pacific Railroad at the North Rim.

NPS director Stephen Mather strongly opposed Davol's plans because the tram was "too costly, and damaging to the canyon's ecology," as well as because Mather believed that national parks were not for "amusements" and "contraptions." Davol's plan was rejected by the NPS when Franklin Lane was replaced as Secretary of the Interior by John Payne in October 1920.

See also: Bass, William and Ada

Further Reading

Lago, Don. 2015. *Grand Canyon: A History of a Natural World and National Park*. Reno: University of Nevada Press.

TRANS-CANYON TELEPHONE LINE

The first trans-canyon communication system at GCNP was a magneto handset system that linked the North Rim to the South Rim in 1922. Within a year, the system included 59 miles of wire. As tourism at GCNP increased, GCNP officials realized that they needed a more reliable system of communication in the park. That system would be the Trans-Canyon Telephone Line.

In 1935, CCC workers completed the Trans-Canyon Telephone Line, which linked the North and South Rims. (Other projects completed by the CCC included replacing the boardwalk between Kolb Studio and Verkamp's with stone and pavement, replacing a fence along the South Rim with a masonry wall, and burying overhead power lines that were installed in 1903). The 25-mile-long (later reduced to 18 mile) line of Copperweld wire, which was part of Mountain Bell, was supported by 592 galvanized metal poles. Those poles, which roughly followed the North Kaibab Trail and Bright Angel Trail, remain visible in GCNP at places such as Clear Creek and on the north side of the Colorado River between the Colorado River and Phantom Ranch. Poles and plaques commemorating the line are near Phantom Ranch and atop the rim wall on the South Rim near the east end of Thunderbird Lodge.

See also: Civilian Conservation Corps

Further Reading

Ford, Erin. 2016. Out of the past: Constructing the Trans-Canyon Telephone Line. Grand Canyon News. Retrieved February 4, 2017, at https://www.grandcanyonnews.com/news/2016/nov/22/out-past-constructing-trans-canyon-telephone-line/

TRANSEPT TRAIL AND BRIDLE TRAIL

Transept Trail is a 1.9-mile (one-way) easy-to-follow, well maintained trail on the North Rim that follows the edge of Transept Canyon, a major tributary of Bright Angel Creek. The trail connects Grand Canyon Lodge to the General Store and the 82-site North Rim Campground (the only GCNP campground on the North Rim; additional sites are in the adjacent Kaibab National Forest). Portions of the trail take you through a forest of quaking aspen, fir, and ponderosa pine that cover four minor draws along the rim. Transept Canyon, which is on river right near River Mile 87.5, was named in 1882 by Clarence Dutton, who described it as "one of the finest and perhaps the most picturesque of the gorges in the whole Kaibab front . . . far grander than Yosemite" (Dutton, 1882).

The nearby Bridle Trail, which parallels Highway 67, connects North Kaibab Trailhead with Grand Canyon Lodge. Leashed pets and bicycles are allowed on this trail. Water is available at the campground and lodge.

See also: North Rim

Further Reading

Aitchison, Stewart. 2008. *Grand Canyon's North Rim and Beyond: A Guide to the North Rim and the Arizona Strip*. Grand Canyon, AZ: Grand Canyon Association.

Dutton, Clarence E. 1882. *Tertiary History of the Grand Canyon*. Washington, DC: U.S. Government Printing Office. (Reprinted 2001 by the University of Arizona Press, with an introduction by Wallace Stegner and a new foreword by Stephen J. Pyne.)

TUSAYAN MUSEUM AND RUINS

Tusayan Museum and Ruins, which is 3 miles west of Desert View Watchtower on Desert View Drive, is the only easily accessible and extensive archaeological site within GCNP. The ruin is one of more than 100 Anasazi sites near the south side of Grand Canyon and one of more than 3,700 documented archaeological sites at Grand Canyon and GCNP. The ancient structures here are the remnants of an ancient village from about 1185 to 1230 CE and consist of eight single-story rooms and two kivas (i.e., ceremonial rooms) that housed 20–30 people. One of the kivas offers an exceptional view of San Francisco Peaks, which is 48 miles away. The site, which was abandoned by 1225, was inhabited from about 500–1220 CE. Soon thereafter, the society withered and then either died or moved east to live with Zunis, Hopis, Acomas, or other Native Americans. People who lived here were among the last at Grand Canyon after 1,200 years of living in the area. (*Tusayan* is a phonetic version of the Hopi word meaning "country of isolated buttes.")

The ruins here were excavated by Emile W. Haury and Harold S. Gladwin in 1930. Helping with the work was Winifred MacCurdy of Pasadena, California, who donated $5,000 for a museum at the site. The Tusayan Museum and Ruin, which opened in 1932, houses displays of early Pueblo culture and includes a 0.1-mile, self-guided trail that examines the kiva, plaza, and storage rooms. The trail is paved and wheelchair accessible, but a short side loop to a prehistoric farming site is not. When GCNP opened, this site was called the Wayside Museum of Archeology. The museum here includes a bookstore and exhibits about the area's history. Some of the oldest artifacts from Grand Canyon are at Tusayan Museum and Ruins.

See also: Thematic Essays section: Grand Canyon's First People

Further Reading

Anderson, Michael. 2001. *Along the Rim: A Guide to Grand Canyon's South Rim from Hermit's Rest to Desert View*. Grand Canyon, AZ: Grand Canyon Association.

U

UNCLE JIM TRAIL

Uncle Jim Trail is a looped, 4.6-mile (round-trip) rim-top trail that begins at the eastern edge of the North Kaibab Trailhead and winds through a fire-damaged alpine forest of spruce, fir, and aspen to the beautiful Uncle Jim Point (previously called Naji Point) that is visible on river right from near River Mile 87. At the trail's intersection with the less used Ken Patrick Trail, hikers turn right and follow the trail into a wooded wash. In this cool, moist area, snow often persists until late May. The trail, which includes no water, then climbs and splits again. Both paths here lead to the same viewpoint; the left path is flatter and passes close to the rim, while the right path has several descents and climbs that offer no views of the canyon. The trail ends in towering ponderosa pine from which hikers can see Bright Angel Canyon, Walhalla Plateau, North Kaibab Trail's switchbacks into Roaring Springs Canyon, and Deva, Brahma, and Zoroaster Temples. Most of the Colorado River gorge is hidden by Bright Angel Point in the west and by the Walhalla Plateau in the east. An NPS information board near the trailhead offers information for hikers and detailed maps of the area. There is no water along the Uncle Jim Trail.

Uncle Jim Trail is named for James T. "Uncle Jim" Owens (c.1848–1936), a prospector and game warden at the North Rim's Grand Canyon Game Preserve in the early 1900s. Jones became famous for using trained dogs to find and kill hundreds of mountain lions and other predators, with the misguided belief that he was protecting the local population of deer. In 1906, Owens brought bison to the Kaibab Plateau; these were the progenitors of the herd that now lives near the entrance to the North Rim. In 1921, Owens—who is the popular character "Owen" in Marguerite Henry's *Brighty of the Grand Canyon* (1953)—proposed building a tourist camp at the North Rim's Cape Royal, but his plan was rejected by NPS director Stephen Mather.

See also: Cattalo, Buffalo, and Pearl Gray; North Rim

Further Reading

Aitchison, Stewart. 2008. *Grand Canyon's North Rim and Beyond: A Guide to the North Rim and the Arizona Strip.* Grand Canyon, AZ: Grand Canyon Association.

UNITED STATES GEOLOGICAL SURVEY

On March 3, 1879, President Rutherford B. Hayes signed legislation consolidating the Powell, King, Hayden, and Wheeler Surveys into the United States Geological

Survey (USGS), the world's first government-funded agency devoted to science. Despite the importance of USGS, the legislation that created the agency was just over one page long (An Act, 1879). The creation of USGS began the systematic mapping and surveying of the region beyond the 100th meridian. Since its creation, USGS has focused most of its work on geology and geologic mapping, assessment of water and other natural resources, and topographic mapping.

The first director of USGS was geologist-adventurer Clarence R. King. One of King's first acts was to assign the completion of John Wesley Powell's work in Grand Canyon to Clarence E. Dutton. King resigned soon thereafter to enter the mining industry. In 1881, Powell became the survey's second director, a position that he held until 1894.

See also: Thematic Essays section: John Wesley Powell; Dutton, Clarence

Further Reading

An Act Making Appropriations for Sundry Civil Expenses of the Government for the Fiscal Year Ending June Thirtieth, Eighteen Hundred and Eighty, and for Other Purposes. 1879. 45th Congress, 3rd session. *U.S. Statues at Large*, 20, 394–395.

United States Geological Survey. Retrieved July 4, 2017, at https://www.usgs.gov/

V

VERKAMP, JOHN

Ohio native John George Verkamp (1877–1944) came to Grand Canyon in 1898 and began selling souvenirs and Native American crafts from a tent he rented near Bright Angel Hotel at the South Rim. Verkamp was an agent for the Babbitt Brothers Trading Company, which raised cattle near Flagstaff and provided start-up money for several Grand Canyon prospectors and other entrepreneurs. Verkamp had grown up in Cincinnati, Ohio, across the street from the Babbitt family; three of Verkamp's sisters had married Babbitt brothers. By 1921, the Babbitt brothers had opened a general store (part of their Babbitt Brothers chain) near the train depot in Grand Canyon Village, but it later closed after a fire.

In 1897, John George Verkamp began selling curios and Native American arts and crafts from this rim-side tent near Bright Angel Camp. After leaving Grand Canyon because of poor sales, Verkamp later returned, and in 1906 opened Verkamp's Curios in Grand Canyon Village. This was the first souvenir shop at Grand Canyon, and by 2008 was the oldest family-owned business in the National Park Service (NPS). The Verkamp family sold their building to the NPS in 2008, and today their former store (and home) is Verkamp's Visitor Center. (Grand Canyon National Park Museum Collection)

Verkamp's store was the first curio shop at Grand Canyon. Business was slow, however, so he sold his inventory and moved to Flagstaff. After the arrival of the Atchison, Topeka and Santa Fe Railway at the South Rim in 1901, visitation at Grand Canyon increased, and in 1905 Verkamp returned to the South Rim to sell souvenirs and Native American crafts from a tent on an acre of land near the El Tovar Hotel. On January 31, 1906, Verkamp opened Verkamp's Curios at its current site, about 100 feet from the South Rim. (For decades, however, Verkamp listed his store in the telephone directory as "Verkamps Souvenir

Shop.") After the first day of business at his permanent store, Verkamp wrote in his ledger, "$4.98. A good day." For comparison, in 1961, revenues exceeded $180,000 and in 1976 exceeded $780,000.

Verkamp initially hired managers to run his souvenir store at Grand Canyon as he pursued other businesses such as ranching and prospecting. However, when Verkamp lost or closed those businesses during the Great Depression, he moved with his wife Catherine and four children into the second-floor space of his building at Grand Canyon. Verkamp's building, which does not blend in well with the area, collected water on its roof that the Verkamps used for drinking, washing, and cooking. The building's first floor included Verkamp's store and storage rooms.

Verkamp's store was one of the few businesses in Grand Canyon Village to survive the Great Depression. Verkamp competed with Hopi House, which was owned by the Fred Harvey Company. Verkamp managed the store until he died in 1944, thus technically ending his concession in GCNP. NPS wanted to replace Verkamp's building with a Fred Harvey Hotel and casino, but Verkamp's wife convinced NPS to renew Verkamp's concession.

Verkamp and his descendants helped develop the South Rim. For example, John George "Jack" Verkamp Jr. (1916–2001) worked with Senator Barry Goldwater to obtain money to build Shrine of the Ages by Pioneer Cemetery, and Margaret Mary "Peggy" Verkamp (1913–1989; the daughter of John and Catherine Verkamp) worked with Senator Carl Hayden to obtain money for Grand Canyon High School. The Verkamps were active in a variety of civic groups, including Rotary Club and Boy Scouts, and strongly supported the local library. The Verkamp family lived in the Verkamp's Curios building until 1978. By 2008, Verkamp's was the oldest family-owned business in the entire NPS.

Historic Plaques at Grand Canyon

Many places at Grand Canyon are decorated with historic markers. For example, iconic sites such as El Tovar Hotel, Hermits Rest, and Lookout Studio display plaques denoting their designation as National Historic Landmarks. These designations originated with the passage of the Historic Sites Act of 1935. The National Historic Landmarks at Grand Canyon previously listed were recognized in 1987. The most recently designated National Historic Landmark at Grand Canyon is the Grand Canyon TWA–United Airlines Aviation Accident Site, which was made a landmark in 2014.

In front of Verkamp's Visitor Center (formerly Verkamp's Curios) is a small plaque designating the site as being on the National Register of Historic Places. This register, which was the first broadly-based national historic preservation program in the United States, was created by the National Historic Preservation Act in 1966. However, as per an agreement between the Verkamp family and governmental officials, the plaque states that the site was added to the register in 1906, which was 60 years before the register originated.

Although a nonfamily member began managing the store in 1995, the Verkamps retained ownership of the store until 2008, when they sold it to GCNP. This was the last echo of the businesses established by Grand Canyon's pioneers. Today, the building houses a small museum and Verkamp's Visitor Center. (As a condition of the sale, the Verkamp family can stay in the upstairs quarters when they visit the canyon.) The Center often displays famed artist Louis Akin's (1868–1913) *Evening—Grand Canyon* (1911), which is Akin's largest painting of a landscape. Verkamp's Visitor Center is the only visitor center in GCNP on the canyon's rim.

Unlike many historic sites in GCNP, which were listed on the National Register of Historic Places in 1987, Verkamp's was listed on the Register in 1974. Today, a smaller-than-usual plaque denoting this designation stands in front of their store. When it was installed, it was backdated to 1906.

See also: Grand Canyon Village

Further Reading

Anderson, Michael. 2001. *Along the Rim: A Guide to Grand Canyon's South Rim from Hermit's Rest to Desert View.* Grand Canyon, AZ: Grand Canyon Association.

W

WALHALLA OVERLOOK AND WALHALLA GLADES PUEBLO

Walhalla Overlook (elevation 7,998 feet), which is 16 miles south of Point Imperial and 21.4 miles from Grand Canyon Lodge, was the summer home of the Anasazi Indians about 900 years ago. The overlook, which is a peninsula surrounded on three sides by Grand Canyon, is slightly lower than most other places on the North Rim; as a result, updrafts of warm air from the inner canyon melt winter snow earlier at Walhalla Overlook than at other areas along the North Rim, thereby making the area a good place for ancestral people to farm. Archaeologists have uncovered more than 100 sites of farming on this plateau, most of them close to the rim. These sites usually consist of small, one-room structures and terraces or of gardens where the natives grew squash, corn, and beans. The larger rooms at these sites were probably living-quarters. Farming here could have resulted from a shortage of land for the growing population living on the 300-acre Unkar Delta along the river below. From Walhalla Overlook, visitors can see Desert View Watchtower (8.5 miles away on the South Rim), Vishnu Temple, and the Colorado River as it flows around Unkar Delta.

Unkar Delta

Unkar Delta is a large, terraced slope along the Colorado River at River Mile 72.5. The area gets its name from the Paiute word *unkar,* which means "red creek" or "red stone." Unkar Delta was the site of the largest known prehistoric settlement along the Colorado River in Grand Canyon. Ancestral Puebloans, notably the Anasazi Indians, lived at the Unkar Delta for at least three centuries until around 1150 CE. (Hopi Indians refer to these ancestors as *Hisasinom.*) Unkar Delta, which is one of the largest archaeological sites in GCNP, is best viewed along a 0.8-mile loop; you can also see Unkar Delta from Lipan Point, from the Unkar Creek Overlook on the south side of the river, and from Cape Royal by looking through Angels Window toward the curve in the Colorado River. Terraces such as Unkar Delta are rare in the canyon's Inner Gorge.

Ruins of the Walhalla Glades Pueblo, which were excavated in 1967–1968, are just across the road and about 350 feet northwest of the overlook. This 900-year-old dwelling built by ancestral Puebloan people (also known as Kayenta Anasazi) had at least nine rooms and was occupied by five generations of people between 1050 and 1150 CE. Cape Royal Road ends about 2 miles past Walhalla Overlook.

See also: Thematic Essays section: Grand Canyon's First People

Further Reading

Grand Canyon National Park. 2001. *Walhalla Glades*. Grand Canyon, AZ: Grand Canyon Association.

WATER AT GRAND CANYON

Accessible water suitable for drinking has always been scarce at Grand Canyon. When GCNP was created in 1919, the Atchison, Topeka and Santa Fe Railway was bringing 60,000–250,000 gallons of water (in water cars carrying 12,500 gallons each) per day to Grand Canyon Village, where it was used and stored in giant towers and tanks. In its early days, the railway also carried water to farmers and ranchers near Grand Canyon at cost.

By 1931, GCNP began pumping water—as much as 576,000 gallons per day—from Roaring Springs to Indian Garden and from there to the South Rim. This more reliable source of water stimulated development at Grand Canyon. Railway water tanks were used "as needed" until the late 1960s, when the Bright Angel Suspension Bridge was completed and carried the water pipeline across the Colorado River, thereby completing the trans-canyon pipeline. This pipeline was designed to last for 40 years, so it was not surprising that by 2015 the ageing, 16-mile pipeline was leaking and even breaking several times per year. For example, in 2017, GCNP had to replace more than 380 feet of the pipeline to the North Rim after it was damaged by rockslides. Repair and replacement of the trans-canyon pipeline are projected to cost $160 million. In the meantime, water from Roaring Springs continues to be pumped to both rims, where it is used and stored (e.g., at the North Rim, the water is stored in four million-gallon tanks). GCNP is also considering using water from Bright Angel Creek as its source of water.

Beginning in 2012, NPS banned disposable bottles of water from being sold in the park. This policy resulted from a study showing that empty plastic bottles accounted for 20 percent of GCNP's trash and 30 percent of its recyclables. GCNP then spent more than $250,000 to install 10 water stations in the park so that visitors could fill their own bottles at no cost. (This water is not from the Colorado River but from Roaring Springs, which is about 5 miles down North Kaibab Trail.) However, in 2017, NPS lifted the ban on the sale of disposable water bottles "to expand hydration options" of visitors.

See also: North Kaibab Trail

Further Reading

National Park Service. 2017. The plastic water bottle reduction in national parks. Retrieved June 3, 2017, at https://www.nps.gov/commercialservices/greenline_water_bottle.htm

WHITE, JAMES

James White (1837–1927) was a prospector who, on September 7, 1867, was pulled from a makeshift cottonwood raft on the Colorado River below the Mormon town of Callville, Nevada (below Pearce Ferry, now submerged in Lake Mead). The delirious, haggard, scantily clothed White claimed that he had been separated from his two prospecting friends during an attack by hostile Indians; one of his friends, Charles Baker, was killed during the attack, and the other, George Strole, died five days later. White, who claimed that he survived by lashing himself to his raft, obtained food by trading his two pistols to Native Americans for dog hindquarters. A letter dated September 26, 1867, to his brother Joshua (in Kenosha, Wisconsin) was White's only written account of his voyage. Throughout his life, White never wavered on his fascinating, but vague, story.

White claimed that he had floated through Grand Canyon in 14 days. If White's claims are true, he was the first person to float, albeit unintentionally, through Grand Canyon. (John Wesley Powell would not float through Grand Canyon until 24 months later.) Grand Canyon pioneer William Bass and a few other people believed White, but most others—including Grand Canyon river running pioneers John Wesley Powell, Julius Stone, and Robert Stanton, who interviewed White at White's home in Trinidad, Colorado—did not. (These people, however, had a vested interest in White's claim: Their expeditions through Grand Canyon were greatly diminished if an unprepared prospector could survive the trip on a makeshift raft.) Stanton hired a stenographer to record the interview but then edited and expanded the transcript significantly. Stanton dismissed White's claims as "wild stories." Grand Canyon river runner Frederick Dellenbaugh, who had boated the Colorado River with John Wesley Powell, claimed, "Just where [White] entered the river is of course impossible to decide, but that he never came through the Grand Canyon is as certain as anything can be. His story reveals an absolute ignorance of the river and its walls throughout its whole course he pretended to traverse" (Berger, 2007). White, who never met Powell, never profited from his story.

The 89-year-old White died in 1927. He is commemorated at Grand Canyon by Whites Butte on river left near River Mile 95.5.

See also: Thematic Essays section: John Wesley Powell; Stanton, Robert

Further Reading

Adams, Eilean. 2001. *Hell or High Water: James White's Disputed Passage through Grand Canyon*. Logan: Utah State University Press.

Berger, Todd R. 2007. *It Happened at Grand Canyon*. Guilford, CT: TwoDot Publishers.

Berger, Todd R., ed. 2008. *Reflections of Grand Canyon Historians: Ideas, Arguments, and First-Person Accounts*. Grand Canyon, AZ: Grand Canyon Association.

WIDFORSS, GUNNAR

Gunnar Mauritz "Weedy" Widforss (1879–1934) was a Swedish American artist who became famous for his watercolor images of wilderness. After meeting NPS director Stephen Mather in 1922 at Yosemite, Widforss began regular trips to GCNP, where he often stayed with the Kolb brothers (Ellsworth and Emery) or in a Fred Harvey Company dormitory. (Emery Kolb loaned Widforss some money in exchange for three paintings that continue to hang in Kolb Studio.) Unlike artists such as Thomas Moran (e.g., *Chasm of the Colorado*) and F. W. von Egloffstein (e.g., *Big Cañon at Diamond River*), Widforss stressed technical accuracy over sentimental or romanticized depictions of the American West. He usually painted while outdoors because he believed it "is much better to sit out in the middle of the scene."

Widforss produced more than 200 paintings and sketches of Grand Canyon, all of which focused on the canyon itself; no humans appear in any of the paintings. Widforss sold his paintings in the lobby of the El Tovar Hotel, where one continues to hang in the hotel's sitting room on the second floor. Widforss, whose favorite scenes included the North Rim's shady groves of aspens, never built or maintained a studio on the canyon's rim. From the early 1920s until his death in 1934 on the South Rim, Widforss considered Grand Canyon his home. Today, Widforss's paintings hang in Bright Angel Lodge, El Tovar Hotel, Kolb Studio, and elsewhere in GCNP, but not at the North Rim, where he did most of his work.

In 1934, after being diagnosed with a serious heart condition, Widforss returned to the South Rim to gather his belongings and move to a lower altitude.

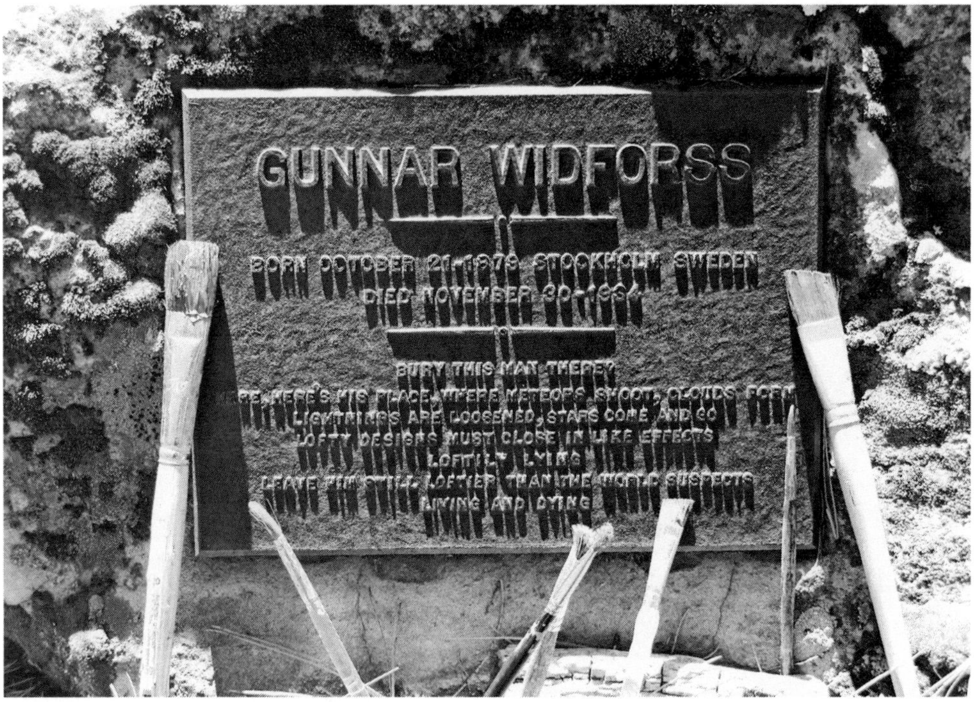

Gunnar Widforss was a Swedish artist who painted many renowned paintings of Grand Canyon, several of which continue to be displayed in Grand Canyon Village. Unlike those of Thomas Moran, Widforss's paintings were technically accurate and never included humans. At the South Rim's Pioneer Cemetery, admirers often leave paintbrushes at Widforss's grave. (Randy Moore)

On November 30, the 55-year-old Widforss died of a heart attack while driving to dinner at El Tovar Hotel. Following a funeral service in El Tovar's music room, Widforss—who was often referred to as "the painter of the national parks"—was buried in the South Rim's Pioneer Cemetery. Widforss rests beneath a quote from Robert Browning's *A Grammarian's Funeral* (1855), a poem that probes the virtues of experiencing and contemplating life. Admiring visitors often leave paintbrushes at Widforss's grave.

Widforss's estate included more than 150 finished and unfinished paintings. In 1972, 72 of these paintings—many of which were of Grand Canyon—were featured in a popular exhibit at the National Gallery of Art in Washington, DC.

In GCNP, Widforss is honored by the North Rim's Widforss Trail. The trailhead for this pleasant, plateau-top 9.7-mile (round-trip) hike is at a parking lot 0.25 mile west of Arizona Highway 67 at the edge of Harvey Meadow, which was once a staging area for cross-canyon mule trips. The mildly undulating, maintained trail winds along the top of the Kaibab Plateau through an aspen and pine forest and along the rim of Transept Canyon to a somewhat restricted viewpoint above the head of Haunted Canyon. (In 1902, USGS topographer François Matthes declared that Transept Canyon was grander than Yosemite Valley.) For an even better view, hikers can continue to a Widforss Point, a rocky, 7,900-foot-high summit where there is a small picnic facility and great views of Manu Temple and Buddha Temple. Brochures describing Widforss Trail are sometimes available at the trailhead. There is no water along Widforss Trail.

See also: Pioneer Cemetery and Other Gravesites

Further Reading

Aitchison, Stewart. 2008. *Grand Canyon's North Rim and Beyond: A Guide to the North Rim and the Arizona Strip*. Grand Canyon, AZ: Grand Canyon Association.

Grand Canyon National Park. 2001. *Widforss Trail*. Grand Canyon, AZ: Grand Canyon Association.

WOOLLEY, EDWIN DILWORTH "UNCLE DEE"

Tourism at the North Rim was pioneered by Kanab resident, Mormon leader, entrepreneur, and hunter Edwin Dilworth "Uncle Dee" Woolley (1845–1920). After seeing the North Rim, Woolley was convinced that "this is one of the wonders of the world! People will come from all quarters of the globe and will pay great sums of money to gaze upon what we behold!" (Brunner, 2017). In 1903, Woolley commenced his plans to cash in on the area by forming, with four other investors, the Grand Canyon Transportation Company. The company hired David Rust, Woolley's son-in-law, to build a trail from the North Rim to the river. Rust, who offered tours from Kanab to Bright Angel Point, finished the trail four years later (i.e., in 1907), and the trail was known as Bright Angel Trail (now the Old Bright Angel Trail) or the Woolley Trail, since it connected at the Colorado River with the Bright Angel Trail from the South Rim. The Old Bright Angel trail began about 4 miles down today's Ken Patrick Trail from the North Kaibab Trailhead and connected with what is today's North Kaibab Trail just past Roaring Springs. Parts of the trail, which was the first direct route from the North Rim to the Colorado River, were later incorporated into today's North Kaibab Trail. In 1907, Woolley and Rust also built a cabin that is today known as Uncle Dee Woolley Cabin near the trailhead. These were the first tourism facilities at the North Rim. In 1923, Rust also became the first commercial boatman in Glen Canyon.

At the river, during 1906–1907, Rust installed a gasoline-engine powered Aerial Tramway to cross the river and constructed a 2-acre tent camp near the tram called Rust's Camp. In 1907, Rose Evans and Lida Bilveal became the first tourists to cross the Colorado River in the tramway car.

Because snow made the North Rim unappealing to tourists in winter, neither Woolley nor Rust lived full-time at Grand Canyon. Rust and Woolley had several other jobs; for example, Rust taught school in Fredonia, edited the *Kanab County Standard* newspaper, and was a member of the state legislature. As a result, neither Rust nor Woolley worked at the camp much, and many people who visited the camp from the South Rim assumed that the rundown camp had been abandoned. However, in 1913, Theodore Roosevelt visited the rustic camp while on a hunting trip; thereafter, the camp was known as Roosevelt's Camp. Today, Rust's camp is known as Phantom Ranch.

After Woolley noted the economic boom produced by the arrival of the railroad at the South Rim in 1901, he began lobbying Utah Pacific Railroad to build a spur to the North Rim. In 1905, Woolley hosted North Rim visits by Utah senator (and future U.S. Supreme Court justice) George Sutherland (1862–1942) and Utah congressman Joseph Howell (1857–1918), as well as NPS director Stephen Mather in 1920, but never convinced railroad officials to bring the railroad to the North Rim.

Woolley and his brother E. G. Woolley Jr. then arranged—with much publicity—for two cars to be driven from Salt Lake City to the North Rim. There were no gas stations near the North Rim; to ensure that the cars had enough fuel, gasoline was carried by wagon and stashed along the dirt road that became Highway 89 and 67 leading to the North Rim.

Rust stopped giving tours in 1915. Although Rust and his business partners never sold their interests in the company, Woolley's company had disappeared by 1919. Woolley is commemorated by Woolley Point and Woolley Butte, both visible to the southeast from Point Imperial.

See also: North Kaibab Trail; Railroad; Roosevelt Point; Tramways

Further Reading

Brunner, Dixie. 2017. Tourism history of the area. Retrieved June 21, 2017, at http://www.kanabguide.com/articledetails.cfm?recordID=27

WYLIE WAY CAMPS

William Wallace Wylie (c.1848–1936) was born in Iowa and, when he turned 30, moved to Montana to be a school administrator. There, in 1880, Wylie visited Yellowstone National Park and was so impressed that he wrote a guide to the park and began giving tours to his colleagues and visitors. By 1883, seven-day tours cost $35. On every night of these tours, Wylie had to set up all of the facilities—tents, the kitchen, dining room, and shelters—and then tear them down the next morning.

As business improved and larger groups joined Wylie's tours, the demands of setting up all of the facilities every night became too cumbersome, prompting Wylie to request permission from the park to establish a camp that he could leave standing in one place. Yellowstone agreed, thereby making Wylie the pioneer of "permanent camps" in national parks. Wylie and his wife Mary Ann Wilson Wylie (1855–1928) called their company Wylie Way Camping Services. By 1905, Wylie had sold his company, but his family remained involved in running his camps.

Meanwhile, at the South Rim, the Atchison, Topeka and Santa Fe Railway had a virtual monopoly on concessions. In 1917, hoping to duplicate Santa Fe's success, the Union Pacific Railroad asked Wylie to open camps at Zion and Grand Canyon. William, who managed the camp at Zion, delegated management of the Grand Canyon camp to his daughter Elizabeth Wylie McKee. In 1917, Elizabeth—with her husband Thomas, 10-year-old son Bobby, and a staff of local Mormon teenagers—set up a 25-site Wylie Way Camp about 300 feet northwest of today's Grand Canyon Lodge. The next year, Bobby and a burro named Brighty began delivering water to the camp.

In 1922, Thomas McKee published an article titled "Brighty, Free Citizen: How the Sagacious Hermit Donkey of the Grand Canyon Maintained His Liberty for Thirty Years" in *Sunset* magazine. Thirty years later, Marguerite Henry—famous for writing stories about horses and other animals—found McKee's article in a public library and used it as the basis for her famous book, *Brighty of the Grand Canyon*. Brighty, who lived in Grand Canyon from around 1892 to 1922, became an icon of the North Rim.

The Wylie Way Camp, which opened almost 30 years after the first businesses opened at the South Rim, was open from June to October, and visitors—who paid $6 per night—could tour Cape Royal, Point Sublime, and other sites along the North Rim. Elizabeth's camp competed with that of Arizona Strip residents Aldus "Blondie" Jensen and his wife Melissa, who took up residence in and began offering mule rides into Grand Canyon and rim-side tours from Uncle Dee Woolley's cabin at the North Rim. When her father retired in 1924, Elizabeth became the first woman to own a Grand Canyon concession. Elizabeth wanted to expand and improve her North Rim camp, but her many requests for a long-term contract were denied. As a result, she operated the camp from 1917 to 1927 with a series of one-year contracts.

In 1927, NPS—wanting to transfer concessions at the North Rim to the Union Pacific Railroad—solicited bids for the first permanent concession at the North Rim. The solicitation required expensive improvements to the property that Elizabeth could not afford, so she had little choice but to sell the property. As a result, the railroad (and its subsidiary, Utah Parks Company) took over the property and soon opened Grand Canyon Lodge. The last of Elizabeth's camps at the North Rim were transferred to the North Rim's current campground, and the last of her tent platforms—which once housed 40 guests—was removed in 1946. No railroad line ever reached the North Rim.

See also: Brighty, the North Rim Burro; North Rim

Further Reading

Aitchison, Stewart. 2008. *Grand Canyon's North Rim and Beyond: A Guide to the North Rim and the Arizona Strip*. Grand Canyon, AZ: Grand Canyon Association.

Leavengood, Betty. 2014, *Grand Canyon Women: Lives Shaped by Landscape*. Grand Canyon, AZ: Grand Canyon Association.

Y

YAVAPAI POINT

An easy (but often crowded) walk west from Mather Point brings you to Yavapai Observation Point (elevation 7,082 feet), which includes the Yavapai Observation Station. The site is named for a Paiute word meaning "sun people." This museum, which was the first museum and interpretive center on the South Rim, is made of Kaibab limestone and ponderosa pine. Most of the funding for the building came from the American Association of Museums, and the Laura Spelman Rockefeller Memorial Carnegie Institution of Washington (now known as the Carnegie Institution for Science). Funds also came from geologists David White and Charles Walcott, as well as *The Brooklyn Eagle*, a newspaper that once enjoyed one of the largest circulations in the United States. The observation station was completed in 1928 to get tourists to see "the real things outside" and tell "the story of Grand Canyon . . . with telescopes and exhibits." Yavapai Observation Station was designed by Herbert Maier, who had built structures in Yellowstone and Yosemite. Yavapai Observation Point was the first museum and formal interpretive structure in GCNP. Although much of GCNP closed during World War II, Yavapai Observation Point remained open.

Yavapai Observation Station was originally an open-air building called Yavapai Point Trailside Museum. The building, which included telescopes and fixed-in-place binoculars through which tourists could see structures in Grand Canyon, resembles Hermits Rest and Lookout Studio (all designed by architect Mary Colter). It was designed to fit into, rather than dominate, the area. Yavapai Observation Station, which was renovated in 1978 and 1990, is now GCNP's main geologic interpretive center; inside, you'll find exhibits, a viewing area, and a bookstore and gift shop operated by Grand Canyon Association. Yavapai also hosts talks by NPS rangers. When it rains on the South Rim, Yavapai Observation Point is crammed with visitors.

Yavapai Point, which is named for Native Americans of central Arizona, is 3.1 miles (as the condor flies) from Phantom Ranch. It offers a dramatic 180° panorama of the inner canyon, including the Colorado River (at the end of Pipe Creek), Plateau Point (immediately west of the river), and the Black Bridge near Phantom Ranch, which is 4,574 feet below Yavapai Point. Upriver are Wotans Throne and Vishnu Temple near the North Rim's Cape Royal; lush vegetation marks Indian Garden on the Tonto Platform. On the North Rim is Trinity Creek, which flows below Shiva Temple and joins the river near Hopi Point. North of Yavapai Point is Bright Angel Creek, and in the east is Clear Creek, which extends toward Walhalla Plateau near Cape Royal.

See also: McKee, Edwin "Eddie"

Further Reading

Anderson, Michael. 2001. *Along the Rim: A Guide to Grand Canyon's South Rim from Hermit's Rest to Desert View.* Grand Canyon, AZ: Grand Canyon Association.

Appendix: Geologic Time Scale

Hadean Eon: 4,600–4,000 million years ago (MYA)
Archaean Eon: 4,000–2,500 MYA
Proterozoic Eon: 2,500–541 MYA
Phanerozoic Eon: 541 MYA–present

- Paleozoic Era: 541–252 MYA
  - Cambrian Period: 541–485 MYA
  - Ordovician Period: 485–443 MYA
  - Silurian Period: 443–419 MYA
  - Devonian Period: 419–359 MYA
  - Carboniferous Period: 359–299 MYA
  - Permian Period: 299–252 MYA
- Mesozoic Era: 252–66 MYA
  - Triassic Period: 252–201 MYA
  - Jurassic Period: 201–145 MYA
  - Cretaceous Period: 145–66 MYA
- Cenozoic Era: 66 MYA–present
  - Paleogene Period: 66–23 MYA
    - Paleocene Epoch: 66–56 MYA
    - Eocene Epoch: 56–34 MYA
    - Oligocene Epoch: 34–23 MYA
  - Neogene Period: 23–2.6 MYA
    - Miocene Epoch: 23–5.3 MYA
    - Pliocene Epoch: 5.3–2.6 MYA
  - Quaternary Period: 2.6 MYA–present
    - Pleistocene Epoch: 2.6 MYA–11,700 years ago
    - Holocene Epoch: 11,700 years ago–present

Source: International Chronostratigraphic Chart, v 2017/02. Retrieved January 22, 2018, at www.stratigraphy.org/ICSchart/ChronostratChart2017-02.jpg

Glossary

Anasazi ("ancient ones"): The Native Americans who occupied Grand Canyon before 800 BCE and who flourished there by 1100 CE. Anasazi are ancestors of several of today's tribes of Native Americans living at and near Grand Canyon (e.g., Hopi, Zuni, and Pueblo).

Basalt: A dark-colored, fine-grained igneous rock made of magnesium- and iron-rich minerals. Basalt forms during lava flows. The Cardenas Basalt of Grand Canyon is a basalt.

Cairn: A stack of rocks designating a trail or route.

Carbonate: A compound of carbon and oxygen, typically found as limestone.

CCC: *See* Civilian Conservation Corps.

Cenotaph: Monument to someone buried elsewhere.

cfs: Cubic feet per second, the USGS measure of stream flow. The number of cfs is the cubic feet of water flowing by a given point in one second. A cubic foot of water, which includes 7.48 gallons, is slightly larger than a basketball.

Civilian Conservation Corps (CCC): A governmental agency that employed single, unemployed, young (18- to 25-year-old) men to work between 1933 and 1942. The CCC worked at GCNP from 1933 to 1942 on trails, buildings, walls, the Trans-Canyon Telephone Line, and other projects.

Colorado Plateau: A large, uplifted area of high desert spanning more than 130,000 square miles of Utah, Arizona, New Mexico, and Colorado having flat Paleozoic and Mesozoic rocks, cliffs, and canyons. The Colorado Plateau is the largest plateau in North America. (For comparison, the largest plateau in the world is the Tibetan Plateau, which covers 970,000 square miles at altitudes up to 16,000 feet.) The Colorado Plateau, which includes GCNP, is characterized by thick, flat layers of sedimentary rocks that have been uplifted thousands of feet above sea level in such a way that the sediments remained nearly horizontal. Most of the Colorado Plateau is an arid desert (with a few scattered forests) that is public land.

Colorado River: A river that arises in the Colorado Rockies and flows more than 1,400 miles to the Gulf of California. In Spanish, its name means "reddish," referring to the sediments carried in the river's water. Today, because Glen Canyon Dam removes most of the river's sediments, the Colorado River is usually clear. The Colorado River, which includes more than 40 large dams (and more than 40 smaller dams), is one of the most politicized, litigated, and controversial rivers in the world.

Continental drift: The discovery that continents move throughout geological time.

Corridor Trails: The Bright Angel, North Kaibab, and South Kaibab Trails of GCNP. These are the most popular trails into and out of Grand Canyon.

Cross-bedding: A place or surface in sedimentary rocks in which internal layers have been deposited obliquely to the predominant orientation of the bedding. Cross-bedding is common in the Coconino Sandstone and Supai Group sandstones in Grand Canyon.

Debris flow: A mass of loose sand, rocks, soil, mud, water, and air moving down a slope under the influence of gravity. In Grand Canyon, debris flows produced by heavy rains can carry boulders into the Colorado River, where they often form rapids.

Dory: A small, rocker-shaped paddleboat made of wood or fiberglass. Dories, which ride like a roller coaster through rapids, are the most stable and maneuverable of all keel-less boats; they also provide the most exciting rides through Grand Canyon's rapids.

Erosion: Chemical (e.g., oxidation) and physical processes (e.g., rain, wind, snow, ice) that loosen, degrade, and displace materials on Earth's surface. In Grand Canyon, water is the primary force producing erosion.

Fault: A fracture in Earth's surface where movement has occurred. An example of a fault at Grand Canyon is Bright Angel Fault.

Formation: A group of rock layers easily distinguished from adjacent strata. An example of a formation at Grand Canyon is the Kaibab Formation.

Geologic time: The 4.6-billion-year timescale used by scientists to describe events in Earth's history (*see* Appendix).

Gneiss: a common metamorphic rock similar to granite but having bands of light- and dark-colored minerals that have segregated into distinct layers. Elves Chasm Gneiss, which is 1.84-billion-years old, are the oldest rocks in Grand Canyon.

Grand Canyon Supergroup: The set of tilted Precambrian beds exposed at the bottom of some parts of Grand Canyon.

Granite: A common, coarse-grained, light-colored, hard, intrusive, igneous rock made of feldspar and quartz.

Great Unconformity: An immense gap in the geological record between the igneous and metamorphic rocks of the Inner Gorge and the overlying Paleozoic rocks in Grand Canyon. This geologic boundary occurs between the dark, tilted Precambrian rocks and the overlying, brownish Tapeats Sandstone. More than a billion years worth of eroded rocks are missing here.

Igneous: Rock formed when magma cools, either at Earth's surface (extrusive) or below ground (intrusive). Granite is an igneous rock.

Kaibab: A Paiute word that means "mountain lying down." The Kaibab National Forest surrounds Grand Canyon, fine-grained Kaibab Limestone forms both rims of Grand Canyon, and the Kaibab Plateau is along the north side of the eastern Grand Canyon.

Kiva: A Puebloan place of worship, commonly entered from above by a ladder.

Lava: Molten rock that erupts and cools at Earth's surface.

Left, right: In river usage, left and right are as seen when looking downstream.

Limestone: A common sedimentary rock made primarily of calcium carbonate (calcite, $CaCO_3$), often containing marine fossils.

Marble Canyon: The relatively straight, northernmost section of Grand Canyon stretching from Lees Ferry to the confluence of the Little Colorado River (River Mile 61.5). John Wesley Powell named this area for its prominent Redwall Limestone which, when polished by the Colorado River, resembles marble.

Metamorphic: Rock transformed from preexisting rock by intense heat and pressure from preexisting rock. Unlike igneous rock, metamorphic rock never melted. Vishnu schist is the dark-colored, metamorphic rock at the bottom of Grand Canyon.

Monocline: A bend or fold in Earth's crust that results in identical beds being found at different elevations on either side of the monocline. The East Kaibab Monocline in eastern Grand Canyon elevates strata in the west by almost 3,000 feet, thereby forming the Kaibab and Coconino Plateaus.

Mule: A sterile offspring of a female horse and a male donkey. Mules are often larger, stronger, healthier, and more cautious than either of their parents.

Paleozoic: The era referring to the period of time between 541 and 252 million years ago.

Pangaea: the supercontinent that formed during the Pennsylvanian and lasted through the Jurassic.

Petroglyph: An image scratched or carved into rock, made by using another rock.

Plate tectonics: The theory that Earth's crust consists of movable, rigid plates that can join or separate over geologic time. The separation, collision, and subduction of these plates account for most of the large-scale features of Earth's surface and geologic events (e.g., earthquakes, uplifts, and mountain ranges).

Precambrian: The time prior to the Paleozoic Era, from 4.6 billion years ago to 545 million years ago.

Quartz: A hard, clear mineral made of silica.

Rapid: A turbulent, fast moving part of a river, containing large waves and obstacles. Constricted river channels increase the velocity of the current, an uneven riverbed increases the turbulence, and a drop in elevation increases the velocity and turbulence of a rapid. In Grand Canyon, virtually all of the more than 160 rapids are formed by debris flows from side canyons. Rapids account for about 9 percent of the total distance of the Colorado River in GCNP.

Riffle: A small rapid, usually having few hazards other than running aground.

River miles: The traditional way to locate points within Grand Canyon, measured as miles downstream from Lees Ferry. For example, Redwall Cavern is at River Mile 33, meaning that it is 33 miles downstream from Lees Ferry.

Sandstone: A common sedimentary rock containing large amounts of quartz sand grains, with small amounts of clay and silt. These materials are cemented together by small mounts of silica or calcite.

Schist: Coarse-grained, metamorphic rock without the layering of minerals that typifies gneiss. Vishnu Schist is a prominent schist in the Inner Gorge of Grand Canyon.

Sediment: Material transported and deposited by water, wind, ice, or gravity. Sedimentary rock is formed by the accumulation and cementing of sediment.

Sedimentary rocks: Rocks formed by the accumulation and consolidation of loose particles or grains that have been cemented together. Examples of sedimentary rock are shale, limestone, and sandstone.

Shale: A common clay-like, fine-grained sedimentary rock made primarily of clay and silt grains. Shale, which is also called mudstone, readily splits into thin layers. In Grand Canyon, an example of shale is Bright Angel Shale.

Sipapu: The Hopi word for the spring (atop a 40-foot travertine dome) where the Hopi and all animals emerged from the underworld. The kiva-like Sipapu is just south of Marble Canyon at the bottom of Grand Canyon near the confluence of the Little Colorado and Colorado Rivers. (The entrance through the roof of a Hopi kiva symbolizes the Sipapu.)

Strata: Layers of sedimentary rock.

Tectonic: Referring to any process that affects Earth's tectonic plates.

Travertine: A type of limestone deposited by springs or as a deposit in caves. In GCNP, Travertine Canyons are on river left at River Miles 95.5 and 229.

Tributary: A river or stream that flows into a larger river or lake. The more than 600 tributaries of the Colorado River in Grand Canyon usually occur on or near faults.

Unconformity: A surface where there is a gap in the geologic record, where a rock layer overlies another that is much older. Unconformities are caused by erosion of previous layers, nondeposition, or both. In an angular unconformity, older strata dip steeper than the younger (i.e., overlying) strata.

Uplift: The elevation of part of Earth's crust. The Colorado Plateau was formed by uplift.

Upper Granite Gorge: The 41-mile-long gorge stretching from the bottom of Hance Rapid to below Elves Chasm.

Bibliography

Adams, Eilean. 2001. *Hell or High Water: James White's Disputed Passage through Grand Canyon*. Logan: Utah State University Press.

Aitchison, Stewart. 2008. *Grand Canyon's North Rim and Beyond: A Guide to the North Rim and the Arizona Strip*. Grand Canyon, AZ: Grand Canyon Association.

Anderson, Michael F., ed. 2005. *A Gathering of Grand Canyon Historians: Ideas, Arguments, and First-Person Accounts*. Grand Canyon, AZ: Grand Canyon Association.

Anderson, Michael. 2001. *Along the Rim: A Guide to Grand Canyon's South Rim from Hermit's Rest to Desert View*. Grand Canyon, AZ: Grand Canyon Association.

Anderson, Michael. 1998. *Living at the Edge: Explorers, Exploiters and Settlers of the Grand Canyon Region*. Grand Canyon, AZ: Grand Canyon Association.

Anderson, Michael F. 2000. *Polishing the Jewel: An Administrative History of Grand Canyon National Park*. Grand Canyon, AZ: Grand Canyon Association.

Audretsch, Robert W. 2012. *Images of America: Grand Canyon's Phantom Ranch*. Charleston, SC: Arcadia Publishers.

Austin, Steven A., ed. 1995. *Grand Canyon: Monument to Catastrophe*. Santee, CA: Institute for Creation Research.

Avery, Valeen Tippetts. 1981. *Free Running: Charlie Spencer and His Most Remarkable Water Project*. Flagstaff, AZ: Flagstaff Corral of Westerners International.

Belknap, Buzz, et al. 2017. *Grand Canyon River Guide*. Evergreen, CO: Westwater Books.

Berger, Todd R. 2007. *It Happened at Grand Canyon*. Guilford, CT: TwoDot Publishers.

Berger, Todd R., ed. 2008. *Reflections of Grand Canyon Historians: Ideas, Arguments, and First-Person Accounts*. Grand Canyon, AZ: Grand Canyon Association.

Boyer, Diane E., and Webb, Robert H. 2007. *Damming Grand Canyon: The 1923 USGS Colorado River Expedition*. Logan: Utah State University Press.

Brian, Nancy. 2005. *River to Rim: A Guide to Place Names along the Colorado River in Grand Canyon from Lake Powell to Lake Mead*. Second edition. Flagstaff, AZ: Earthquest Press.

Brooks, Juanita. 1992 (1962). *John Doyle Lee: Zealot, Pioneer Builder, Scapegoat*. Logan: Utah State University Press.

Brownold, Tom, and Dimock, Brad. 2010. *The Grandest Ride*. Tucson, AZ: Rio Nuevo Publishers.

Butchart, Harvey. 1998. *Grand Canyon Treks: 12,000 Miles through the Grand Canyon*. Bishop, CA: Spotted Dog Press.

Coder, Christopher. 2000. *An Introduction to Grand Canyon Prehistory*. Flagstaff, AZ: Grand Canyon Association.

Crumbo, Kim. 1994. *A River Runner's Guide to the History of Grand Canyon*. Boulder, CO: Johnson Books.

Dobyns, Henry F. 1971. *The Havasupai People*. Phoenix: Indian Tribal Series.

Dobyns, Henry, and Euler, Robert. 1971. *The Hopi People*. Phoenix: Indian Tribal Series.

Dobyns, Henry F., and Euler, Robert C. 1976. *The Walapai People*. Phoenix: Indian Tribal Series.

Dutton, Clarence E. 1882. *Tertiary History of the Grand Canyon*. Washington, DC: U.S. Government Printing Office.

Eddy, Clyde L. 1929. *Down the World's Most Dangerous River.* New York: Frederick A. Stokes.

Fletcher, Colin. 1967. *The Man Who Walked through Time*. New York: Random House.

Fradkin, Philip A. 1981. *A River No More: The Colorado River and the West*. New York: Alfred A. Knopf.

Ghiglieri, Michael P., and Myers, Thomas M. 2011. *Over the Edge: Death in Grand Canyon*. Flagstaff, AZ: Puma Press.

Grattan, Virginia L. 1992. *Mary Colter: Builder upon the Red Earth*. Grand Canyon, AZ: Grand Canyon Association.

Hefley, Flood. 2013. *Grand Canyon Nankoweap: Hiking the Notorious Horsethief Trail*. Boulder, CO: Johnson Books.

Henry, Marguerite. 1953. *Brighty of the Grand Canyon*. New York: Rand McNally.

Hill, Carol, et al., eds. 2016. *The Grand Canyon, Monument to an Ancient Earth: Can Noah's Flood Explain Grand Canyon?* Grand Rapids, MI: Kregel Publications.

Hirst, Stephen. 2007. *I Am the Grand Canyon: The Story of the Havasupai People,* Third edition. Flagstaff, AZ: Grand Canyon Association.

Hirst, Stephen, ed. 2016. *We Call the Canyon Home: American Indians of the Grand Canyon Region*. Grand Canyon, AZ: Grand Canyon Association.

Hubbard, Samuel L. 1924. *The Doheny Scientific Expedition to the Hava Supai Canyon, Northern Arizona, October and November, 1924*. Oakland, CA: Oakland Museum.

Hughes, Donald J. 1978. *In the House of Stone and Light: A Human History of the Grand Canyon*. Flagstaff, AZ: Grand Canyon Natural History Association.

Ingram, Jeff. 2003. *Hijacking a River: A Political History of the Colorado River in the Grand Canyon*. Flagstaff, AZ: Vishnu Temple Press.

Kaiser, James. 2011. *Grand Canon: The Complete Guide*. Fourth edition. Chicago: Destination Press.

Kolb, Ellsworth. 1914. *Through the Grand Canyon from Wyoming to Mexico*. New York: Macmillan.

Lago, Don. 2014. *Canyon of Dreams: Stories from Grand Canyon History*. Salt Lake City: University of Utah Press.

Lavender, David. 1985. *River Runners of the Grand Canyon*. Tucson: University of Arizona Press.

Leavengood, Betty. 2014, *Grand Canyon Women: Lives Shaped by Landscape*. Grand Canyon, AZ: Grand Canyon Association.

Loving, Nancy J., and Bean, Tom. 2001. *Along the Rim: A Road Guide to the South Rim of Grand Canyon*. Grand Canyon, AZ: Grand Canyon Association.

Marston, Otis Reed "Dock." 2014. *From Powell to Power: A Recounting of the First One Hundred River Runners through the Grand Canyon*. (Edited by Tom Martin.) Flagstaff, AZ: Vishnu Temple Press.

Martin, Tom. 2012. *Big Water, Little Boats: Moulty Fulmer and the First Grand Canyon Dory on the Last of the Wild Colorado River.* Flagstaff, AZ: Vishnu Temple Press.

Montgomery, David R. 2012. *The Rocks Don't Lie: A Geologist Investigates Noah's Flood*. New York: W. W. Norton.

Morehouse, Barbara J. 1996. *A Place Called Grand Canyon: Contested Geographies*. Tucson: University of Arizona Press.

Neumann, Mark. 1999. *On the Rim: Looking for the Grand Canyon*. Minneapolis: University of Minnesota Press.

Pearson, Byron E. 2001. *Still the Wild River Runs: Congress, the Sierra Club, and the Fight to Save Grand Canyon*. Tucson: University of Arizona Press.

Poling-Kempes, Lesley. 1989. *The Harvey Girls: Women Who Opened the West*. New York: Paragon House.

Powell, John Wesley. 1895. *Canyons of the Colorado*. New York: Flood and Vincent.

Pyne, Stephen J. 1998. *How the Canyon Became Grand*. New York: Penguin Books.

Quartaroli, Richard D., ed. 2013. *A Rendezvous of Grand Canyon Historians: Ideas, Arguments, and First-Person Accounts*. Flagstaff, AZ: Grand Canyon Historical Society.

Reilly, P. T. 1999. *Lee's Ferry: From Mormon Crossing to National Park*. Logan: Utah State University Press.

Reilly, P. T. 1987. *Norman Nevills: Whitewater Man of the West*. Salt Lake City: Utah Historical Society.

Richmond, Al. 1995. *Cowboys, Miners, President and Kings: The Story of the Grand Canyon Railway*. Williams, AZ: Grand Canyon Railway.

Richmond, Al, and Pearsall, Marc. 2007. *The Grand Canyon Railway: Sixty Years in Color*. Second edition. Flagstaff, AZ: Grand Canyon Railway.

Schwartz, Douglas. 1989. *On the Edge of Splendor: Exploring Grand Canyon's Human Past*. Santa Fe, NM: School of American Research Press.

Shankland, Robert. 1951. *Steve Mather of the National Parks*. New York: Knopf.

Shepherd, Jeffery P. 2010. *We Are an Indian Nation: A History of the Hualapai People*. Tucson: University of Arizona Press.

Shields, Kenneth, Jr. 2000. *The Grand Canyon: Native People and Early Visitors*. Charleston, SC: Arcadia Publishing.

Silverthorn, Suzanne, and Chiang, I-Ting. 2014. *Grand Canyon National Park: Past and Present*. Atglen, PA: Schiffer Publishing.

Spamer, Earle E. 2017. *The Grand Canon: A Worldwide Bibliography of the Grand Canyon and Lower Colorado River Regions in the United States and Mexico*. Greatly expanded and revised third edition, 1535–2017. Philadelphia: Raven's Perch Media.

Stampoulos, Linda L. 2004. *Visiting the Grand Canyon: Views of Early Tourism*. Charleston, SC: Arcadia Publishing.

Stenger, Wallace. 1954. *Beyond the Hundredth Meridian: John Wesley Powell and the Second Opening of the West*. Boston: Houghton Mifflin.

Stevens, Larry. 2013. *The Colorado River in Grand Canyon: A River Runner's Map and Guide to Its Natural and Human History*. Flagstaff, AZ: Grand Canyon Wildlands Council.

Suran, William C. 1991. *The Kolb Brothers of Grand Canyon*. Grand Canyon, AZ: Grand Canyon Natural History Association.

Swanson, Frederick H. 2007. *David Rust: A Life in the Canyons*. Salt Lake City: University of Utah Press.

Teal, Louise. 1994. *Breaking into the Current: Boatwomen of the Grand Canyon*. Tucson: University of Arizona Press.

Thybony, Scott. 2001. *Phantom Ranch*. Grand Canyon, AZ: Grand Canyon Association.

Thybony, Scott. 2007. *The Incredible Grand Canyon: Cliffhangers and Curiosities from America's Greatest Canyon*. Grand Canyon, AZ: Grand Canyon Association.

Wilkins, Thurman. 1998. *Thomas Moran: Artist of the Mountains*. Revised edition. Norman: University of Oklahoma Press.

Witt, Greg. 2010. *Exploring Havasupai: A Guide to the Heart of the Grand Canyon*. Birmingham, AL: Menasha Ridge Press.

Worster, Donald. 2001. *A River Running West: The Life of John Wesley Powell*. Oxford: Oxford University Press.

Index

Page numbers in *italics* indicate photos or figures, and page numbers followed by *t* indicate tables. Page numbers in **bold** indicate the location of main entries.

About the Authors

Randy Moore, PhD, is H.T. Morse Distinguished Professor at the University of Minnesota in Minneapolis and St. Paul, Minnesota. He has published several books, including *Dinosaurs by the Decade*, *Chronology of the Evolution–Creationism Controversy*, and *Arguing for Evolution: An Encyclopedia for Understanding Science* (all with ABC-CLIO). He is a member of several professional societies at Grand Canyon (e.g., Grand Canyon Historical Society) and has won a variety of awards from organizations such as the National Association of Biology Teachers, the National Center for Science Education, the Carnegie Foundation, and the National Science Teachers Association.

Kara Felicia Witt, PhD, is a psychologist, researcher, writer, and Grand Canyon enthusiast who has visited and studied Grand Canyon for many years. She has investigated virtually all parts of Grand Canyon, ranging from Supai and Phantom Ranch to sites along both rims. This is Kara's first published book.